Manolis Wallace, Marios Angelides, Phivos Mylonas (Eds.)

Advances in Semantic Media Adaptation and Personalization

Studies in Computational Intelligence, Volume 93

Editor-in-chief
Prof. Janusz Kacprzyk
Systems Research Institute
Polish Academy of Sciences
ul. Newelska 6
01-447 Warsaw
Poland
E-mail: kacprzyk@ibspan.waw.pl

Further volumes of this series can be found on our homepage: springer.com

Vol. 67. Vassilis G. Kaburlasos and Gerhard X. Ritter (Eds.)
Computational Intelligence Based on Lattice Theory, 2007
ISBN 978-3-540-72686-9

Vol. 68. Cipriano Galindo, Juan-Antonio Fernández-Madrigal and Javier Gonzalez
A Multi-Hierarchical Symbolic Model of the Environment for Improving Mobile Robot Operation, 2007
ISBN 978-3-540-72688-3

Vol. 69. Falko Dressler and Iacopo Carreras (Eds.)
Advances in Biologically Inspired Information Systems: Models, Methods, and Tools, 2007
ISBN 978-3-540-72692-0

Vol. 70. Javaan Singh Chahl, Lakhmi C. Jain, Akiko Mizutani and Mika Sato-Ilic (Eds.)
Innovations in Intelligent Machines-1, 2007
ISBN 978-3-540-72695-1

Vol. 71. Norio Baba, Lakhmi C. Jain and Hisashi Handa (Eds.)
Advanced Intelligent Paradigms in Computer Games, 2007
ISBN 978-3-540-72704-0

Vol. 72. Raymond S.T. Lee and Vincenzo Loia (Eds.)
Computation Intelligence for Agent-based Systems, 2007
ISBN 978-3-540-73175-7

Vol. 73. Petra Perner (Ed.)
Case-Based Reasoning on Images and Signals, 2008
ISBN 978-3-540-73178-8

Vol. 74. Robert Schaefer
Foundation of Global Genetic Optimization, 2007
ISBN 978-3-540-73191-7

Vol. 75. Crina Grosan, Ajith Abraham and Hisao Ishibuchi (Eds.)
Hybrid Evolutionary Algorithms, 2007
ISBN 978-3-540-73296-9

Vol. 76. Subhas Chandra Mukhopadhyay and Gourab Sen Gupta (Eds.)
Autonomous Robots and Agents, 2007
ISBN 978-3-540-73423-9

Vol. 77. Barbara Hammer and Pascal Hitzler (Eds.)
Perspectives of Neural-Symbolic Integration, 2007
ISBN 978-3-540-73953-1

Vol. 78. Costin Badica and Marcin Paprzycki (Eds.)
Intelligent and Distributed Computing, 2008
ISBN 978-3-540-74929-5

Vol. 79. Xing Cai and T.-C. Jim Yeh (Eds.)
Quantitative Information Fusion for Hydrological Sciences, 2008
ISBN 978-3-540-75383-4

Vol. 80. Joachim Diederich
Rule Extraction from Support Vector Machines, 2008
ISBN 978-3-540-75389-6

Vol. 81. K. Sridharan
Robotic Exploration and Landmark Determination, 2008
ISBN 978-3-540-75393-3

Vol. 82. Ajith Abraham, Crina Grosan and Witold Pedrycz (Eds.)
Engineering Evolutionary Intelligent Systems, 2008
ISBN 978-3-540-75395-7

Vol. 83. Bhanu Prasad and S.R.M. Prasanna (Eds.)
Speech, Audio, Image and Biomedical Signal Processing using Neural Networks, 2008
ISBN 978-3-540-75397-1

Vol. 84. Marek R. Ogiela and Ryszard Tadeusiewicz
Modern Computational Intelligence Methods for the Interpretation of Medical Images, 2008
ISBN 978-3-540-75399-5

Vol. 85. Arpad Kelemen, Ajith Abraham and Yulan Liang (Eds.)
Computational Intelligence in Medical Informatics, 2008
ISBN 978-3-540-75766-5

Vol. 86. Zbigniew Les and Mogdalena Les
Shape Understanding Systems, 2008
ISBN 978-3-540-75768-9

Vol. 87. Yuri Avramenko and Andrzej Kraslawski
Case Based Design, 2008
ISBN 978-3-540-75705-4

Vol. 88. Tina Yu, David Davis, Cem Baydar and Rajkumar Roy (Eds.)
Evolutionary Computation in Practice, 2008
ISBN 978-3-540-75770-2

Vol. 89. Ito Takayuki, Hattori Hiromitsu, Zhang Minjie and Matsuo Tokuro (Eds.)
Rational, Robust, Secure, 2008
ISBN 978-3-540-76281-2

Vol. 90. Simone Marinai and Hiromichi Fujisawa (Eds.)
Machine Learning in Document Analysis and Recognition, 2008
ISBN 978-3-540-76279-9

Vol. 91. Horst Bunke, Kandel Abraham and Last Mark (Eds.)
Applied Pattern Recognition, 2008
ISBN 978-3-540-76830-2

Vol. 92. Ang Yang, Yin Shan and Lam Thu Bui (Eds.)
Success in Evolutionary Computation, 2008
ISBN 978-3-540-76285-0

Vol. 93. Manolis Wallace, Marios Angelides and Phivos Mylonas (Eds.)
Advances in Semantic Media Adaptation and Personalization, 2008
ISBN 978-3-540-76359-8

Manolis Wallace
Marios C. Angelides
Phivos Mylonas
(Eds.)

Advances in Semantic Media Adaptation and Personalization

With 148 Figures and 20 Tables

Springer

Dr. Manolis Wallace
Department of Computer Science and Technology
University of Peloponnese
Terma Karaiskaki
22 100, Tripoli
Greece
wallace@uop.gr

Professor Marios C. Angelides
School of Information Systems, Computing and Mathematics
Brunel University, Uxbridge
Middlesex UB8 3PH
United Kingdom
marios.angelides@brunel.ac.uk

Dr. Phivos Mylonas
Image, Video and Multimedia Laboratory
Department of Electrical and Computer Engineering
School of Electrical Engineering
National Technical University of Athens
Iroon Ploytechneioy 9
157 80, Zographoy Campus, Athens
Greece
fmylonas@image.ntua.gr

ISBN 978-3-540-76359-8 e-ISBN 978-3-540-76361-1

Studies in Computational Intelligence ISSN 1860-949X

Library of Congress Control Number: 2007941931

Cover Design: Deblik, Berlin, Germany

Printed on acid-free paper

9 8 7 6 5 4 3 2 1

springer.com

Contents

Dynamic Linking of Web Resources: Customisation and Personalisation
Yeliz Yesilada, Sean Bechhofer, and Bernard Horan 1
1 Introduction 1
2 COHSE 5
2.1 Deploying COHSE 6
3 Portals 8
3.1 Architecture 9
3.2 COHSE Portlet 12
4 Personalisation 14
4.1 User Tasks on the Web 17
5 Related Work 20
6 Summary and Discussion 21
References 22

A Multi-Purpose Ontology-Based Approach for Personalised Content Filtering and Retrieval
Iván Cantador, Miriam Fernández, David Vallet, Pablo Castells, Jérôme Picault, and Myriam Ribière 25
1 Introduction 25
2 Ontology-Based Personalisation for Content Retrieval 27
3 Adaptation of Semantic User Preferences 30
3.1 Automatic Creation and Update of Semantic Preferences 31
3.2 Exploitation of Semantic Links Between Concepts for Preferences Learning 33
4 Contextual Personalisation 36
4.1 Context Representation 36
4.2 Evaluation of Personalisation in Context 37
5 Augmented Social Networking and Collaborative Filtering 40
5.1 Semantic Group Profiling 40
5.2 Semantic Social Networking 43

5.3 Semantic Profile Expansion for Collaborative Group Profiling ... 46
6 Conclusions ... 48
References ... 49

Analysing the Behaviour of Students in Learning Management Systems with Respect to Learning Styles
Sabine Graf and Kinshuk ... 53
1 Introduction ... 53
2 Felder–Silverman Learning Style Model ... 55
3 Investigated Patterns of Behaviour ... 56
4 Design of the Study ... 57
4.1 Description of the Course ... 57
4.2 Instrument for Identifying Learning Styles ... 58
4.3 Extensions in Moodle for Tracking Students' Behaviour ... 58
5 Results ... 59
5.1 Behaviour Vs. Learning Style Preferences ... 60
5.2 Correlations Between Behaviour and Learning Style Preferences ... 65
6 Benefits ... 68
7 Related Work ... 70
8 Conclusion and Future Work ... 71
References ... 72

Improving Image Retrieval Using Semantic Resources
Adrian Popescu, Gregory Grefenstette, and Pierre-Alain Moellic ... 75
1 Introduction ... 75
2 Related Work ... 77
3 Ontological Issues ... 79
3.1 Hierarchies ... 79
3.2 Picturable Objects ... 80
3.3 Automatic Ontology Building ... 81
4 Visual Ontology Construction ... 82
4.1 The WordNet Nouns Hierarchy ... 82
4.2 Image Database Constitution ... 83
4.3 Image Retrieval Phase ... 84
4.4 Image Clustering ... 88
5 Limits ... 90
5.1 Image Processing Limitations ... 90
5.2 WordNet Coverage ... 90
5.3 Complex Queries ... 91
5.4 Quality of the Semantic Structure ... 91
6 Evaluation ... 92
6.1 Experimental Settings ... 92
6.2 Results ... 92

7 Perspectives ... 93
8 Conclusions ... 94
References ... 95

Implementing an Internet Image Search Service Based on the AJAX Web Programming Techniques
I. Anagnostopoulos, C. Anagnostopoulos, I. Psoroulas, and I. Maglogiannis ... 97
1 Introduction ... 97
1.1 Internet Programming with Asynchronous Javascript and XML ... 98
2 System Architecture ... 98
2.1 The User-Interface (UI) ... 99
2.2 The Query-Transmitter ... 101
2.3 The Results-Collector ... 102
2.4 The Local Repository and XML Handler ... 103
2.5 The Meta-Results Handler ... 103
3 Retrieving Images with Human Appearance on the Web ... 105
3.1 A Case Study ... 105
3.2 The Meta-Search Engine Performance ... 109
3.3 Asynchronous–Synchronous Analysis Performance ... 111
4 Conclusions: Future Considerations ... 112
References ... 115

A Rule-Based Adaption Model for Ontology-Based Personalization
Thanh Tran, Philipp Cimiano, and Anupriya Ankolekar ... 117
1 Introduction ... 117
2 Overview of Adaptation Approaches ... 118
2.1 State-of-the-Art ... 119
2.2 On the Use of Ontologies and Rules for Adaptation ... 119
2.3 ODAS - A Domain Ontology for Adaptive Hypermedia Systems ... 121
3 A Rule-Based Adaptation Model ... 122
3.1 Applying Rules-Based Adaptation to a Portal ... 122
3.2 A Formal Definition of the Adaptation Model ... 125
3.3 ODAS-Based Instantiations of the Adaptation Model ... 126
4 Implementation and Discussion ... 131
5 Related Work ... 132
6 Conclusion ... 133
References ... 134

Management and Processing of Personalized Annotations in Image Retrieval Systems
Johanna Vompras and Stefan Conrad ... 137
1 Motivation ... 137
2 Image Annotation Process ... 139

3 Multi-level Annotation Model 140
3.1 Subjectivity in Image Annotations 142
3.2 Components of the Annotation Analysis Framework 143
3.3 Graph Representation for Multi-level Annotations 144
3.4 Relations Between Terms of Annotation 148
3.5 Extending the Probabilistic Annotation by Multi-level Annotations 149
3.6 Example of Context-Based Similarity 151
4 Related Work 152
5 Summary and Future Work 153
References 153

An Incremental Framework for Adapting the Hypermedia Structure of Multimedia Documents
Sébastien Laborie and Jérôme Euzenat 157
1 Introduction 157
2 Multimedia Document Specification 158
3 Semantic Adaptation of Multimedia Documents 159
4 Adaptation of the Hypermedia Dimension 162
5 Adaptation Based on the Hypermedia Structure 165
5.1 Generating the Multimedia Document Sub-Specifications According to the Hypermedia Structure 166
5.2 Adapting All Sub-Specifications at Once 166
5.3 Adapting Sub-Specifications Step by Step 168
6 Implementation 169
7 Application to Concrete Multimedia Documents 170
7.1 Concrete Multimedia Documents Adaptation 171
7.2 Prototype 172
8 Related Work 173
9 Conclusion 174
References 175

Improving the Adaptiveness of an e-Learning System
Ioannis Hatzilygeroudis, Constantinos Koutsojannis, and Nikolaos Papachristou 177
1 Introduction 177
1.1 Web-Based Intelligent Education Systems 178
1.2 Medical Education 178
1.3 Cognitive Flexibility Theory 179
1.4 The Motivation 180
2 The COFALE Environment 181
3 Extending COFALE 182
3.1 Domain Knowledge 182
3.2 Test Creation 182
3.3 Student Assessment 185
4 Architecture and Implementation Aspects 188

5 System Evaluation ... 190
6 Related Work ... 192
7 Conclusions ... 194
References ... 195

The Context Oriented Architecture: An Augmentation of Context Awareness and Reactivity into Web Services
Aisha Mohamed-Salama Elsafty, Sherif Gamal Aly, and Ahmed Sameh ... 199
1 Introduction ... 199
2 Related Work ... 201
2.1 Context Defined ... 201
2.2 OWL-S: The Semantic Web Service Ontology Language ... 203
2.3 Context Awareness Infrastructures ... 204
2.4 Context Aware Web Services ... 204
3 The Context Oriented Architecture ... 206
3.1 The Negotiator ... 208
3.2 The Context Sensors ... 208
3.3 The Monitor ... 208
3.4 The Context Response ... 208
3.5 The Response Manager ... 209
3.6 The Component Registry ... 209
3.7 The Context Repository ... 209
3.8 Usage Scenario ... 210
4 OWL-S Extension ... 211
4.1 Profile Ontology Extension ... 212
4.2 Monitors Attribute ... 212
4.3 Uses Attribute ... 212
4.4 Responds Attribute ... 213
4.5 Process Model Extension ... 213
4.6 Exposes Attribute in the Process Model ... 214
5 Results and Analysis: The File Storage Scenario ... 215
5.1 Scenario Context ... 215
5.2 Scenario Formalization in the Framework ... 216
5.3 Semantic Model ... 216
5.4 Interaction Model ... 216
6 Conclusion ... 218
References ... 219

Knowledge-Assisted Analysis of Video for Content-Adaptive Coding and Transmission
Vasileios Mezaris, Nikolaos Thomos, Nikolaos V. Boulgouris, and Ioannis Kompatsiaris ... 221
1 Introduction ... 221
2 Domain Knowledge ... 224

3 Knowledge-Assisted Video Analysis 226
3.1 Color and Motion Initial Segmentation 226
3.2 Low-level Descriptors Extraction 227
3.3 Genetic Algorithm 228
4 Content-Adaptive Coding and Transmission 230
4.1 Basic Problem Formulation 230
4.2 Channel Rate Optimization for Wireless Transmission Scenarios 232
5 Experimental Results 235
6 Conclusions 238
References 238

Mobile Photography within a Social Context
Kris Luyten, Ruben Thysv, Kristof Thys, and Karin Coninx 241
1 Introduction 241
2 Related Work 244
3 Case Study 245
4 Using PhotoFOAF for Social Photo Management 247
4.1 Photo Production Phase 247
4.2 Photo Consumption Phase 248
5 Determining Social Relevance 250
6 Embedded Metadata 254
7 System architecture 255
8 Social Network Visualization Experiments 256
9 Conclusions and Future Work 258
References 259

An Adaptive Semantic Framework to Support Multiple User Viewpoints over Multiple Databases
Stefan Poslad and Landong Zuo 261
1 Introduction 261
1.1 Information Heterogeneity 262
1.2 Outline of the Chapter 263
2 Related Work 263
2.1 Logical Data Views vs. User Views 263
2.2 Integrating Semantics, Rules, Logic and Databases 265
3 A Multiple Viewpoint Semantic Framework 267
3.1 Viewpoint Conceptualisation 269
3.2 Semantic Mapping 270
3.3 The Process to Project a Viewpoint 271
3.4 Modelling Individual Users 273
3.5 View-Based Query Answering and Result Adjustment 274
3.6 Applying Preference and Rules in Query Answering 277
3.7 User-Centre Query Answering 278
4 Implementation and Validation 280

5 Conclusions and Further Work 282
References 283

User Preferences for Access to Textual Information: Model, Tools and Experiments
Thibault Roy and Stéphane Ferrari 285
1 Introduction 285
2 Motivations 286
2.1 Textual Information and Users' Point of View 286
2.2 Visual and Interactive Tools for Access to Textual Information 287
2.3 Our Approach 289
3 Models and Tools 289
3.1 LUCIA: A Model for Representing User's Knowledge on Domains 289
3.2 User-Centred Tools 291
4 Experiment 1: Accessing Information 294
4.1 Context and Materials 294
4.2 Results and Discussion 295
5 Experiment 2: Conceptual Metaphors 298
5.1 Constraints on the Model for Metaphor Characterisation 298
5.2 Map and Texts "Metaphoricity" 300
6 Conclusion 302
References 303

Modeling the Reusable Content of Adaptive Web-Based Applications Using an Ontology
Mária Bieliková and Michal Moravčík 307
1 Introduction 307
2 Related Work 309
2.1 Content Representation 309
2.2 Content Transformation 310
3 Development of Reusable Content 311
3.1 The Method for Domain Modeling 312
3.2 Importing Existing Domain Model Ontologies 314
3.3 Delivering Ontology Content into Existing Adaptive Application 315
4 Ontology Models of the Content 316
4.1 Core Ontology of Content Model 316
4.2 Domain and User Models Interconnection 321
5 Meta-Model of the Adaptive Application Content 322
6 Evaluation of Proposed Method 322
7 Conclusions 325
References 325

Indexing and Browsing of Color Images: Design Considerations
Christos Diou, Nikos Batalas, and Anastasios Delopoulos 329
1 Introduction 329
2 Color Descriptors and Similarity Measures 331
 2.1 Histogram and Dominant Color 331
 2.2 Distance Metrics 332
3 Indexing and Dimensionality 334
 3.1 Limits of *kd*-Tree Effectiveness 335
4 Point Configurations 335
 4.1 MDS and FastMap 337
 4.2 Browsing 337
5 Efficiency of FastMap Configurations 338
6 Conclusions 343
References 345

Generic Personal Assistance Architecture for Mobile Environments
Gerald Bieber 347
1 Introduction 347
2 Related Work 348
 2.1 Personal Task Model 350
3 Personal Assistant 352
 3.1 Goal Description 353
 3.2 Situation Influence by Task Execution 354
 3.3 Task Generation 355
 3.4 Situation Dependent Task Suggestion 357
 3.5 Control Circuit 358
4 Personal Assistance Architecture 358
5 Proof of Concept-Applications 360
 5.1 xGuide: Electronic Exhibition Guide 360
 5.2 Applied Task Definition 361
 5.3 StepMan 361
 5.4 Personal Assistant DiaTrace 363
 5.5 Mobile Personal Trainer 365
6 Outcome Analysis 366
7 Conclusion and Further Work 367
References 367

Foreword

Different media types have converged under the realm of multimedia, or simply media. Consequently, modern media systems are expected to handle all possible media types and formats seamlessly and efficiently, taking into account not only the means of transmission and presentation but also the tastes and preferences of the user. Combined with the emergence of advanced network infrastructures which enable fast and reliable storage and retrieval and end-to-end transmission of large volumes of multimedia content and the development of a plethora of terminal devices for media presentation, has lead to immense opportunities for media content and service development and provision.

Although systems succeed in considering and adapting to diverse networking infrastructures and terminal devices, not enough consideration is granted to the user. Although presentation preferences, such as video quality, can be addressed by techniques as simple as low level content adaptation, users' preferences at a semantic level are more difficult to understand, model and satisfy. The realization of notions of context and contextualization at a semantic level is expected to aid significantly in this area.

Realizing the growing importance of semantic adaptation and personalization of media, the editors of this book decided in the summer of 2006 to bring together researchers and practitioners of the field to discuss the state of the art and explore emerging exciting developments. The meeting that took place in Athens in December 2006, the first International Workshop on Semantic Media Adaptation and Personalization (SMAP 2006), managed to trigger an unexpected amount of interest: 42 high quality submissions were received from 20 countries across Europe, North America and Asia. The quality of the workshop submissions was so high that the organizers were compelled to seek further means of dissemination. This volume comprises extended versions of selected papers presented at the workshop. Two rounds of reviews have assured the quality and depth of analysis of all the chapters.

Following the success of the first event, the SMAP initiative looks forward to a second event, this time in London during December 2007 and will feature, additionally to the main workshop, a Doctoral Consortium for

Ph.D. candidates specializing in the field. As with SMAP 2006, SMAP 2007 will be supported by the IEEE Computer Society, which publishes the proceedings of the Workshop, the IEEE Computational Intelligence Society, the IEEE Systems, Man and Cybernetics Society and the IEEE-SMC Technical Committee on Intelligent Internet Systems, which provide technical sponsorship and, importantly, the community from which a great deal of the SMAP participants are drawn.

In life you can only do this much without help from others. Thus, the editors of this book and founders of the SMAP initiative and workshop series would like to express their gratitude to many people for their help and support. Unable to list all here, we would like to thank Nicoletta Georgiou from the University of Indianapolis Athens, Stefanos Kollias and Marios Phinikettos from the National Technical University of Athens, John Lee from the IEEE Systems, Man and Cybernetics Society TC on Intelligent Internet Systems and Gary Greenwood from IEEE Computational Intelligence Society. Our thanks also go to our students who volunteered to help during the event and last, but certainly not least, to all those who embraced the SMAP initiative and help in its growth.

M. Wallace
M.C. Angelides
Ph. Mylonas

Dynamic Linking of Web Resources: Customisation and Personalisation

Yeliz Yesilada[1], Sean Bechhofer[1], and Bernard Horan[2]

[1] University of Manchester, Manchester, UK http://www.manchester.ac.uk/cs, sean.bechhofer@manchester.ac.uk, yeliz.yesilada@manchester.ac.uk
[2] Sun Microsystems Laboratories, Menlo Park, CA 94025, USA http://research.sun.com, bernard.horan@sun.com

Summary. Conceptual Open Hypermedia Service (COHSE) provides a framework that integrates a knowledge service and the open hypermedia link service to dynamically link Web documents via knowledge resources (e.g., ontologies or controlled vocabularies). The Web can be considered as a *closed* hypermedia system – Links on the Web are unidirectional, embedded, difficult to author and maintain. With a Semantic Web architecture COHSE addresses these limitations by dynamically creating, multi-headed links on third party documents by integrating third party knowledge resources and third party services. Therefore *open-ness* is a key aspect of COHSE. This chapter first presents how COHSE architecture is reengineered to support customisation and to create an *adaptable* open hypermedia system where the user explicitly provides information about himself. It then presents how this architecture is deployed in a portal and discusses how this portal architecture can be extended to turn COHSE from an adaptable system to an *adaptive* system where system implicitly infers some information about the user.

1 Introduction

Links on the Web are typically embedded within pages. Although this is a simple and scalable approach, there are a number of limitations:

- Links can only be *unidirectional*.
- *Ownership* is required to place an anchor in a document.
- Documents need to be written in a particular *format* to support the addition of links (e.g., HTML or PDF).
- There is a weight of *maintenance* in creating and updating links [14].

Open Hypermedia Systems (OHS) address these limitations by viewing links as "first class citizens" [24]. By using the existing infrastructure of the Web, COHSE (Conceptual Open Hypermedia Service) marries the work on OHS and Semantic Web which aims to improve machine processing on the Web.

Y. Yesilada et al.: *Dynamic Linking of Web Resources: Customisation and Personalisation*, Studies in Computational Intelligence (SCI) **93**, 1–24 (2008)
www.springerlink.com

Thus, COHSE aims to combine the advances of both fields; including *semantics* for machine-processing can also be used to improve the linking and navigation of Web pages intended for human end-users. Through the use of different domain ontologies or vocabularies, COHSE provides different hypertext views of Web resources. For instance, [5] presents how the COHSE system is used in Bioinformatics to link biology resources using the Gene Ontology[1] and [9] presents how Sun's Java tutorial pages are augmented using a Java and programming ontology.

In this chapter as a motivating scenario, we consider the problems of Web site maintenance and presentation that a large organisation like Sun Microsystems faces. Sun provides a number of different Web sites that are targeted at different user communities. For example, Sun Store (`store.sun.com`) provides access to the Sun product catalogue; SunSolve (`sunsolve.sun.com`) provides access to a troubleshooting knowledge base; and a third site (`docs.sun.com`) provides further documentation about Sun products. None of these sites is explicitly cross linked, however – managing associations between these sites is a challenge, potentially crossing organisational boundaries. The sites do share subject content, in that they all contain information relevant to the products and objects that Sun manufactures and sells. It would be of great benefit to the users if, for example, when browsing the product catalogue links to relevant documentation could be provided.

Our solution is to make use of a Unified Product Taxonomy (UPT) that describes the products and services that Sun manufactures and sells. The UPT is a controlled vocabulary created for cataloguing products – we are able to repurpose it here and use to drive the linking process, identifying sources and targets. In conjunction with the COHSE infrastructure and services searching Sun content, we are able to support dynamic linking which is illustrated in Fig. 1: (a) this shows the original page; (b) COHSE identifies and highlights concepts (e.g., Sun fire E4900 server, etc.) and (c) when the user clicks on a highlighted term "Ultrasparc iv", COHSE creates a box that includes a set of links (e.g., to the relevant documentation) along with a short description that comes from the UPT.

Although we are describing a particular problem here, the solution is generic, and can be applied to other domains – simply by substituting alternative knowledge resources and target services. For example, NELI (National Electronic Library of Infection)[2] is a collection of Web sites (e.g., Bugs and Drugs, Training in infection, etc.) forming a digital library of materials related to infectious diseases. As with the Sun scenario, although these sites present relevant information, there are very few cross links between them and other relevant sites on the Web. Although the Sun scenario illustrates the problem of linking within the same organisation, in this case we are talking about the problem of linking across organisations – how do we link sites that belong

[1] GO, `http://www.geneontology.org/`.

[2] NELI, `http://www.neli.org.uk/`.

(c)

Fig. 1. COHSE in action: (**a**) a screenshot of the original page, (**b**) COHSE identifies and highlights terms and (**c**) when the user clicks on an highlighted term, a link box is displayed

Table 1. NeLI users and example targets from the UK

User group	Question	Targets
Family doctor (GP)	Tuberculosis drugs and side effects?	British National Formulary (BNF), Other NeLI pages
Clinicians	Tuberculosis treatment guidelines, X-ray examinations?	NeLI pages, BNF, Public Health Observatories (PHO)
Molecular biologists	Drug resistant tuberculosis species?	PubMed, Wikipedia, PHO
General public	What is tuberculosis?	Health Protection Agency (HPA), NHS direct

to different organisations? To address this, we propose to use the COHSE infrastructure along with controlled vocabularies and medical ontologies such as the GALEN ontology[3] (see Fig. 6) to dynamically add links to NELI pages.

In both of these motivating scenarios, an additional constraint is that the sites provide content targeting a particular user group – overlaying links appropriate for a particular user context is important if we are to repurpose the content effectively. For example, NELI has a range of user groups; Table 1 shows four different user groups that might come into the NELI Web site each wanting to answer different questions relating to a particular disease. For example, if a user from the general public is looking for information about "Tuberculosis", it would be good to suggest pages that give broad overview of the disease. Whereas if a family doctor is looking for information about "Tuberculosis", it may be more appropriate to suggest pages that provide information about the drugs, treatments and local hospital's facilities related to that disease. Similarly the Sun sites are also accessed by a range of user groups, for example experienced engineers or non-technical customers. Knowing more about these user groups and overlaying customised and personalised links would be a great benefit to these users; we will be able to repurpose the content based on the user's context to provide personalised custom links.

Supporting customisation and personalisation has a number of requirements including identifying users and being able to store preferences persistently. The previous implementations of COHSE which include a proxy and a browser plug-in provide some customisation, however the available options are limited. Furthermore, with these implementations it was almost impossible to support personalisation techniques such as collaborative filtering [34]. To meet these requirements, we have moved to a portal-based implementation. A portal provides a framework to aggregate content from different sources, and supports storage of user profiles, customisation and personalisation [16].

In this chapter, we first present the original COHSE architecture and discuss its limitations (Sect. 2). We explain how this architecture is extended

[3] The *GALEN* ontology defines medical terms and surgical procedures.

and deployed in a portal (Sect. 3). We then discuss how this new architecture can be used to support personalisation (Sect. 4). Finally, we present some related work (Sect. 5) and complete this chapter with a summary and discussion (Sect. 6).

2 COHSE

COHSE's architecture is described elsewhere [9] – we provide a brief overview here. COHSE is composed of the following three components.

The *Knowledge Service (KS)* supports interaction with knowledge resources such as OWL[4] ontologies and SKOS[5] vocabularies. It provides services such as:

- Mappings between concepts and lexical labels (synonyms, e.g., skos:alt Label)
- Information about specialisation (narrower, e.g., skos:narrower) and generalisation of concepts (broader, e.g., skos:broader)
- Access to related concepts (e.g., skos:related)
- Descriptions of concepts (e.g., skos:scopeNote)

The service has a simple HTTP interface and can host third party knowledge resources (accessed via URLs).

The *Resource Service (RS)* manages and stores annotations of resources. An annotation represents a mapping between a concept and a resource fragment (e.g., a paragraph) that may be produced by parties other than the document owners or users of the COHSE system [17]. The annotations are used in two ways:

- A concept to resource mapping provides candidate *targets* for links – given a concept it will return a list of candidate link targets.
- A resource to concept mapping provides candidate *source* anchors for links – given a resource it will return a list of candidate link sources.

Therefore, the RS is used for both source anchor identification and discovery of possible target anchors. Various techniques are used to generate annotations; currently, we use lexical matching algorithms (e.g., regular expression matching), manual annotation with our annotation tool [9], external third party services (such as Google Web Service (WS) or Amazon WS) and the GATE (A General Architecture for Text Engineering)[6] framework with extended ontological services [9]. These give us flexibility to manage various sets of annotations.

[4] Web Ontology Language (OWL), `http://www.w3.org/2004/OWL/`.
[5] Simple Knowledge Organisation Systems (SKOS), `http://www.w3.org/2004/02/skos/`.
[6] GATE, `http://gate.ac.uk/`.

The *DLS agent* is responsible for modifying pages and adding links depending on information provided by the KS and RS. It processes documents and inserts link source anchors into the document along with the possible link targets. The DLS agent works roughly as follows:

1. The agent contacts the KS to get a collection of relevant lexical terms, for example in the Sun scenario, it will get all the lexical terms from the UPT.
2. The requested document is processed by looking for occurrences of these terms. Any that are found provide potential link sources (see Fig. 1b).
3. For the identified link sources, the DLS agent queries the RM to determine appropriate targets; it also uses the knowledge about the term encoded in the knowledge resource to enhance the suggested link targets – using narrower, broader and related terms to suggest more resources. For example, in Fig. 1 a list of resources are suggested for the term "Ultrasparc iv" along with a list of resources for related terms such as "Sun Fire E4900 Server", etc. Knowledge about the relationship between these terms comes from the knowledge resource used, in this case the UPT. The DLS agent thus uses the KS and RS in co-ordination to enhance linking opportunities (see Fig. 1).

COHSE has a loosely coupled architecture; it supports easy interaction of third party knowledge resources, third party services to add links dynamically to third party documents. Therefore *open-ness* is a key aspect of COHSE.

2.1 Deploying COHSE

Based on the client–server architecture of the Web, various Open Hypermedia Systems (OHS) have utilised different techniques to augment Web pages with extra links [24]. Extra links are either displayed alongside the page or inserted into the page. COHSE uses the latter approach as this is familiar to most users [1]. However, this requires documents to be modified. Regarding the fundamental client–server architecture of the Web this can be done at three different places. Each approach has pros and cons, particularly in relation to different levels of customisation support:

(1) Server Side: A specialised server side application can be used to add links to documents before delivery to the client. This might be an extension to a Content Management System (CMS) or a standalone server side application. Since the documents do not need to be updated on the client side, no end user installation of specialist browsing software is required. Additionally, on the server side, more than one document can be processed (i.e., batch processing), as opposed to client-side approach where one document can be accessed at a time. Although dynamically generated sites can be better handled by a server-side application, each site requires its processing machinery. Moreover, one needs ownership of the pages to add links. This means that with this approach there still be a clear distinction between author and reader. One

of the criteria for referring to a hypermedia system as *open* is to have no such artificial distinction between author and reader [24]. Although pages can be linked dynamically with a server side system, unless the user plays a role in this, it will not be a true *open* system. Server side processing also means that COHSE links (i.e., multi-headed extra links) will only be available for that particular site. The user will not have a continuous browsing experience; when they traverse to other sites COHSE links will not be available. Finally regarding customisation and personalisation, limited support can be provided for customisation as the server needs to keep track of users and store their preferences persistently.

(2) Intermediary: A proxy server acts as an intermediary between a client and server and can be used to add links to documents. The proxy rewrites documents, inserting links where appropriate. This approach also potentially opens up the use of COHSE to alternative platforms such as WAP and mobile phones as appropriate transformations an be used for different clients. As with server side deployment, no end user installation of specialist browsing software is required. However, the use of a proxy may introduce processing delays and the need to re-write URLs to ensure that users continue to browse via the proxy. Moreover users may need to modify their browser's configuration which can be inconvenient and might not be allowed in some Intranets. To overcome this, COHSE employed a rewriting proxy which allows clients to make use of the proxy without adjusting their browser settings. The Rewriting proxy takes requests in the form `http://proxy_url/http://page_url` where `proxy_url` is the URL of the rewriting proxy and `page_url` is the required page. Although this addresses the problem of configuring browsers, it becomes hard to read URLs; When a user moves his mouse over a link to read the target URL, he will see a long and confusing URL. Although customisation can be easier with a proxy, storing users' preferences persistently can be troublesome. To address this problem, one of the COHSE implementations support customisation per IP address. If more than one user accesses this proxy from the same IP address, which may be the case if a firewall is in use, one users' configuration will be changed by the other. And if a user accesses this proxy from different IP addresses, then he will not be able to transfer and use his settings.

(3) Client Side: A plug-in or browser extension can be used to add links to documents. This is easy to prototype because the underlying development framework of the browser (e.g., Mozilla) can be used. The implementation is then tightly coupled to a specific browser and typically to a specific version. For example, although Firefox is a continuation of Mozilla, a Mozilla extension cannot be installed to Firefox, leading to issues with maintenance. Similarly, Mozilla extensions cannot be installed to Internet Explorer, requiring multiple implementations for different browsers. A client side extension also requires local installation and this could be an issue if the user does not have control over the settings of the browser. However, with a browser extension, the browsers' internal engine can be used to parse the documents helping to overcome issues with mal-formed documents. Additionally, storing

user preferences persistently is easier with client side implementations compared to others, as the underlying preferences structure of the browser can be used. However, maintenance and communication of these preferences with external components has to be performed in an ad-hoc manner. Use of a plug-in may also make it more difficult to share information between users which can be a problem in supporting some personalisation techniques such as collaborative filtering.

COHSE has, in the past, been deployed as a proxy and as a suite of plug-ins to Mozilla [9]. However, as discussed above these client-side and proxy implementations have limitations, in particular with respect to the customisation and personalisation. Furthermore, for large organisations such as Sun these implementations proved to be inappropriate; an extension installation can be difficult and using a proxy is not easy due to the internal firewalls used. Therefore, in order to address these limitations, we are proposing to use a portal to deploy COHSE.

3 Portals

A portal is defined as "a Web application that *commonly* provides personalisation, single sign-on and content aggregation from different sources". A portal uses a number of pluggable user interface components called *portlets* that provide a presentation layer and enable aggregation of content from different sources. From a user's perspective, a portlet is a window in a portal that provides a specific service (e.g., a calendar). From an application perspective, a portlet is a software component written in Java, managed by a portlet container, which handles user requests and generates dynamic contents.

JSR168[7] standardises how portlets are to be developed by addressing preferences (storing customisation preferences persistently), user information, security, etc. It allows portlets adhering to JSR168 to be deployed to any portal that conforms to the specifications by enabling interoperability among the portals and the portlets. The COHSE DLS agent has been re-implemented as a portlet to exploit the advantages of portals.

- *Customisation*: With a portlet, different options can be declared and stored persistently. JSR168 specifies a method for handling persistent preferences and most portals that conform to JSR168 provide an implementation of that. Therefore, using an existing portal provides us an advantage of using an existing infrastructure to support customisation and user profiling.
- *Single Sign-on*: This feature enables single sign on across a range of applications. Users need to enter their details (i.e, user name, password) once and that information will be used to authenticate users to other resources that will be accessed through the portal. Since the COHSE DLS

[7] JSR168, `http://www.jcp.org/en/jsr/detail?id=168`.

agent relies on external services including the RS and KS, this can be an advantage in passing user authentication information to these services.

- *Aggregation*: This is the process of integrating content from different sources within a portal. A portal COHSE DLS agent can be used along with diverse types of information from a variety of sources (e.g., search engines). For example, with Grid community portals[8] which provide a means to access different grid services, the DLS agent can be provided as another service.
- *Authentication*: Portals provide a mechanism to deal with authentication issues which means an existing development can be reused without implementing a new one.
- *User Profiles/Authorisation*: Most portals accommodate different roles and user groups. This feature gives us flexibility to support customisation in different ways. For example, different user groups can have different customisation options depending on their role, for instance "administrators" would be able to customise supporting services (e.g., KS) but not "end-users". Moreover having user groups allows us to provide *personalised links* [33]; certain types of links would be visible to a specific group of users, rather than to every user or different user groups would get different type of information about extra COHSE links. For example, novice users in a particular domain would get definitions of the identified concepts but experienced users would not.

3.1 Architecture

In order to deploy COHSE as a portlet in a portal, the original architecture is extended. Figure 2 (1–3) shows the components that contribute to the identification of resources (i.e., annotation producers) and Fig. 3 (5–8) highlights the components that use those resources (i.e., annotation consumers). Fig. 2 (4) is the KS which is the main mediating component that enables the communication between these two parties.

(1) Document Object Model (DOM). Both annotation producers and consumers use the DOM to process documents. Tidy is used at both ends of the architecture in order transform input documents from HTML format to XHTML and to create a DOM representation of those documents. According to Chen et al. [10], few pages on the Web are valid and conform to standards. When documents are not well-formed, a lot of technical problems can occur in using the DOM. For instance, different DOM parsers can generate different DOM trees if the document is not well-formed. Therefore, Tidy[9] is used to ensure that both consumers and producers access the same DOM representation.

---

[8] For some examples, see `http://www.grids.ac.uk/`.

[9] Tidy, `http://tidy.sourceforge.net/`.

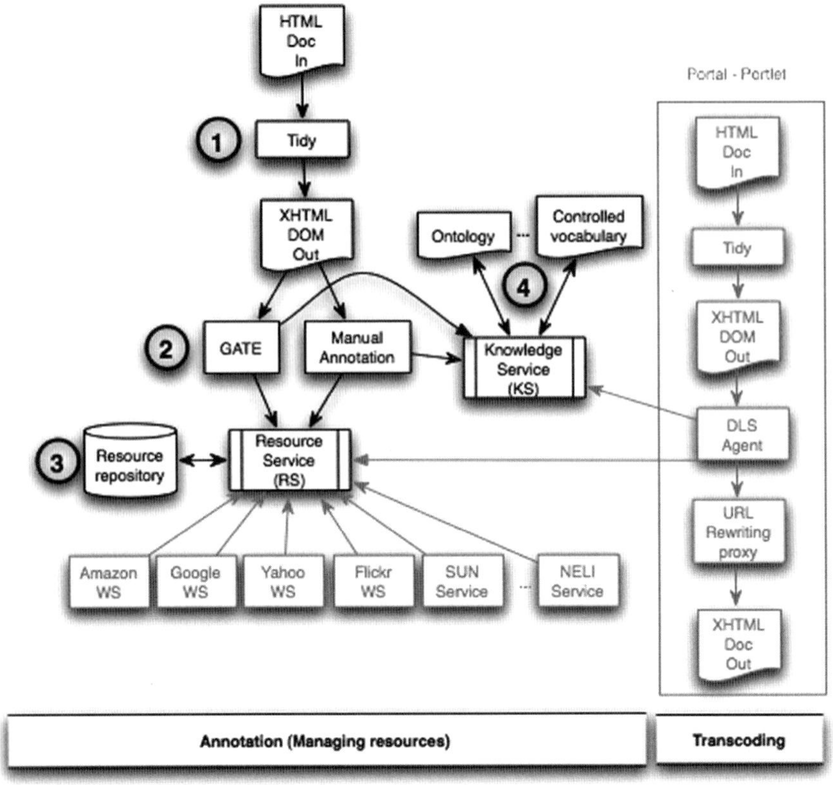

Fig. 2. The COHSE architecture – annotation producers

(2) Advance Processing. The RS identifies resources for dynamic linking in two ways: by using a resource repository which is populated with advance processing techniques and by using different online services dynamically. For advance processing, we currently use manual annotation techniques and GATE to populate our resource repository. GATE's ANNIE[10] pipeline is extended with additional components to process documents using OWL ontologies. A crawler is used along with the extended GATE components to annotate Web sites. Advance processing has a number of pros and cons. For example, it supports batch processing; a group of documents can be processed and annotated together. This gives an opportunity to create better associations between processed documents. But the robustness of the referencing mechanism becomes crucial.

(3) Resource Repository and XPointer. In the central repository, identified resources are described and stored using XPointers. XPointer is based on XPath but provides finer granularity that can be used to point to sections of

[10] ANNIE, `http://gate.ac.uk/ie/annie.html`.

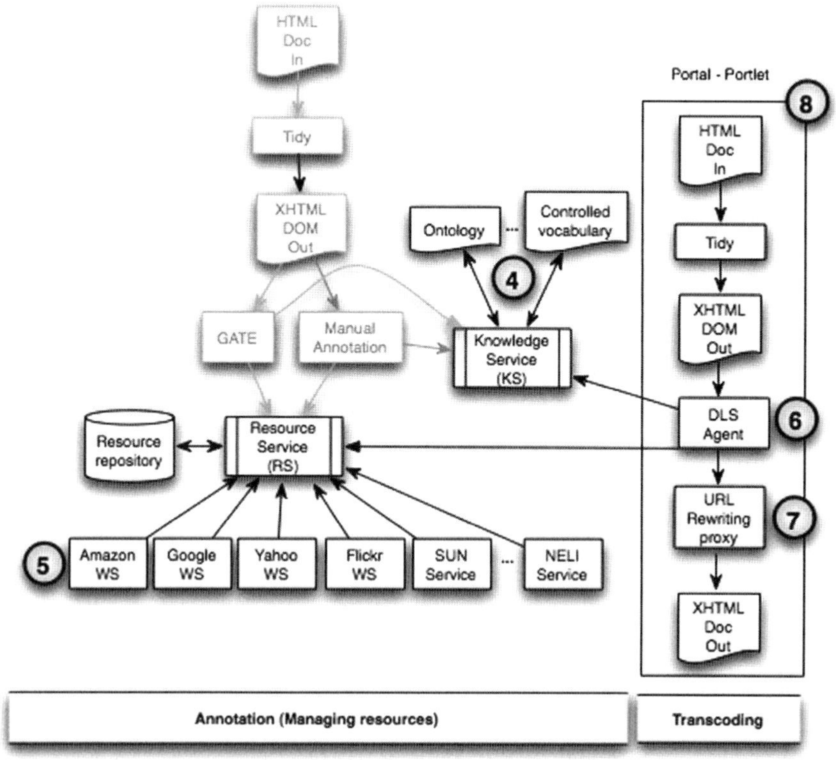

Fig. 3. The COHSE architecture – annotation consumers

an XML document such as paragraphs. Using Tidy to transform HTML pages into XHTML allows us to use XPointer as a referencing language with HTML pages. XPointer gives us a common representation for storing annotations and in principle should allow the use of multiple annotation techniques and tools.

(4) Dynamic Processing. With emerging WS standards such as SOAP, more services are becoming available. In COHSE, we currently use Google (`code.google.com/`), Amazon (`aws.amazon.com`), Sun search engine (`search.sun.com`), Yahoo (`developer.yahoo.com`), Flickr (`www.flickr.com/services/`), BBC News (`backstage.bbc.co.uk/`), MeshPubMed (`www.meshpubmed.org/`) and GoPubMed (`www.gopubmed.org/`) to identify target resources. In contrast to manual annotation and GATE, resources identified by these WSs are not stored in our central repository. Wrapper components for these services are invoked when extra COHSE link targets are requested from the RS. We have also created specialised wrappers, for example our NELI service uses Google to search the NELI Web site. New WSs can easily be included in our set of services (e.g., Ebay WS). Such a flexible architecture allows for easy extension, and ensures that our system meets one of the

important criteria for being a true *open* system [24]. *(5) Knowledge Resources and "Openness"*. The KS is a kind of bridge between annotation producers and consumers. The combination of a variety of resource manager components (i.e., WSs and advanced processing techniques) and knowledge resources provide us a rich set of domain specific resources. Again, being able to use different knowledge resources contributes to an *open* system that does not impose a particular domain model.

3.2 COHSE Portlet

Figure 3 (part 8) illustrates the architecture of the DLS portlet. The DLS portlet conforms to JSR168 so it can be easily deployed to any JSR168 compliant portal (e.g., Sun Portal Server[11]). Figure 4a shows the portlet deployed using an open source portal called Liferay.[12] This portlet stores preferences persistently and allows users to use COHSE without the need for local installation. It currently supports three major groups of customisation options: (1) users can specify the location of KS and can choose which controlled vocabulary or ontology to use; (2) users can specify the location of RS and can choose which services to use to get link targets; (3) users can customise the visualisation of COHSE links (e.g., extra links can be added as an icon, or terms can be highlighted, etc.) and can customise the visualisation of the created link boxes. Figures 5 and 4b show three different ways of presenting a link box with the same set of links. The current COHSE portlet provides a number of pre-defined link box visualisation styles, however more techniques

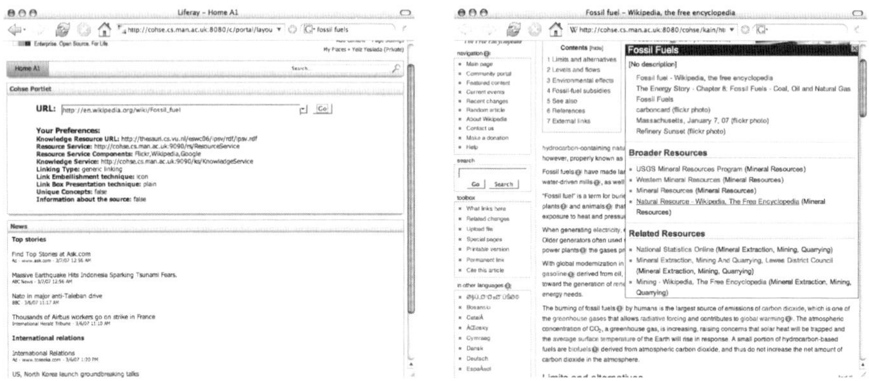

(a) (b)

Fig. 4. The COHSE DLS Portlet: (**a**) deployed in the Liferay portal along with other portlets and (**b**) in action; links are represented by small icons added to the document. When a user clicks on an icon, a link box is displayed

[11] Sun Portal Server, `http://www.sun.com/software/`.

[12] Liferay, `http://www.liferay.com/`.

(a) (b)

Fig. 5. Different visualisation styles of a link box – same set of links presented differently

can easily be introduced and as discussed in the following section depending on the user's profile, the system can make automated decisions about how to display a link box.

The portlet works as follows: the user first enters the URL of a page (Fig. 4a). The DLS agent then retrieves the page. Extra links are added, based on the user's preferences and information provided by the KS and RS. A URL rewriting proxy then rewrites all URLs to ensure that the user will continue to browse pages through the portlet, and finally the portlet returns the transcoded page in a new window (Fig. 4b). Although the COHSE DLS agent is implemented as a portlet, it does not fit in the true spirit of a portal approach; rather than returning a fragment of a document for aggregation into the portal, COHSE actually returns a complete document. We have essentially used the portal to provide the infrastructure supporting single sign-on, persistent storage of user profiles, and authentication (see Sect. 3).

Figure 6 shows an example from NELI scenario; the first screenshot (a) shows the original BBC page that talks about an outbreak of Polio. COHSE, by using relevant target services and knowledge resources, provides links to relevant resources that had not been included by the original author of the document. Such resources include links to information about the Polio virus, its effect on humans, vaccination information, and also geographical information about the local area. Note that when the COHSE DLS agent processes a document, highlights terms and returns it to the user, the content of the link boxes is not actually created. To generate link boxes, COHSE uses AJAX (Asynchronous Javascript and XML) technology; when the user clicks on a COHSE link, a request is made to our portlet and the portlet creates a link box dynamically, *on-demand*.

With this architecture, COHSE adds an extra layer to Web pages that provides a different hypertext experience to the user based on his preferences.

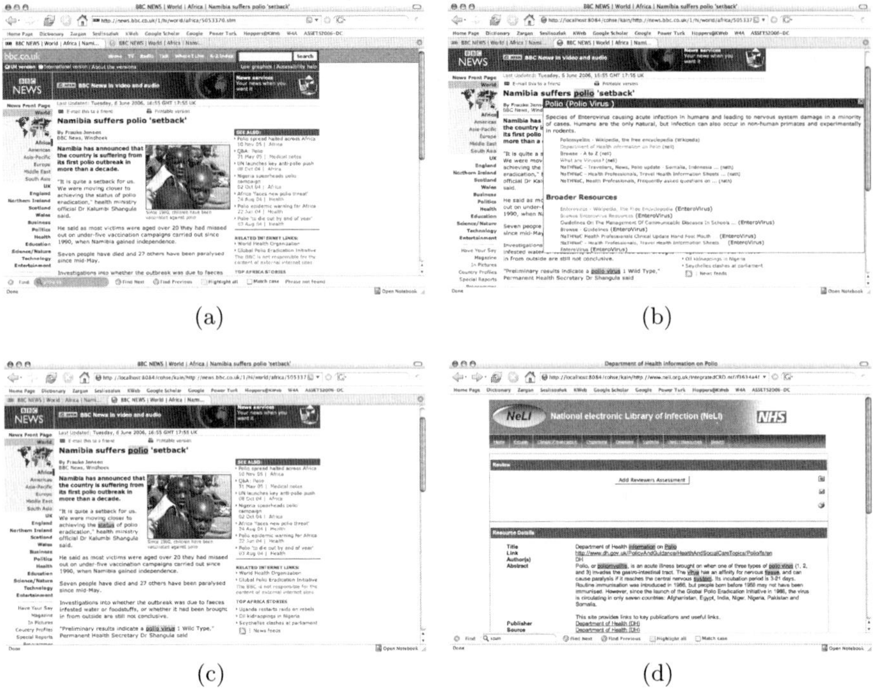

Fig. 6. Dynamic linking in action

For example, Fig. 1 uses the UPT ontology and Sun services, Fig. 6 uses a medical controlled vocabulary and health related services, and Fig. 4 uses a controlled vocabulary that describes water pollution and services such as Wikipedia, etc. These figures also illustrate different customisation options supported by this portlet; different link visualisation techniques (colour, e.g., Fig. 6 vs icon, e.g., Fig. 4) and different ways of structuring link boxes (plain, e.g., Fig. 4 vs group by concepts, e.g., Fig. 5).

Although with our current implementation we have not fully exploited the advantages of portals, this portlet provides the basis to be extended to support more sophisticated usage of user profiles and history. For instance, *personalisation* can be supported where the system makes decisions based on some observation of user behaviour [16].

4 Personalisation

Existing research suggest that *disorientation* and *cognitive overhead* are the two main problems that users experience when trying to navigate within hypertext systems. Cognitive overhead is defined as "additional effort and concentration necessary to maintain several tasks or trails at one time" [31].

This can happen when users are given a large number of choices and need to make decisions about which one to follow. Disorientation and/or cognitive overhead are sometimes associated with the number of links in a hypermedia system. For example, Tsandilas and schraefel [32] indicate that as the number of links increases, reading and learning performance decreases. These results suggest that users could benefit from adapted and personalised COHSE links where extra links added by COHSE will be designed and created to better meet the users' needs.

Adaptive hypermedia systems aim to reduce the gap between user interests and information by providing navigational support according to a user model [32]. A user model can be created in two ways: *explicitly* where the user tells the system directly about his preferences and *implicitly* where the system infers information about the user. According to Brusilovsky et al. [7], a system is *adaptable* if the user model is created *explicitly* and a system is *adaptive* if the user model is created *implicitly*. Looking at these definitions, COHSE deployed in a portal can be considered as an adaptable system since the user provides information about himself and the links are dynamically created based on this information. However, with this architecture, we can take it one step further, have an hybrid approach and create a system which is between an adaptable system that supports customisation and an adaptive system that supports personalisation.

Brusilovsky et al. [7] suggest that there are two high level processes that are required to realise an adaptive system (Fig. 7): *user-modelling* and *adaptation decision making*. These can be easily encoded as two additional components to our portlet. In fact an AJAX approach helps us to capture users' interaction easily [3]. We plan to create a user model that includes information

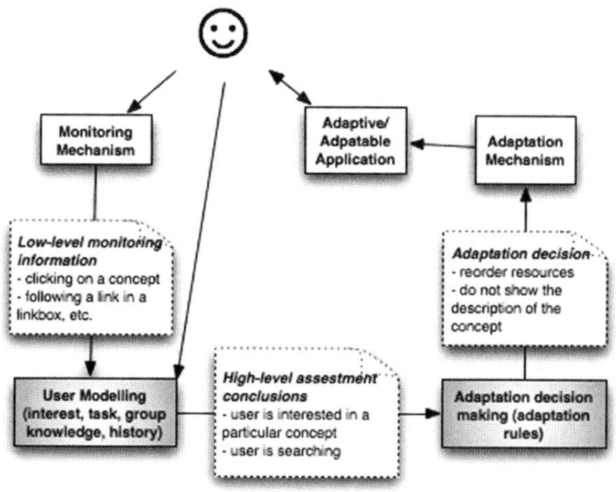

Fig. 7. Adaptation decomposed (adapted from [3])

about the user's roles/groups (e.g., doctors, patients), browsing history (e.g., concepts browsed, links navigated) and experience (e.g., professional, novice, etc.). Having such information about the user, we will be able to support the following five groups of adaptation techniques.

(1) Content. Our aim here is to improve the efficiency of the content provided in a link box. A link box is composed of the following five components: (1) concept title, (2) concept description, (3) a list of related resources, (4) a list of broader resources and (5) a list of specific resources. By default a link box is created for all users with all of these, however based on a user model different combinations can be provided (e.g., only description, only broader resources). For instance, in the NELI scenario if the user is a medical doctor then the description of the term can be unnecessary so a link box can be created without a description.

(2) Navigation. Brusilovsky et al. [7] identify five types of adaptive navigation support which are (1) direct guidance, (2) adaptive ordering, (3) hiding and (4) adaptive annotation. Different adaptation techniques can be provided for each of these. Zellweger et al. [36] indicate that annotation of links (augmented with extra information) can reduce the number of visited nodes thus further reduces the learning time. Thus, we can use different techniques to add additional information to the suggested links in a link box (e.g., provide summaries, use different colours, etc.). For instance, in the NELI scenario, for the general public a very short description of the identified terms could be provided as a tooltip.

(3) Presentation. Our aim is to improve the visualisation of the links and link boxes. According to Dalal et al. [13], an unstructured long list of links can cause cognitive overhead and disorientation. These can be addressed by better structuring and organising link boxes to assist navigation, browsing and comprehension [13]. Broadly speaking, presentational adaptation can be provided regarding the (1) structure (e.g., list, tabular, menu, etc.), (2) ordering of links in a link box (e.g., use history, alphabetical, etc.) and (3) interactivity (a variety of interaction techniques can be used). For example, if the user's current task is "browsing", then the interaction supported by link boxes (i.e., click and go model) can be quite efficient; the user will get it when he wants it and will close it when he finishes. However, if the user has a very well-defined goal, i.e., If the user is "searching", then the "click and go" model might not be that efficient and in that case link boxes can be presented as tooltips. These obviously require information about the user's task. For this we can either create stereotype tasks so that the user can choose or we can infer the user's task from his browsing history. For instance, in the Sun scenario, if a user keeps using the search box in a session, then we can infer that the current task is "searching" and can adapt the COHSE link box interactivity accordingly (for further discussion on users' tasks, see the following section).

(4) Browsing History. The user's browsing history can be recorded and used to adapt both presentation and navigation aspects. For example, conventional browsers make a distinction between visited and unvisited links.

Through use of the user's history we can keep track of the concepts that have been visited and highlight them differently. This can also be useful for orientation as users will be able to find out where they have been. For example, in the NELI scenario (see Fig. 6) if the user clicks on the COHSE link "PolioVirus" she will get the list of resources suggested for the concept "EntroVirus" (due to the *rdfs:SubClassOf* relationship between the terms). If the user subsequently visits the "EntroVirus" concept, she will also get the resources for the concept "PolioVirus" but with rendering hints (e.g., use of icons) indicating that the concept has already been seen.

(5) Collaborative Filtering. Collaborative filtering is the idea of filtering information for a user based on a collection of user profiles [34]. One common usage of this is to make recommendation which is popularised by Amazon; "users who bought X also bought Y". For COHSE, it can be a useful technique to help users decide which of the suggested concepts and links are more relevant to them to read and follow. In particular, this can be quite efficient when the knowledge model used is large. For collaborative filtering, we can make use of the user's browsing history and the groups or roles to make recommendations to users. For example, when user A in group B for concept C has visited link D, then for another user in the same group we can suggest link D in our recommended *popular* list of links. Similarly, we can create *popular* list of concepts. For example, if we observe that a number of users from the same group, with the same ontology and in the same page, traverses a particular concept, then we can infer that, that particular concept is quite *interesting* for that particular ontology and page and that can be included in a possible *hot* list of concepts. For instance, in the Sun scenario, if most of the "IT managers" click on the term "Sun Fire E4900 Server" on Fig. 1, then we can put this term into *hot list* for that particular user group. Support for this kind of reasoning, requires storage of information about the user's browsing history and groups/roles, etc. As with (4) above, information regarding history and choices is being recorded with respect to the concepts in the ontology.

4.1 User Tasks on the Web

User modelling is a crucial part of personalisation. Depending on the system, different granularity can be used [6], here we focus on users' tasks – "why is the user using the system and what does the user actually want to achieve?". We have investigated a number of studies that aim to understand what kind of tasks users perform on the Web and the behaviours or tactics used to achieve those tasks. With this investigation, our aim is to understand how COHSE can support these tasks and we aim to investigate how task-based personalisation can be supported by COHSE.

In the literature, the two fundamental paradigms that are discussed for finding information on the Web are browsing and searching [20, 25, 28]. The basic distinction between these two is the object of search (the information need) and systematicity of tactics used. With *searching* the user is looking for something specific whereas with *browsing* the user does not have a well-defined task. Looking at these simple definitions, broadly we can say that COHSE supports browsing. Studies on the browsing paradigm take this view one step further and propose three different kinds of browsing again based on the users' goal or specificity of the object searched; it can be either very well-defined, not very well defined and not defined at all [2, 12, 19, 23, 25]. Based on these, we can say that COHSE supports the latter two.

There are also relevant information retrieval paradigms that can be used to describe what kind of tasks COHSE supports. These paradigms include exploratory search [35], berrypicking [4] and information foraging theory [29]. These paradigms are relevant to each other in the sense that they all propose that the users' queries continually shift. They propose that accessing and learning new information may yield new queries. According to information foraging theory, users typically forage information on the Web by moving from page to page along hyperlinks. Therefore by adding extra hyperlinks, COHSE improves the potential exploration of the environment. It provides extra paths by using the terms in an ontology that enable users to better explore their environment. According to Marchionini [26], learning and investigation activities are parts of exploratory search. Looking at the tasks involved with these two activities, the way COHSE adds extra links and link boxes, we can say that COHSE supports these tasks. Marchionini [26] highlights that exploratory search is more concerned with *recall* (maximising the number of possibly relevant objects that are retrieved) than *precision* (minimising the number of possibly irrelevant objects that are retrieved). This is a good distinction between what COHSE does and what search engines do. We can say that COHSE aims to improve the *recall* and search engines aim to improve the *precision*. When a Web page is accessed, COHSE aims to add and return as many relevant links as possible by using the knowledge encoded in a domain ontology (e.g., using broader and narrower relationships, etc.).

If the user's task is very well defined (i.e., *searching*) then we can say that COHSE does not support such tasks very well. However, the COHSE approach could be used to better support *searching* tasks. When users are looking for something specific they can easily use search engines to search for their specific query. However, when they visit the pages that the search engines return, they move to a new context and lose their search context. Although Google's toolbar allows users to highlight searched terms in the visited page after the search, they do not get links to other pages relevant to their search query from that visited page. Through the use of COHSE, links could be dynamically added to that visited page based on the keywords searched to allow users to continue to browse within the context of their original search.

There are also a number of user studies that seek to understand what users do on the Web. The results of these studies are summarised in Table 2. The tasks that COHSE can support are also highlighted. If we look at the tasks summarised in Table 2 and the other information seeking paradigms, most of the tasks discussed are based on the specificity of the search goal.

Table 2. User Tasks – highlighted tasks are potentially supported by COHSE

Byrne et al. 1999 [8]	Morrison et al. 2001 [27]	Choo et al. 2000 [11]	Sellen et al. 2002 [30]	Heath et al. 2005 [18]	Kellar et al. 2006 [22]
Locate on page	Find	Formal search	Finding	Locating	Fact finding
Use information	Collect	Informal search	Information gathering	Exploring	Information gathering
Go to page	Monitor	Conditioned viewing	Browsing	Monitoring	Just browsing
Provide information	Explore	Undirected viewing	Transacting	Grazing	Transaction
Configure browser			Communication	Sharing	Other
React to environment			Housekeeping	Notifying	
				Asserting	
				Discussing	
				Evaluating	
				Arranging	
				Transacting	

This literature survey not only helps us understand what kind of tasks COHSE can support but it also helps us confirm that it is not enough to only study the click-stream to draw some conclusions about the user's goals and motivation [30]. Therefore, we can say that the fully automatic task based personalisation will be almost impossible without users' intervention. Another problem with capturing information about user's task is that it changes quite often: it changes from session to session and often can change several times within a single working session.

5 Related Work

In an *open* hypermedia such as Microcosm the reader is given the same access as the author. With respect to this, the Web is considered as a *closed* hypermedia and a number of systems such as DLS are introduced to address this limitation [24]. These systems allow readers to create links (or other hypermedia structures) on top of arbitrary Web pages and share these links with others. Compared to these, COHSE marries the work on open hypermedia and SW by using the existing infrastructure of the Web.

Semantic annotation is the process of semantically enriching Web pages by adding metadata so that they are not only understandable by humans but also by machines [17]. Manual annotation approaches allow users to define annotations by hand such as our COHSE annotator [9]. Automatic annotation approaches focus on extracting annotations automatically by using different techniques such natural language processing techniques (e.g., our GATE crawler). Some tools are introduced to annotate pages when they are designed and created (e.g., SMORE[13]) which is important for bringing in the annotation process to the early stage of design. There are also systems that use RDF (Resource Description Framework) such as Annotea [21] to create and store annotations. Although Annotea shows an exemplar way of acquiring annotations, we cannot use those annotations in COHSE as the annotations are not performed in a controlled way, i.e., by using an ontology.

Semantic annotations also play an important role in bioinformatics research. For example, Entrez[14] provides access to the PubMed literature resource where articles are marked up with terms from a medical controlled vocabulary called MeSH. The Gene Ontology has been used in an extension to PubMed called GOPubMed[15] to provide searches based on Gene Ontology terms. Whilst in GOPubMed an ontology is used to organise documents, there is currently no actual linking of documents.

Another Semantic Web application similar to COHSE is MAGPIE [15]. It is implemented as a client-side plug-in to a browser, so as a system it has the

[13] SMORE, `http://owl.mindswap.org`.

[14] Entrez, `http://www.ncbi.nlm.nih.gov/entrez/`.

[15] GOPubMed, `http://www.gopubmed.org/`.

limitations that we have discussed in Sect. 2, for example it requires a local installation.

Systems like Google's Autolink also provide dynamic linking functionality, but tied to a limited number of types of data. Vibrant Media's IntelliTXT[16] gives a mechanism where advertising messages are linked to keywords, but this is a server-side process, whereas our infrastructure can enhance third-party resources.

There are also some emerging technologies that are introduced to overcome the limitations of the embedded links on the Web such as XLink.[17] Although XLink is a powerful language for representing links that offers a lot of improvements, it is not widely used or supported by existing browsers.

6 Summary and Discussion

In this chapter, we have first presented an extended and modified architecture of COHSE and then presented how this architecture is deployed in a portal. This architecture makes a clear distinction between the annotation producers and consumers and demonstrates how knowledge resources such as ontologies or controlled vocabularies can play an important role in mediating between these two parties. We have also presented how COHSE introduces a new hypertext layer on top of existing Web pages by using different knowledge resources and services. The portal approach also allows easy customisation of COHSE and thus shows how customised links or essentially hypertexts can be created with third party resources.

The proposed portal approach has a number of advantages; people can use COHSE without requiring installation or local storage of their preferences. This approach opens up the usage of COHSE in large organisations such as Sun Microsystems. We have also discussed how our portal architecture can be extended to turn COHSE from an *adaptable* system where the user *explicitly* provides information about himself to an *adaptive* system where the system *implicitly* infers some information about the user. We have identified five groups of techniques that can be introduced to turn COHSE into an adaptive system: (a) content adaptation techniques, (b) navigation adaptation techniques, (c) presentation adaptation techniques, (d) browsing history based adaptation techniques and (e) collaborative filtering. To support these techniques, we have also discussed a model that includes information about user's roles/groups, browsing history and experience. Regarding the user model, we have also discussed a number of studies from literature that aim to understand user's tasks on the Web. Table 2 summarises some of these studies and highlights tasks that are potentially supported by COHSE. Finally, we have discussed how the portal architecture can be extended to support these

[16] IntelliTXT, `http://www.intellitxt.com/`.
[17] XLink, `http://www.w3.org/TR/xlink/`.

techniques. Future work will focus on encoding these techniques into our portal implementation and performing a user evaluation of both adaptable and adaptive COHSE.

Acknowledgments

Sun Microsystems Laboratories supported this research, whose support we are pleased to acknowledge. We would also like to acknowledge Robert Stevens, Simon Jupp and the Sealife team for providing the NELI use case and the related materials.

References

1. K.M. Anderson. Integrating open hypermedia systems with the world wide web. In *Proceedings of the Eighth ACM Conference on Hypertext*, pages 157–166. ACM, 1997.
2. S.M. Apted. General purposive browsing. *Library Association Record*, 73(12): 228–230, 1971.
3. R. Atterer, M. Wnuk, and A. Schmidt. Knowing the user's every move – user activity tracking for website usability evaluation and implicit interaction. In *Proceedings of the 15th International Conference on World Wide Web*, 2006.
4. M. Bates. The design of browsing and berrypicking techniques for the online search interface. *Online Review*, 13(5):407–424, 1989.
5. S. Bechhofer, R. Stevens, and P. Lord. Ontology driven dynamic linking of biology resources. *Journal of Web Semantics*, 4(3):155–163, 2006.
6. P. Brusilovsky. Adaptive navigation support: From adaptive hypermedia to the adaptive web and beyond. *Psychnology*, 2(1), 2004.
7. P. Brusilovsky, C. Karagiannidis, and D. Sampson. Layered evaluation of adaptive learning systems. *International Journal of Continuing Engineering Education and Lifelong learning*, 14(4/5), 2004.
8. M.D. Byrne, B.E. John, N.S. Wehrle, and D.C. Crow. The tangled web we wove: A taskonomy of www use. In *CHI '99: Proceedings of the SIGCHI conference on Human factors in computing systems*, pages 544–551, 1999.
9. L. Carr, S. Kampa, W. Hall, S. Bechhofer, and C. Goble. *Handbook on Ontologies*, chapter COHSE: Conceptual Open Hypermedia Service, pages 193–209. Springer, Berlin Heidelberg New York, 2004.
10. S. Chen, D. Hong, and V.Y. Shen. An experimental study on validation problems with existing html webpages. In *Proceedings of International Conference on Internet Computing (ICOMP'05)*, pages 373–379, 2005.
11. C.W. Choo, B. Detlor, and D. Turnbull. Information seeking on the web: An integrated model of browsing and searching. *First Monday (online)*, 5(2), 2000.
12. J.F. Cove and B.C. Walsh. Online text retrieval via browsing. *Information Processing & Management*, 24(1):31–37, 1998.
13. N.P. Dalal, Z. Quible, and K. Wyatt. Cognitive design of home pages: An experimental study of comprehension on the world wide web. *Information Processing and Management: An International Journal*, 36(4):607–621, 2000.

14. H. Davis. To embed or not to embed. *Communications of the ACM*, 38(8):108–109, 1995.
15. M. Dzbor, E. Motta, and J. Domingue. Opening up magpie via semantic web services. In *Proceedings of the Third International Semantic Web Conference*, volume 3298 of *Lecture Notes in Computer Science*. Springer, Berlin Heidelberg New York, 2004.
16. N. Ferguson, S. Schmoller, and N. Smith. Personalisation in presentation services. Technical report, The Joint Information Systems Committee, 2004.
17. S. Handschuh and S. Staab. *Annotation for the Semantic Web*, volume 96 of *Frontiers in Artificial Intelligence and Applications*. IOS, Amsterdam, 2003.
18. T. Heath, M. Dzbor, and E. Motta. Supporting user tasks and context: Challenges for semantic web research. In *ESWC2005 Workshop on End-user Aspects of the Semantic Web (UserSWeb)*, Heraklion, Crete, 2005.
19. S. Herner. Browsing. *Encyclopedia of Library and Information Science*, 3:408–415, 1970.
20. S. Jul and G.W. Furnas. Navigation in electronic worlds: A CHI 97 workshop. *ACM SIGCHI Bulletin*, 29(4):44–49, 1997.
21. J. Kahan and M. Koivunen. Annotea: An open RDF infrastructure for shared web annotations. In *Proceedings of the 10th International World Wide Web Conference*, pages 623–632, 2001.
22. M. Kellar, C. Watters, and M. Shepherd. The impact of task on the usage of web browser navigation mechanisms. In *Proceedings of the 2006 conference on Graphics interface*, pages 235–242, Toronto, ON, Canada, 2006. Canadian Information Processing Society.
23. M.M. Levine. An essay on browsing. *Reference Quarterly*, 93:35–36, 1969.
24. D. Lowe and W. Hall. *Hypermedia and the Web: An Engineering Approach*. Wiley, New York, 1998.
25. G. Marchionini. *Information Seeking in Electronic Environments*. Cambridge Series on Human-Computer Interaction. Cambridge University Press, Cambridge, 1995.
26. G. Marchionini. Exploratory search: From finding to understanding. *Communications of the ACM*, 49(4):41–46, 2006.
27. J.B. Morrison, P. Pirolli, and S.K. Card. A taxonomic analysis of what world wide web activities significantly impact people's decisions and actions. In *CHI '01 extended abstracts on Human factors in computing systems*, pages 163–164, 2001.
28. C. Olston and E.H. Chi. ScentTrails: Integrating browsing and searching on the web. *ACM Transactions Computer–Human Interaction*, 10(3):177–197, 2003.
29. P. Pirolli and S.K. Cart. Information foraging. *Psychological Review*, 106(4), 1999.
30. A.J. Sellen, R. Murphy, and K.L. Shaw. How knowledge workers use the web. In *CHI '02: Proceedings of the SIGCHI conference on Human factors in computing systems*, pages 227–234, 2002.
31. M. Thuering, J. Hannemann, and J. Haake. Hypermedia and cognition: Designing for comprehension. *Communications of the ACM*, 38(8):57–66, 1995.
32. T. Tsandilas and M.C. Schraefel. User-controlled link adaptation. In *Proceedings of the fourteenth ACM conference on Hypertext and hypermedia*, pages 152–160. ACM, 2003.
33. F. Vitali and M. Bieber. Hypermedia on the web: What will it take? *ACM Computing Surveys*, 31(4es):31, 1999.

34. J. Wang, A.P. de Vries, and M.J.T. Reinders. Unifying user-based and item-based collaborative filtering approaches by similarity fusion. In *Proceedings of the 29th Annual International ACM SIGIR Conference on Research and Development on Information Retrieval (SIGIR06)*, 2006.
35. R.W. White, B. Kules, S.M. Drucker, and M.C. Schraefel. Introduction – exploratory search. *Communications of the ACM*, 49(4):36–39, 2006.
36. P.T. Zellweger, S.H. Regli, J.D. Mackinlay, and B. Chang. The impact of fluid documents on reading and browsing: An observational study. In *Proceedings of the SIGCHI Conference on Human Factors in Computing Systems*, pages 249–256. ACM, 2000.

A Multi-Purpose Ontology-Based Approach for Personalised Content Filtering and Retrieval

Iván Cantador[1], Miriam Fernández[1], David Vallet[1], Pablo Castells[1], Jérôme Picault[2], and Myriam Ribière[2]

[1] Escuela Politécnica Superior, Universidad Autónoma de Madrid, 28048 Madrid, Spain, ivan.cantador@uam.es, miriam.fernandez@uam.es, david.vallet@uam.es, pablo.castells@uam.es

[2] Motorola Labs, Parc Les Algorithmes, St. Aubin, 91193 Gif-sur-Yvette, France, jerome.picault@motorola.com, myriam.ribiere@motorola.com

Summary. Personalised multimedia access aims at enhancing the retrieval process by complementing explicit user requests with implicit user preferences. We propose and discuss the benefits of the introduction of ontologies for an enhanced representation of the relevant knowledge about the user, the context, and the domain of discourse, as a means to enable improvements in the retrieval process and the performance of adaptive capabilities. We develop our proposal by describing techniques in several areas that exemplify the exploitation of the richness and power of formal and explicit semantics descriptions, and the improvements therein. In addition, we discuss how those explicit semantics can be learnt automatically from the analysis of the content consumed by a user, determining which concepts appear to be significant for the user's interest representation. The introduction of new preferences on the user profile should correspond to heuristics that provide a trade-off between consistency and persistence of the user's implicit interests.

1 Introduction

Personalised multimedia access aims at enhancing the retrieval process by complementing explicit user requests with implicit user preferences, to better meet individual user needs [9]. Automatic user modelling and personalisation has been a thriving area of research for nearly two decades, gaining significant presence in commercial applications around the mid-90s. Popular online services such as Google [1, 37] or Amazon [26, 32] are nowadays exploiting some personalisation features, in particular to improve their content retrieval systems. Even if these systems have the merit of having been deployed at a large scale, they rely on rather simple models, which may often be inaccurate or still provide results that do not completely match

I. Cantador et al.: *A Multi-Purpose Ontology-Based Approach for Personalised Content Filtering and Retrieval*, Studies in Computational Intelligence (SCI) **93**, 25–51 (2008)
www.springerlink.com

users' expectations. Indeed, personalising a content retrieval system involves considerable complexity, mainly because finding implicit evidence of user needs and interests through their behaviour is not an easy task. This difficulty is often considerably increased by an imprecise and vague representation of the semantics involved in user actions and system responses, which makes it even more difficult to properly pair user interests and content descriptions. The ambiguity of terms used in this representation, the unclear relationships between them, their heterogeneity, especially in current ever-growing large-scale networked environments such as the WWW, often constitute a major obstacle for achieving an accurate personalisation, e.g. when comparing user preferences to content items, or users among themselves.

In this chapter we argue for the introduction of ontologies [17] as an enhanced representation of the relevant knowledge about the domain of discourse, about users, about contextual conditions, involved in the retrieval process, as a means to enable significant improvements in the performance of adaptive content retrieval services. We illustrate our point by describing the development of advanced features and enhancements in specific areas related to personalisation where the ontology-based approach shows its benefit, including:

- Basic personalised content search and browsing based on user preferences;
- Learning semantic user preferences over time;
- Dynamic contextualisation of user preferences;
- Dynamic augmented social networking and collaborative filtering.

Domain ontologies and rich knowledge bases play a key role in the models and techniques that we propose in the above areas, as will be described in the sequel. The approaches presented in this chapter share and exploit a common representation framework, thus obtaining multiple benefits from a shared single ontology-rooted grounding. Furthermore, it will be shown that modular semantic processing strategies, such as inference, graph processing, or clustering, over networked ontology concepts, may be reused and combined to serve multiple purposes.

The rest of the chapter is organized as follows. Section 2 introduces the basic approach for the ontology-oriented representation of semantic user preferences, and its application to personalised content search and retrieval. Following this, Sect. 3 explains how these semantic user preferences may evolve automatically over time. Then Sect. 4 describes an approach for the dynamic contextualisation of semantic user preferences, and Sect. 5 shows the extension of the techniques described in previous sections to multi-user environments, based on collaborative personalisation strategies. Finally, some conclusions are given in Sect. 6.

2 Ontology-Based Personalisation for Content Retrieval

A very significant body of research in the last two decades has paid attention to the problem of personalising information access and delivery, commonly addressed under such names as information filtering, collaborative filtering, or personalised information retrieval, with variations in approach and perspective [16,18,20]. Formulated and approached from different angles, the problem has been a major research topic in the information retrieval, user modelling, and machine learning fields. The research activity in this area has been paralleled by a comparable interest towards making such techniques commercially profitable [1,26,32,37].

Most of the proposed techniques in this area keep and process long records of accessed documents by each user, in order to infer potential preferences for new documents (e.g. by finding similarities between documents, or between users). The data handled by these techniques have been rather low-level and simple: document IDs, text keywords and topic categories at most [21,28]. In contrast, the recent proposals and achievements towards the enrichment of text and multimedia content by formal, ontology-based, semantic descriptions open new opportunities for improvement in the personalisation field from a new, richer representational level [4,9]. We see indeed the introduction of ontology-based technology in the area of personalisation as a promising research direction [14]. Ontologies enable the formalisation of user preferences in a common underlying, interoperable representation, whereby user interests can be matched to content meaning at a higher level, suitable for conceptual reasoning.

An ontology-based representation is richer, more precise, and less ambiguous than a keyword-based model. It provides an adequate grounding for the representation of coarse user interests (e.g. for sports, cinema, finance, baseball, sci-fi movies, oil industry) to fine-grained preferences (e.g. for individual items such as a sports team, an actor, a stock value) in a hierarchical way, and can be a key enabler to deal with the subtleties of user preferences. An ontology provides further formal, computer-processable meaning on the concepts (e.g. who is coaching a team, an actor's filmography, financial data on a stock), and makes it available for the personalisation system to take advantage of. Moreover, an ontology-rooted vocabulary can be agreed and shared (or mapped) between different systems, or different modules of the same system, and therefore user preferences, represented this way, can be more easily shared by different players. For instance, a personalisation framework may share a domain ontology with a knowledge-based content analysis tool that extracts semantic metadata from audio/visual content, conforming to the ontology [4]. On this basis, it is easier to build algorithms that match preference to content, through the common domain ontology.

In an ontology-based approach, semantic user preferences may be represented as a vector of weights (numbers from -1 to 1), representing the intensity of the user interest for each concept, where negative values indicative

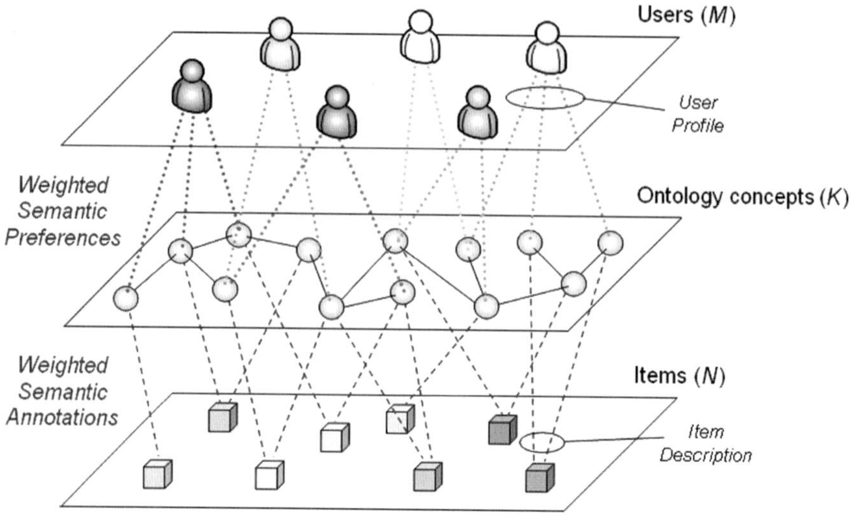

Fig. 1. Association of users and content

of a dislike for that concept [9]. Similarly, content can be described by a set of weighted concepts (values from 0 to 1, indicating the intensity of relation between the content and the concept) in such a way that users can be accurately related through the ontology layer to the content units in the search space (see Fig. 1).

If a content analysis tool identifies, for instance, a cat in a picture, and the user is known to like cats, the personalisation module can make predictions on the potential user interest for the picture by comparing the metadata of the picture, and the preferred concepts in the user profile. Furthermore, ontology standards backed by international consortiums (such as the W3C), and the corresponding available processing tools, support inference mechanisms that can be used to further enhance personalisation, through the middle ontology layer, so that, for instance, a user interested in animals (superclass of cat) is also recommended pictures of cats. Inversely, a user interested in lizards, snakes, and chameleons can be inferred to be interested in reptiles with a certain confidence. Also, a user keen of Sicily can be supposed to like Palermo, through the transitive *locatedIn* relation, assuming that this relation has been seen as relevant for inferring previous underlying user's interests. In fact, it is even possible to express complex preferences based on generic conditions, such as "athletes that have won a gold medal in the Olympic Games".

Based on preference weights, measures of user interest for content units can be computed, with which it is possible to discriminate, prioritize, filter and rank contents (a collection, a catalogue section, a search result) in a personal way.

The basis for the personalisation of content retrieval is the definition of a matching algorithm that provides a *personal relevance measure* (PRM) of a content object d for a particular user u, according to his/her semantic preferences. The measure is computed as a function of the semantic preferences of u and the semantic metadata of d. In this calculation, user preferences and content metadata are seen as two vectors in a K-dimensional vector space, where K is the number of elements in the universe $\mathcal{O}$ of ontology terms, and the coordinates of the vectors are the weights assigned to ontology terms in user preferences and content annotations. The weights represent the intensity of preference by u for each concept, and the degree of importance of each concept in the meaning conveyed by d, respectively. The PRM is thus represented as the algebraic similarity between the user preferences and the content vector. Using the classic Information Retrieval vector-space model [2], this similarity can be measured by the cosine function. Note that in our approach the preference vector plays an equivalent role to the query vector in classic IR. Of course, the accuracy of the PRM comparison is directly dependent on the accuracy of content annotations and the represented user preferences.

Figure 2 represents the similarity between two different items d_1 and d_2, and the semantic preferences of the user u.

The PRM algorithm thus matches two concept-weighted vectors and produces a value between -1 and 1. Values near -1 indicate that the preferences of the user do not match the content metadata (i.e. the two vectors are dissimilar); values near 1 indicate that the user interests do match the content. In cases where annotations are attached to the items in a Boolean way (because the underlying model or system does not provide for a finer weighting), a weight of 1 is assigned by default to all annotations.

Figure 3 shows an example where $\mathcal{O} = \{$*beach*, *construction*, *family*, *vegetation*, *motor*$\}$ is the set of all domain ontology terms (classes and instances). According to her profile, the user is interested in the concepts "*beach*", "*motor*", and "*vegetation*", with different intensity, and has a negative preference for "*construction*". The preference vector for this user is thus $\overrightarrow{u} = \{1.0, -0.7, 0.0, 0.2, 0.5\}$.

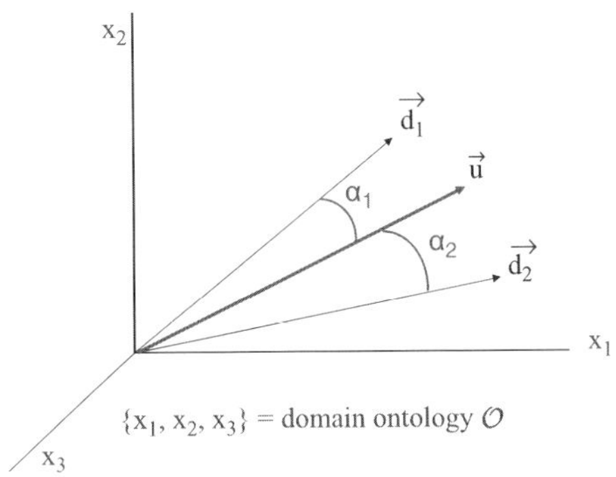

Fig. 2. Visual representation of metadata and preference's vector similarity

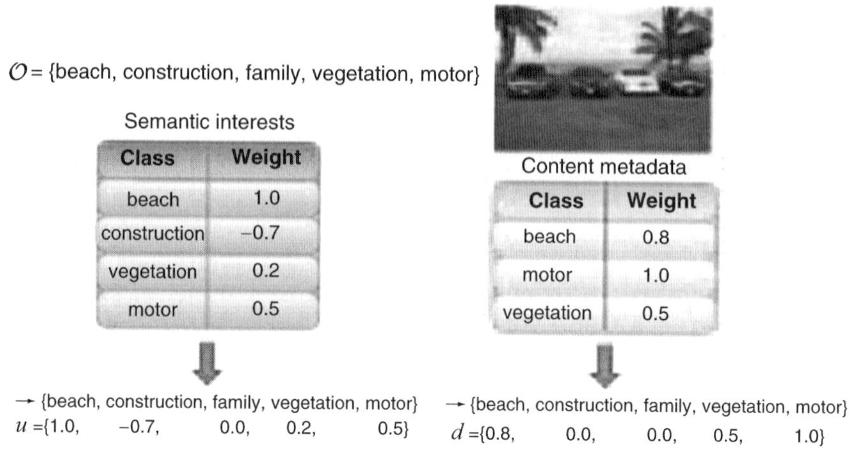

Fig. 3. Example of semantic preferences and metadata matching

A still image is annotated with the concepts "*beach*", "*motor*" and "*vegetation*", with the metadata vector $\overrightarrow{d} = \{0.8, 0.0, 0.0, 0.5, 1.0\}$.

The PRM of the still image for this user is thus $PRM = \cos(\overrightarrow{u}, \overrightarrow{d}) \approx 0.82$[1].

This measure can be combined with the relevance measures computed by user-neutral algorithms (e.g. a search result score), or other user-based algorithms like the ones introduced in the following sections, resulting in a personalised bias on the ranking of search results and/or content recommendations.

3 Adaptation of Semantic User Preferences

In the approach described in the previous section, personalised content retrieval relies on a description of the user preferences as a vector of concept weights. In the simplest approach, these weights can be set by the users themselves. However, this solution presents a number of drawbacks: in such a case, user preferences are often static, therefore do not mimic correctly the natural evolution of users' tastes over time. In addition, setting manually preferences may take time, and requires some knowledge on the structure of the personalisation system, such as the exact meaning of concepts used to describe preferences, or the rating scale for weighting preferences, in order to minimize the introduction of self-induced bias [5]. An alternative to manually setting

---

[1] $\cos(\overrightarrow{x}, \overrightarrow{y}) = \frac{\overrightarrow{x} \cdot \overrightarrow{y}}{|x| \times |y|} = \frac{\sum_{i=1}^{n} (x_i \times y_i)}{\sqrt{\sum_{i=1}^{n} (x_i^2)} \times \sqrt{\sum_{i=1}^{n} (y_i^2)}}$

preferences consist in analysing content retrieved and consumed by the user over a significant period of time to infer implicit associated preferences.

3.1 Automatic Creation and Update of Semantic Preferences

Thus, several methods exposed in various recent research [14, 15, 22] are exploiting collected data such as document representation, user feedback and some other metrics such as the time the user spent to read the document, the length of the document, the mean number of documents read by a user, etc. to create and update user preferences. Although most of these techniques rely on a keyword- or taxonomy-based description of user preferences, they can be applied and expanded for a richer ontology-based representation of user profiles.

Our approach consists in analysing the content consumed by a user to determine from the content annotations which concepts appear and how often they occur during a given time period. These content concepts (i.e. metadata) are compared to the user profile concepts (i.e. preferences). The process is slightly different depending if the content concepts appear or not in the user profile.

The introduction of new preferences on the user profile should correspond to heuristics that propose a trade-off between *consistency* (e.g. if the content consumed by the user deals with diverse, semantically-unrelated concepts or not) and *persistence* (i.e. how stable and recurrent content concepts are) of content metadata in the user's history. We can envision the following situations:

- A concept occurs once and its occurrence is confirmed with time with roughly the same level, this concept can be introduced after a period as a long-term preference of the user;
- A concept occurs once and its occurrence is very high on a short period, and then disappears very quickly. Even if this concept can be considered as a preference during a period of time, it must be removed very fast from the preferences, once the interest of the user is over;
- A concept occurs once but the occurrence is not very high and even if confirmed in time, it does not constitute a significant interest for the user. In that case this concept will never become a user preference;
- A concept occurs and becomes a preference as in the first case, but disappears with time. It must at a certain time be removed from the preferences.

To handle those use-cases, we propose to use a *concept history stack* as a mechanism to store all concepts representing potential user interests (because they occurred in the consumed contents).

The insertion of new preferences is an *off-line process* that uses the history stack to decide whether or not to add new preferences in the user profile. This process can be run periodically, for example at the end of the day, or at any other frequency (which could be determined based on appropriate user studies).

The decision is based on the comparison between a value of *concept occurrence* $C_{occ} = N_{occ}/(D-d)^2$, computed during the off-line process, and a threshold P_{thd} for which a concept candidate of the stack history can become a new preference. Candidate concepts for which $C_{occ} > P_{thd}$, are introduced as new preferences. Their weight into the user profile is initialised to a default, neutral value (for example 0.5).

The removal of concepts from the profile can rely on two mechanisms. First, define a stack size limit: when a new concept is introduced as a possible preference candidate and the limit of the stack is reached, the concept with the lower C_{occ} is removed. Second, define a threshold R_{thd} so that when the value C_{occ} for a concept is under this threshold, the concept must be removed[3].

The computation of C_{occ} is naturally reflecting a *decay factor* – or *gradual forgetting* of preferences – since the occurrence of a concept is divided by the age of the concept in the history stack. This decay factor is sufficient to handle gradual interest changes [24]. A main issue concerns user *shift of interest* [34], since an important interest of one day can potentially create a new preference in the profile that will take a week to disappear from the profile (based on the natural decay factor). Some solutions have been proposed to adapt the size of the time window which is considered for adaptation [25]. We propose to have an additional shorter time window to handle shift of interest, where the occurrence of the concept is followed day by day: if a user is showing a particular interest for a concept during a one day period, this concept will be quickly taken into account in preferences, but if this interest is not confirm the following day, we would like this preference to disappear quickly in the user profile. When this phenomenon is encountered in the process, an additional decay factor is added to the concept.

Those decay factors, are not impacting directly the weight of the user preferences, but are impacting the decision for keeping or not a preference in the profile, so that the weight of a concept still reflect the real user consumption of contents.

However, the adaptation of semantic user preferences does not only consists in adding or removing preferences, but also in updating the concept weights in the user profile, based on the analysis of consumed content. A possible mathematical model to apply for the update of concept weights, influenced by the proposal made by [30] is the following:

[2] Where N_{occ} is the number of time the concept occurs in the set of metadata of a consumed content (considered as relevant for the user); D is the date of the day when the process is launch; and d is the date of first appearance of the concept into the set of metadata of a consumed content.

[3] Values of P_{thd} and R_{thd} have to be determined through experiments with real data.

$$w_{new} = w_{old} + fd * ContentRating * e^{-\beta^* x^* y} * \log \frac{time}{\log length}.$$

The w_{old} factor represents the current weight of the concept. *fd* is the relevance feedback factor given through an analysis of the content consumption; the relevance feedback can take either Boolean value or be multi-valued [6]. *ContentRating* is the rank assigned to the content by the personalised retrieval system; it can use a cosine similarity measure between the content and the user profile, or any other measure implemented to rank content. The $\log \frac{time}{\log length}$ expression incorporates the time spent reading or watching a content item and the length of the content, which operates as the normalizing factor. The $e^{-\beta^* x^* y}$ factor is used to attend the personalised non-linear change of the concept's weight according to user usage history data. x represents the mean number of content that the user is consuming per day; y represents the number of consumed content where the concept appears in the set of metadata. The more content a user consumes per day, for example, the more slowly the weights increase. The β factor is a constant, which takes different values in the two opposite scenarios of consumed/non-consumed content. More precisely, in the case of non-consumed content, the changing rate (i.e. the decreasing rate) should be slower, since a non-consumed content does not constitute an explicit indication for non-interest. On the contrary, in case of consumed content the changing rate (i.e. the increasing rate) should be faster, since a read news item demonstrates a better indication for interest.

3.2 Exploitation of Semantic Links Between Concepts for Preferences Learning

We believe that the method described above can be significantly improved by taking into account the benefits of a domain ontology. We propose to exploit semantic links between concepts to provide two independent mechanisms:

- Interest assumption completion, which adds more potential user interests in the concept history stack, by using hierarchical and semantic relationships between concepts;
- Preference update expansion, which expands the re-weighting of a concept to the subset of correlated concepts.

Interest Assumption Completion

We propose to use semantic relationships between concepts as a mean to enhance the management of the concept history stack. The concept history stack expresses the set of assumptions on user interests. In their simplest form, those assumptions are only relying on the exact occurrence of the keywords or the concepts appearing in the content annotations. We propose to complement those conjectures by deducing additional interests from the semantic relationships expressed in the domain ontologies. Two types of semantic relationships can be considered: (1) inheritance relationships and (2) semantic well-defined relationships (also called concept properties).

In the case of inheritance, the following example can be exploited: a user is interested in "cats", we should deduce just by expansion of the concept "cat", that the user is interested also to some extend in "animals". By doing such assumption, the user will receive information that could not be directly related to cats. This assumption can be true, but it has to be confirmed by the occurrence of other subtypes of "animals". Thus, if a new concept c_{new} appears in a user consumed content, its super-type $c_{supertype}$ is also introduced as a potential interest in the concept history stack, with a pseudo-occurrence[4] value proportional to the occurrence of c: $N_{occ}(c_{supertype}) = \gamma_1 \times N_{occ}(c_{subtype})$, where $\gamma_1 < 1$, has to be determined empirically. This mechanism allows expanding the user preferences based on semantic knowledge of the concept, but avoids making any assumption *directly from the ontology* on user interests. Indeed, by doing this expansion of concepts into the concept history stack, the super-type can only be added into the user preferences when it has been confirmed enough by other sub-concepts so that its pseudo-occurrence reaches the insertion threshold P_{thd}. Following our example, if concepts "dog" and "horse" appear later in other consumed content items, they will confirm the assumption that the user is not only interested in cats, since the occurrence value of the super-type is increased each time one of its subtype appears in the concept history stack.

As ontologies are richer than simple taxonomies, the exploitation of other kinds of semantic relationships than inheritance can bring a significant value. Thus, if a concept c appears in a user consumed content, then we propose to introduce in the concept history stack all related concepts $c_{related}$ such as $\exists p_{j=1..m} \in \mathcal{P}, \exists c_{related} \in \mathcal{O}/p_j(c, c_{related})$, where $\mathcal{O}$ is the set of ontology concepts and $\mathcal{P}$ is the set of ontology properties. In that way, all concepts related directly (through semantic relationships) to a user interest (a concept) can be considered as candidates for becoming user preferences. This is done by setting a value for the pseudo-occurrence of related concepts: $N_{occ}(c_{related}) = \gamma_2 \times N_{occ}(c)$, where $\gamma_2 < 1$ is determined empirically.

By using knowledge represented in the ontology, we can enhance the preliminary acquisition of preferences, by selecting potential interests that are related to concepts appearing in the consumed content. By introducing such additional assumptions in the history stack, with factors to calculate the pseudo-occurrence of deduced concepts, we are just helping the system to acquire more quickly preferences, but we are not making direct assumptions on user preferences. This mechanism avoids risks of errors by always relying on the concept occurrence to confirm a possible interest.

---

[4] Whereas the term "occurrence" refers to concepts that explicitly appear in the content, the term "pseudo-occurrence" applies to concepts that do not explicitly appear in the content, i.e. to concepts that "could have appeared" because of their semantic proximity with concepts present in the document.

Preference Update Expansion

In Sect. 3.1 concepts are learnt in isolation; the updated weight of an existing preference is only based on its own previous value, regardless of other influential related concepts. But, if concepts like "Mars", "mission", "NASA", "spacecraft" appear, they should be counted as more significant than if seen in an isolated way, because there is a semantic recurrence, which may be detected by finding semantic paths between the concepts. The exploitation of the ontology knowledge may enable to take also into consideration in the preference weight update of a given concept the weight of semantically related concepts. For example, we want to update the weight of the concept c in the user preferences, and we know that this concept is semantically linked to at least another concept: $\exists c_{related} \in \mathcal{O}, \exists p_{j=1..n} \in \mathcal{P}/p(c, c_{related})$, then the new weight of c is updated as described in Sect. 3.1, and the new weight for each concept $c_{related}$ related to c can be computed thanks to the following formula: $w_{new}(c_{related}) = w_{old}(c_{related}) + sf_{c,c_{related}} \times w_{new}(c)$, where $w_{new}(c_{related})$ is the new weight of the concept, seen as a related concept to the concept c, $w_{old}(c_{related})$ is the old weight value of the concept, $sf_{c,c_{related}}$ is a semantic factor that depends of the type of semantic link existing between $c_{related}$ and c, and $w_{new}(c)$ is the new weight value for the current concept. It describes the influence (semantic effect) that concepts c has on concept $c_{related}$.

A special care has to be given to the definition of the semantic factor *sf*, which may decrease with the level of semantic proximity between $c_{related}$ and c:

- *Level 1.* $c_{related}$ is part of the definition of c. The relationships to be considered are of any type (transitive, inverse, etc.). One example is the direct concepts that constitute the definition of a given other concepts. For example, direct concepts that relates to "car" are "wheels", "road", etc.
- *Level 2.* $c_{related}$ is related to c by a combination of the same transitive relationship, and $c_{related} \cap c \neq Class$, meaning that they have a super-type in common.
- *Level n.* $c_{related}$ is related to c through the combination of the n same transitive relationship, and $c_{related} \cap c \neq Class$, meaning that they have a super-type in common.

If it is rather intuitive that this semantic factor depends of the semantic proximity of the two concepts, another issue concerns the relative values between two semantic factors sf_{c,c_1} and sf_{c,c_2}, i.e. the problem of assigning different weights to the semantic relations between the concepts of an ontology. Although this approach is still in an early stage, it is thought to be promising and could rely on the attempts that have been made to define a notion of similarity or distance between two concepts, such as in [11, 33], which propose an approach to similarity between concepts dealing with not only atomic concepts with Inclusion relation but also complex concepts with all kinds of semantic relations.

4 Contextual Personalisation

The shallowest consideration is sufficient to notice that human preferences are complex, variable and heterogeneous, and that not all preferences are relevant in every situation [35]. For instance, if a user is consistently looking for some contents in the Formula 1 domain, it would not make much sense that the system prioritizes some Formula 1 picture with a helicopter in the background just because the user happens to have a general interest for aircrafts. In other words, in the context of Formula 1, aircrafts are out of (or at least far from) context. Context is a difficult notion to grasp and capture in a software system, and the elements than can, and have been considered in the literature under the notion of context are manifold: user tasks and goals, computing platform, network conditions, social environment, physical environment, location, time, noise, external events, text around a word, visual context of a graphic region, to mention a few.

Complementarily to the ones mentioned, we propose a particular notion, for its tractability and usefulness in semantic content retrieval: that of semantic runtime context, which we define as the background themes under which user activities occur within a given unit of time. Using this notion, a finer, qualitative, context-sensitive activation of user preferences can be defined. Instead of a uniform level of personalisation, user interests related to the context are prioritized, discarding the preferences that are out of focus. The problems to be addressed include how to represent such context and determine it at runtime, and how the activation of user preferences should be related to it, predicting the drift of user interests over time.

4.1 Context Representation

Our approach is based on a concept-oriented context representation, and the definition of distance measures between context and preferences as the basis for the dynamic selection of relevant preferences [35].

A runtime context is represented (is approximated) in our approach as a set of weighted concepts from the domain ontology. This set is obtained by collecting the concepts that have been involved, directly or indirectly, in the interaction of the user (e.g. issued queries and accessed items) with the system during a retrieval session. The context is built in such a way that the importance of concepts fades away with time (number of user requests back when the concept was referenced) by a decay factor. This simulates a drift of concepts over time, and a general approach towards achieving this follows. Therefore, at each point t in time, context can be represented as a vector $C(t) \in [0,1]^{|\mathcal{O}|}$ of concept weights, where each $x \in \mathcal{O}$ is assigned a weight $C_x(t) \in [0,1]$. This context value may be interpreted as the probability that x is relevant for the current semantic context. Additionally, time is measured by the number of user requests within a session. In our approach, $C(t)$ is built as a cumulative combination of the concepts involved in successive user requests, in such a way that the importance of concepts fades away with time. Right after each user's request, a request vector $Req(t) \in X_{\mathcal{O}}$ is defined. In the next step, an initial context vector $C(t)$ is defined by combining the newly constructed request vector $Req(t)$ from the previous step with the context $C(t-1)$, where the context weights computed in the previous step are automatically reduced by the mentioned decay factor ξ, a real value in [0,1]. The decay factor will define how many *action units* will be considered for context-building, and how fast a concept will be "forgotten" by the system.

Once a context is built, the contextual activation of preferences is achieved by a computation of the semantic similarity between each user preference and the set of concepts in the context. In spirit, the approach consists of finding semantic paths linking preferences to context, where the considered paths are made of existing semantic relations between concepts in the domain ontology. The shorter, stronger, and more numerous such connecting paths, the more in context a preference is considered.

The proposed techniques to find these paths use a form of constrained spreading activation (CSA) strategy [13], similar to that explained in Sect. 3.2 for preference learning. In fact, in our approach a semantic expansion of both user preferences and the context takes place, during which the involved concepts are assigned preference weights and contextual weights, which decay as the expansion grows farther from the initial sets. This process can also be understood as finding a sort of fuzzy semantic intersection between user preferences and the semantic runtime context, where the final computed weight of each concepts represents the degree to which it belongs to each set.

Finally, the perceived effect of contextualisation is that user interests that are out of focus, under a given context, are disregarded, and only those that are in the semantic scope of the ongoing user activity (the "intersection" of user preferences and runtime context) are considered for personalisation. The inclusion or exclusion of preferences is in fact not binary, but ranges on a continuum scale, where the contextual weight of a preference decreases monotonically with the semantic distance between the preference and the context.

Contextualised preferences can be understood as an improved, more precise, dynamic, and reliable representation of user preferences, and as such they can be used directly for the personalised ranking of content items and search results, as described in Sect. 4.1, or they can be input to any system that exploits this information in other ways, such as the one described in Sect. 4.2.

4.2 Evaluation of Personalisation in Context

The contextualisation techniques described in this section have been implemented in an experimental prototype, and tested on a medium-scale corpus. Evaluating personalisation is known to be a difficult and expensive task [31, 36]. In order to measure how much better a retrieval system can perform with the proposed techniques than without them, it is necessary to compare the performance of retrieval (a) without personalisation, (b) with simple personalisation, and (c) with contextual personalisation. The standard evaluation measures from the IR field require the availability of manual content ratings with respect to (i) query relevance, (ii) query relevance and general user preference (i.e. regardless of the task at hand), and (iii) query relevance and specific user preference (i.e. constrained to the context of his/her task).

For this purpose, we have conducted an experiment with real human subjects, focusing on the testing of the retrieval performance. The search space is based on a textual corpus, consisting of 145,316 documents (445 MB) from the CNN web site (http://dmoz.org/News/Online Archives/CNN.com), plus the KIM domain ontology and KB [23], publicly available as part of the KIM Platform, developed by Ontotext Lab, with minor extensions. The Ontology Knowledge Base contains a total of 281 RDF classes, 138 properties, 35,689 in-stances, and 465,848 sentences. The CNN documents are annotated with KB concepts, amounting to over three million annotations in total. The user-neutral retrieval system used for this experiment is a semantic search engine developed by the authors [10]. Human subjects are given three different retrieval tasks, each expressing a specific information need, so that users are given the goal of finding as many documents as possible which fulfil the given needs. The sequence of actions is not fixed but is defined with full freedom by users as they seek to achieve the proposed tasks. A total of 18 subjects were selected for the experiment, all of them being PhD students from the authors' institutions. Three tasks were set up for the experiment, which can be briefly summarized as

1. News about agreements between companies.
2. Presentations of new electronic products.
3. Information about cities hosting a motor sports event.

Each task was tested (a) with contextual personalisation, (b) with simple personalisation, and (c) without personalisation. In order for users not to repeat the same task twice or more, each of the three modes was used with six users (3 modes $\times$ 6 users = 18 tests for each task), in such a way that each user tried each of the three modes a, b, and c, exactly once. This way, each mode is tried exactly 18 times: once for each user, and six times for each task, in such a way that neither mode is harmed or favoured by different task difficulty or user skills. User preferences are obtained manually from the user by asking her to explicitly rate a predefined list of domain concepts at the beginning of the session. The relevant documents for each task are marked beforehand by an expert (a role that we played ourselves), so that users are relieved from providing extensive relevance judgements. However users are encouraged to open the documents that seem more relevant according to their subjective interests, in order to provide the system with more contextual tips. Context information is gathered based on concepts annotating such selected results, and the concepts that are related to the keywords in user queries (using the keyword-concept mapping provided in the KIM KB).

At the end of every task the system asks the user to mark the documents in the final result set as related or unrelated to her particular interests and the search task. For the computation of precision and recall after the experiment logs were collected, the following two simplifications are made for each interactive sequence (i.e. for each task and user):

- The search space is simplified to be the set of all documents that have been returned by the system at some point in the iterative retrieval process for the task conducted by this user.
- The set of relevant documents is taken to be the intersection of the documents in the search space marked as relevant for the task by the expert judgement, and the ones marked by the user according to her particular interests.

Figure 4 shows the results obtained with this setup and methodology. The curve on the left of this figure shows a clear improvement at high precision levels by the contextualisation technique both with respect to simple personalisation and no personalisation. The improvement decreases at higher recall levels. The cut-off precision curve clearly shows a significant performance improvement by the contextual personalisation, especially in the top 10 results. Personalisation alone achieves considerably lower precision on the top documents, showing that the contextualisation techique avoids further false positives which may still occur when user preferences are considered out of context. The mean average precision values for contextual, simple, and no personalisation in this experiment were 0.135, 0.106, and 0.046, respectively, which reflects that our technique globally performs clearly above the two baselines.

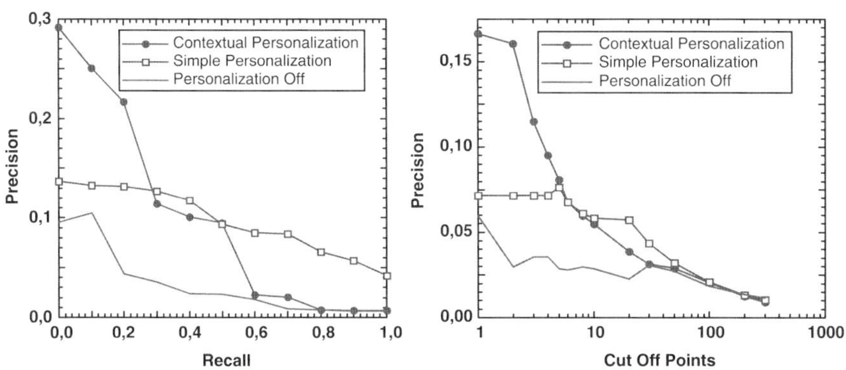

Fig. 4. Comparative performance of personalised search with and without contextualisation tested with 18 subjects on three proposed tasks. The graphics show (**a**) the precision vs. recall curve, and (**b**) the precision at cut-off points. The results are averaged over the set of all users and tasks

Most cases where our technique performed worse were due to a lack of information in the KB, as a result of which the system did not find that certain user preferences were indeed related to the context. Another limitation of our approach is that it assumes that consecutive user queries tend to be related, which does not hold when sudden changes of user focus occur. However, not only the general improvements pay off on average, but the potential performance decay in such cases disappears after two or three queries, since the weight of contextual concepts decreases exponentially as the user keeps interacting with the system.

5 Augmented Social Networking and Collaborative Filtering

When the system perspective is widened to take in contextual aspects of the user, it is often relevant to consider that in most cases the user does not work in isolation. Indeed, the proliferation of virtual communities, computer-supported social networks, and collective interaction (e.g. several users in front of a set-top box), call for further research on group modelling, opening new problems and complexities. A variety of group-based personalisation functionalities can be enabled by combining, comparing, or merging preferences from different users, where the expressive power and inference capabilities supported by ontology-based technologies can act as a fundamental piece towards higher levels of abstraction [7, 8].

5.1 Semantic Group Profiling

Group profiling can be understood under the explicit presence of a priori given user groups, or as an activity that involves the automatic detection of implicit links between users by the system, in order to put users in contact with each other, or to help them benefit from each other's experience. In the first view, collaborative applications may be required to adapt to groups of people who interact with the system. These groups may be quite heterogeneous, e.g. age, gender, intelligence and personality influence on the perception and demands on system outputs that each member of the groups may have. The question that arises is how the system can adapt itself to the group in such a way that each individual benefits from the results.

In [7] we have explored the combination of the ontology-based profiles defined in Sect. 2 to meet this purpose, on a per concept basis, following different strategies from social choice theory [27] for combining multiple individual preferences. In our approach, user profiles are merged to form a shared group profile, so that common content recommendations are generated according to this new profile (see Fig. 5).

With the combination of several profiles using the considered group modelling strategies we seek to establish how humans create an optimal ranked

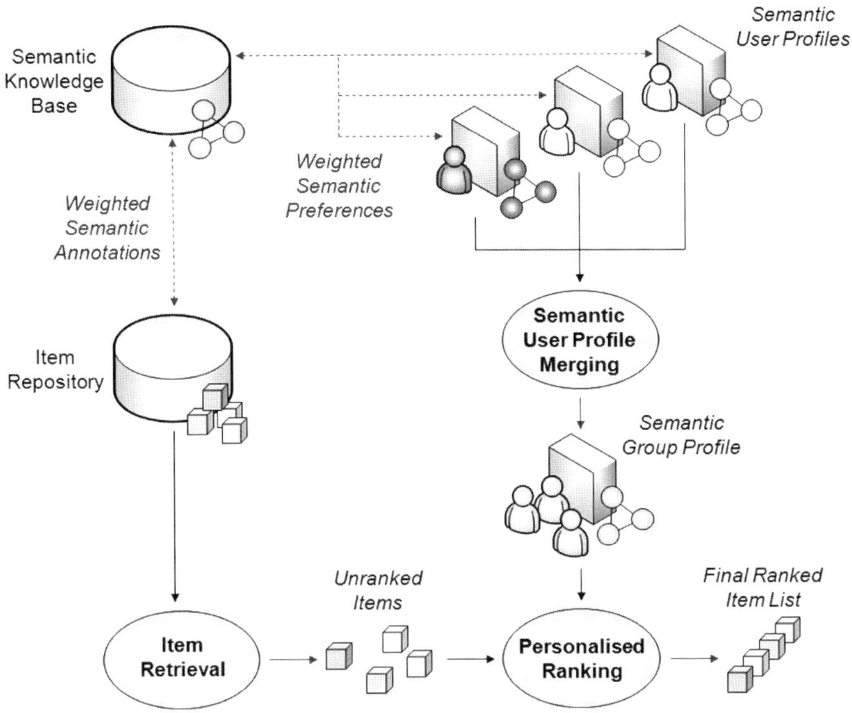

Fig. 5. Group profiling by aggregation of individual user profiles

item list for a group, and how they measure the satisfaction of a given list. Our preliminary experiments have shown that improved results can be obtained from the accuracy and expressivity of the ontology-based representation as proposed in this approach [7], and have exhibited which user profile combination strategies could be appropriate for a collaborative environment.

Specifically, we define a distance that measures the existing difference between two given ranked item lists, and we use this distance to determine which group modelling strategies give rankings closest to those empirically obtained from several subjects.

Consider $\mathcal{D}$ as the set of items stored and retrieved by the system. Let $\tau_{sub} \in [0,1]^N$ be the ranked item list for a given subject and let $\tau_{str} \in [0,1]^N$ be the ranked item list for a specific combination strategy, where N is the number of items stored by the system. Using the notation $\tau(d)$ to refer the position of the item $d \in \mathcal{D}$ in the ranked list τ, we define the distance R between the previous ranked lists as follows:

$$R\left(\tau_{sub}, \tau_{str}\right) = \sum_{n=1}^{N} P\left(n\right) \frac{1}{n} \sum_{d \in \mathcal{D}} \left|\tau_{sub}\left(d\right) - \tau_{str}\left(d\right)\right| \cdot \chi_n\left(d, \tau_{sub}, \tau_{str}\right),$$

where $P(n)$ is the probability that the user stops browsing the ranked item list at position n, and

$$\chi_n(d, \tau_{sub}, \tau_{str}) = \begin{cases} 1 & \text{if } \tau_{str}(d) \le n \text{ and } \tau_{sub}(d) > n \\ 0 & \text{otherwise} \end{cases} .$$

The distance R basically sums the differences between the positions of each item in the subject and strategy ranked lists. Thus, the smaller the distance is, the more similar the lists are. Additionally, and motivated by the fact that in typical information retrieval systems a user usually takes into account only the first n top ranked items, the formula considers more those items that appear before the nth position of the strategy ranking and after the nth position of the subject ranking. The idea here is to penalize those of the top n items in the strategy ranked list that are not relevant for the user.

The scenario of the experiments was the following. A set of 24 pictures was considered. For each picture several semantic-annotations were taken, describing their topics (at least one of *beach*, *construction*, *family*, *vegetation*, and *motor*) and the degrees (real numbers in [0,1]) of appearance these topics have on the picture. Twenty subjects participated in the experiments. They were Ph.D. students from the authors' institutions, and they were asked in all experiments to think about a group of three users with different tastes. In decreasing order of preference (i.e. progressively smaller weights): (a) User_1 liked *beach*, *vegetation*, *motor*, *construction* and *family*, (b) User_2 liked *construction*, *family*, *motor*, *vegetation* and *beach*, and (c) User_3 liked *motor*, *construction*, *vegetation*, *family* and *beach*.

Observing the 24 pictures, and taking into account the preferences of the three users belonging to the group, the subjects were asked to make an ordered list of the pictures. With the obtained lists we measured the distance R with respect to the ranked lists given by the group modelling strategies. Although an approximation to the distribution function for $P(n)$ can be taken, e.g. by interpolation of data from a statistical study, we simplify the model fixing $P(10) = 1$ and $P(n) = 0$ for $n \ne 10$, assuming that users are only interested in those items shown in the screen at first time after a query. Additionally, we also compared the strategies lists with those obtained using semantic user profiles in our personalised retrieval model explained in Sect. 2.

The average results are shown in Fig. 6. Surprisingly, both comparisons resulted quite similar. They agree with the strategies that seem to be more or less adequate for group modelling. From the figure, it can be seen that strategies like "Borda Count" and "Copeland Rule" give lists more similar to those manually created by the subjects, and strategies such as "Average Without Misery" and "Plurality Voting" obtained the greatest distances. For more details, see [7].

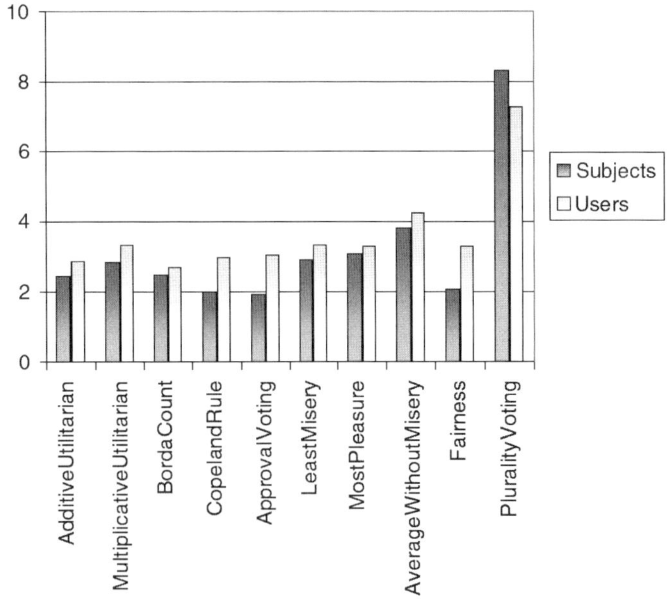

Fig. 6. Average distance R between the ranked lists obtained with the combination strategies, and the lists created by the subjects and the lists retrieved using the individual semantic user profiles

5.2 Semantic Social Networking

Even when explicit groups are not defined, users may take advantage of the experience of other users with common interests, without having to know each other. The issue of finding hidden links between users based on the similarity of their preferences or historic behaviour is not a new idea. In fact, this is the essence of the well-known collaborative recommender systems [3], where items are recommended to a certain user concerning those of his interests shared with other users or according to opinions, comparatives, and ratings of items given by similar users.

However, in typical approaches, the comparison between users and items is done globally, in such a way that partial, but strong and useful similarities may be missed. For instance, two people may have a highly coincident taste in cinema, but a very divergent one in sports. The opinions of these people on movies could be highly valuable for each other, but risk to be ignored by many collaborative recommender systems, because the global similarity between the users might be low.

In recommendation environments there is an underlying need to distinguish different layers within the interests and preferences of the users. Depending on the current context, only a specific subset of the segments (layers) of a user profile should be considered in order to establish his similarities with other people when a recommendation has to be performed. Models of social networks partitioned into different common semantic layers can achieve more accurate and context-sensitive results.

The definition and generation of such models can be facilitated by a more accurate semantic description of user preferences, as supported by ontologies. A multilayered approach to social networking can be developed by dividing user profiles into clusters of cohesive interests, so that several layers of social networks are found. This provides a richer model of interpersonal links, which better represents the way people find common interests in real life.

Taking advantage of the relations between concepts, and the (weighted) preferences of users for the concepts, we have defined a strategy that clusters the semantic space. The obtained clusters are based on the correlation of concepts appearing in the preferences of individual users, representing thus those sets of preferences shared by specific cliques of people [8]. Considering the concept clusters, the user profiles are partitioned by projecting the clusters into the set of preferences of each user (see Fig. 7). Thus, users can be compared on the basis of the resulting subsets of interests, in such a way that several, rather than just one, (weighted) links can be found between two users.

Multilayered social networks are potentially useful for many purposes. For instance, users may share preferences, items, knowledge, and benefit from each other's experience in focused or specialized conceptual areas, even if they have very different profiles as a whole. Such semantic subareas need not be defined manually, as they emerge automatically with our proposed method. Users may be recommended items or direct contacts with other users for different aspects of day-to-day life.

In addition to these possibilities, our two-way space clustering, which finds clusters of users based on the clusters of concepts built in a first pass, offers a reinforced partition of the user space that can be exploited to generate group profiles for sets of related users as explained in Sect. 5.1. These group profiles enable efficient strategies for collaborative recommendation in real-time, by using the merged profiles as representatives of classes of users.

On the other hand, the degree of membership of the obtained sub-profiles to the clusters, and the similarities among them, can be used to define social

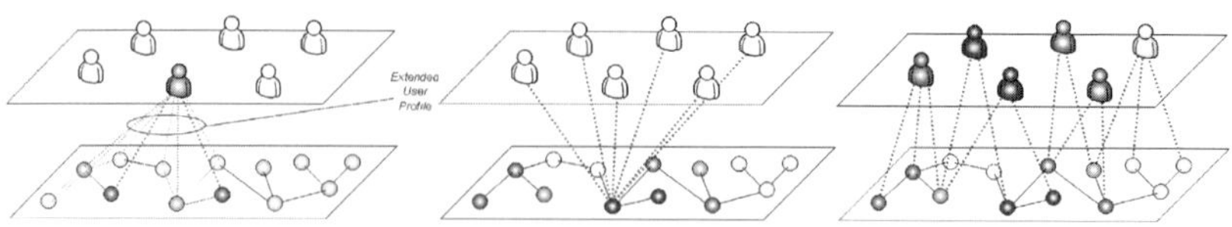

(a) Semantic preference spreading (b) Semantic concept clustering (c) Semantic user clustering

Fig. 7. Multilayer generation of social links between users: (**a**) the initial sets of individual interests are expanded, (**b**) domain concepts are clustered based on the vector space of user preferences, and (**c**) users are clustered in order to identify the closest class to each user links to be exploited by collaborative filtering systems. We report early experiments with real subjects in [8], where the emergent augmented social networks are applied to a variety of collaborative filtering models, showing the feasibility of the clustering strategy.

Specifically, for a user profile $\mathbf{u}_m = (u_{m,1}, u_{m,2}, \ldots, u_{m,K})$, an item vector $\mathbf{d}_n = (d_{n,1}, d_{n,2}, \ldots, d_{n,K})$, and a cluster C_q, we denote by $\mathbf{u}_m^q$ and $\mathbf{d}_n^q$ the projections of the vectors onto cluster C_q, i.e. the kth component of $\mathbf{u}_m^q$ and $\mathbf{d}_n^q$ is $u_{m,k}$ and $d_{n,k}$ respectively if $c_k \in C_q$, and 0 otherwise.

The profile of the user u_m is used to return a unique list. The score of an item d_n is computed as a weighted sum of the indirect preferences based on similarities with other users in each cluster. The sum is weighted by the similarities with the clusters:

$$pref(d_n, u_m) = \sum_q nsim\left(d_n, C_q\right) \sum_i nsim_q\left(u_m, u_i\right) \cdot sim_q(d_n, u_i),$$

where:

$$sim\left(d_n, C_q\right) = \frac{\sum\limits_{c_k \in C_q} d_{n,k}}{\left\|\mathbf{d}_n\right\| \sqrt{|C_q|}}, \quad nsim\left(d_n, C_q\right) = \frac{sim\left(d_n, C_q\right)}{\sum\limits_i sim\left(d_n, C_i\right)}$$

are the single and normalized similarities between the item d_n and the cluster C_q,

$$sim_q\left(u_m, u_i\right) = \cos\left(\mathbf{u}_m^q, \mathbf{u}_i^q\right) = \frac{\mathbf{u}_m^q \cdot \mathbf{u}_i^q}{\left\|\mathbf{u}_m^q\right\| \cdot \left\|\mathbf{u}_i^q\right\|}, \quad nsim_q\left(u_m, u_i\right) = \frac{sim_q\left(u_m, u_i\right)}{\sum\limits_j sim_q\left(u_m, u_j\right)}$$

are the single and normalized similarities at layer q between users u_m and u_i, and:

$$sim_q\left(d_n, u_i\right) = \cos\left(\mathbf{d}_n^q, \mathbf{u}_i^q\right) = \frac{\mathbf{d}_n^q \cdot \mathbf{u}_i^q}{\left\|\mathbf{d}_n^q\right\| \cdot \left\|\mathbf{u}_i^q\right\|}$$

is the similarity at layer q between item d_n and user u_i.

This model, which we have called UP (user profile-based), can be slightly simplified if we only consider the similarities within the cluster C_q for which the user has the highest membership. With this simplification, we present the model UP-q (user profile-based) as follows:

$$pref_q(d_n, u_m) = \sum_i nsim_q(u_m, u_i) \cdot sim_q(d_n, u_i).$$

If the current semantic cluster is well identified for a specific item, we expect to achieve better precision/recall results than those obtained with the overall model.

The experiments with real subjects we conducted in order to evaluate the effectiveness of our proposed social networking and recommender models were very similar to those explained in Sect. 5.1. Again, the set of 24 pictures was defined as the retrieval space. Each picture was annotated with weighted semantic metadata describing what the image depicts, using an ontology including six topics: *animals*, *beach*, *construction*, *family*, *motor* and *vegetation*. The 20 Ph.D. students that participated in the previous experiments were asked to define their weighted preferences for a list of concepts related to the above topics, and present also in the annotations of the pictures. No restriction was imposed on the number of preferred topics and concepts to be selected by the subjects. The generated user profiles showed very different features, observable not only in their joint interests, but also in their complexity. Once the user profiles were created, we run our method. After the execution of the preference spreading procedure, the domain concept space was clustered according to interest similarity.

We evaluated the recommendation models UP and UP-q computing their average precision/recall curves for the users of each of the existing clusters. In this case we calculate the curves at different number of clusters ($Q = 4$, 5, 6). Figure 8 exposes the results.

The UP-q version, which returns ranked lists according to specific clusters, outperforms the UP version, which generates a unique list assembling the contributions of the users in all the clusters. Additionally, for both models, we have plotted with dotted lines the curves obtained without spreading preferences. It can be observed that our clustering strategy performs better when it is combined with the constrained spreading activation algorithm, thus showing preliminary evidence of the importance of extending the profiles before the clustering processes, as it is discussed in Sect. 5.3.

5.3 Semantic Profile Expansion for Collaborative Group Profiling

In real scenarios, user profiles tend to be very scattered, especially in those applications where user profiles have to be manually defined. Users are usually not willing to spend time describing their detailed preferences to the system, even less to assign weights to them, especially if they do not have a clear understanding of the effects and results of this input. On the other hand, applications where an automatic preference learning algorithm is applied tend to recognize the main characteristics of user preferences, thus yielding profiles that may entail a lack of expressivity. To overcome this problem, the semantic preference spreading mechanism described in Sect. 3 has proved highly useful for improving our group profiling techniques as well.

Previous experiments without the semantic spreading feature showed considerably poorer results. The profiles were very simple and the matching between the preferences of different users was low. Typically, the basic user profiles provide a good representative sample of user preferences, but do not reflect the real extent of user interests, which results in low overlaps between the preferences of different users. Therefore, the extension is not only

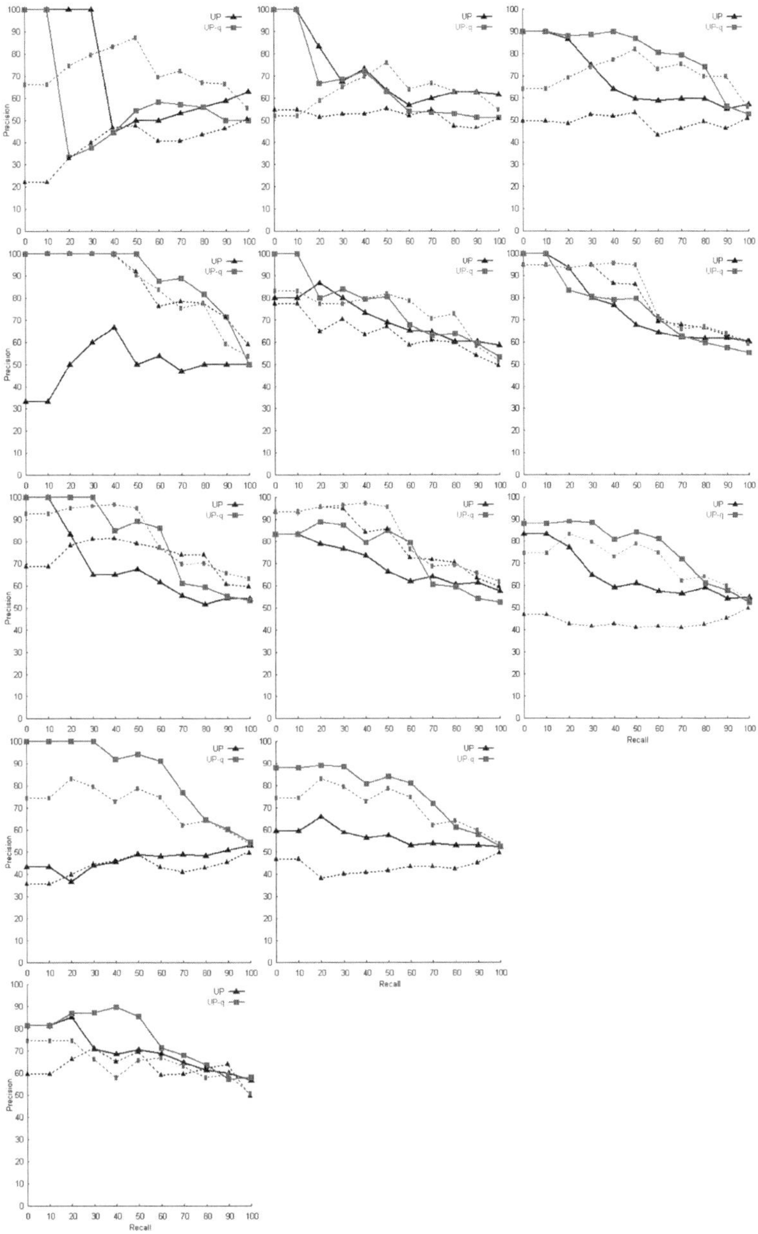

Fig. 8. Average precision vs. recall curves for users assigned to the user clusters obtained with the UP (*black lines*) and UP-q (*gray lines*) models at levels of $Q = 6$ (*graphics on the left*), $Q = 5$ (*graphics in the middle*), and $Q = 4$ (*graphics on the right*) concept clusters. For both models, the *dotted lines* represent the results achieved without semantic preference spreading

important for the performance of individual personalisation, but is essential for the clustering strategy described in Sect. 5.2.

In very open collaborative environments, it is also the case that not only direct evidence of user interests needs to be properly completed in their semantic context, but that they are not directly comparable with the input from other users in its initial form. If the environment is very heterogeneous, the potential disparity of vocabularies and syntax used by different users or subsystems pose an additional barrier for collaborative techniques. One of the major purposes for which ontologies are conceived is that of reflecting or achieving a consensus between different parties in a common knowledge space [17]. Therefore, they provide special-purpose facilities to ensure the required interoperability between semantic user spaces, and match descriptions that are syntactically different but semantically related.

6 Conclusions

Ontology-based technology has been identified as a key tool for moving beyond current state of the art in knowledge-intensive fields, where a precise, formal representation of the involved semantics can make a difference. In this chapter we address the potential of the ontology-based approach in the area of personalisation for content access and retrieval. We have shown ways in which ontology-driven representations can be used to improve the effectiveness of different personalisation techniques, focusing on specific functionalities where the introduction of ontologies enables qualitative enhancements. In the proposed approaches, ontologies are used to model the domain of discourse in terms of which user interests, content meaning, retrieval context, and social relationships, can be described, analysed, and interrelated.

The presented research is being currently continued in the MESH[5] project, in which the proposed ontology-driven mechanisms are combined to provide users with personalised multimedia news retrieval. Automatic personalisation has a particularly high potential in the news domain, because personal preferences play an important role in the way users approach, move and browse across the space of available information. Indeed, there are many situations where users seek information for the sake of it, without the necessary intervention of a specific, immediate goal or task beyond their own personal interest or whim. As such, this project provides a good recipient for further experimenting with the benefits of ontology-driven personalisation, and concurring with the development of a general personalisation infrastructure, supporting user profile construction and evolution, and contextual, semantic preference-based content retrieval functionalities.

The advantages of ontology-driven representations (expressiveness and precision, formal properties, inference capabilities, interoperability) enable

[5] Multimedia sEmantic Syndication for enHanced news services, http://www.mesh-ip.eu

further developments that exploit such capabilities, beyond the ones proposed here, on top of the basic personalisation framework described in this chapter. A trade-off of our proposals is the cost and difficulty of building well-defined ontologies and populating large-scale knowledge bases, which is not addressed here. Recent research on these areas is yielding promising results [23], in a way that any advancement on these problems can be played to the benefit of our proposed achievements.

Acknowledgements

The research leading to this document has received funding from the European Community's Sixth Framework (FP6-027685-MESH), and the Spanish Ministry of Science and Education (TIN2005-06885). However, it reflects only the authors' views, and the European Community is not liable for any use that may be made of the information contained therein.

References

1. Badros, G.J. and Lawrence, S.R. (2005). Methods and systems for personalised network searching. *US Patent Application 20050131866.*
2. Baeza-Yates, R. and Ribeiro-Neto, B. (1999). Modern Information Retrieval. Addison-Wesley, Reading, MA.
3. Balabanovic, M. and Shoham, Y. (1997). Content-based collaborative recommendation. *Communications of the ACM* 40(3), pp. 66–72.
4. Bloehdorn, S., Petridis, K., Saathoff, C., Simou, N., Tzouvaras, V., Avrithis, Y., Handschuh, S., Kompatsiaris, Y., Staab, S., and Strintzis, M. G. (2005). Semantic annotation of images and videos for multimedia. *Proceedings of the Second European Semantic Web Conference* (ESWC 2005), Heraklion, Greece. Springer Verlag Lecture Notes in Computer Science, vol. 3532, pp. 592–607.
5. Block, J. and Funder, D.C. (1986). Social roles and social perception: Individual differences in attribution and "error". *Journal of Personality and Social Psychology* 51, pp. 1200–1207.
6. Bookstein, A. (1983). Outline of a general probabilistic retrieval model. *Journal of Documentation* 39(2), pp. 63–72.
7. Cantador, I., Castells, P., and Vallet, D. (2006). Enriching group profiles with ontologies for knowledge-driven collaborative content retrieval. *Proceedings of the First International Workshop on Semantic Technologies in Collaborative Applications* (STICA 2006), *at the 15th IEEE International Workshops on Enabling Technologies* (WETICE 2006), Manchester, UK, pp. 358–363.
8. Cantador, I. and Castells, P. (2006). Multilayered semantic social network modelling by ontology-based user profiles clustering: application to collaborative filtering. *Proceedings of the 15th International Conference on Knowledge Engineering and Knowledge Management* (EKAW 2006), Podebrady, Czech Republic. Springer Verlag Lectures Notes in Artificial Intelligence, vol. 4248, pp. 334–349.

9. Castells, P., Fernández, M., Vallet, D., Mylonas, P., and Avrithis, Y. (2005). Self-tuning personalised information retrieval in an ontology-based framework. *Proceedings of the First International Workshop on Web Semantics* (SWWS 2005), Agia Napa, Cyprus. Springer Verlag Lecture Notes in Computer Science, vol. 3762, pp. 977–986.
10. Castells, P., Fernández, M., and Vallet, D. (2007). An adaptation of the vector-space model for ontology-based information retrieval. *IEEE Transactions on Knowledge and Data Engineering* 19(2), pp. 261–272.
11. Cross, V. (2004). Fuzzy semantic distance measures between ontological concepts. *IEEE Annual Meeting of the Fuzzy Information*, pp. 635–640.
12. Darragh, J.J., Witten, I.H., and James, M.L. (1990). The reactive keyboard: a predictive typing aid. *IEEE Computer* 23(11), pp. 41–49.
13. Crestani, F. (1997). Application of spreading activation techniques in information retrieval. *Artificial Intelligence Review* 11, pp. 453–482.
14. Gauch, S., Chaffee, J., and Pretschner, A. (2003). Ontology-based personalised search and browsing. *Web Intelligence and Agent Systems Journal* 1(3–4), pp. 219–234.
15. Gauch, S. and Trajkova, J. (2004). Improving ontology-based user profiles, *Proceedings of the Recherche d'Informations Assistee par Ordinateur (RIAO 2004)*, Avignon, France, April 2004, pp. 380–389.
16. Good, N., Schafer, J., Konstan, J., Borchers, A., Sarwar, B., Herlocker, J., and Riedl, J. (1999). Combining collaborative filtering with personal agents for better recommendations. *Proceedings of the 16th National Conference on Artificial Intelligence (AAAI 1999)*, Orlando, FL, pp. 439–446.
17. Gruber, T.R. (1993). A translation approach to portable ontology specification. *Knowledge Acquisition* 5, pp. 199–220.
18. Hanani, U., Shapira, B., and Shoval, P. (2001). Information filtering: overview of issues, research and systems. *User Modeling and User-Adapted Interaction* 11(3), pp. 203–259.
19. Hirsh, H., Basu, C., and Davison, B.D. (2000). Enabling technologies: learning to personalize. *Communications of the ACM* 43(8), pp. 102–106.
20. Hull, D.A. (1998). The TREC-7 filtering track: description and analysis. *Proceedings of the Seventh International Text REtrieval Conference (TREC-7)*, Gaithersburg, Maryland, pp. 33–56.
21. Jeh, G. and Widom, J. (2003). Scaling personalised web search. *Proceedings of the 12th International World Wide Web Conference* (WWW 2003), Budapest, Hungary, pp. 271–279.
22. Kim, H. R. and Chan, P.K. (2003). Learning implicit user interest hierarchy for context in personalisation. *Proceedings of the Eigth International Conference on Intelligent User Interfaces (IUI 2003)*, Miami, FL, pp. 101–108.
23. Kiryakov, A., Popov, B., Terziev, I., Manov, D., and Ognyanoff, D. (2004). Semantic annotation, indexing, and retrieval. *Journal of Web Semantics* 2(1), pp. 47–49.
24. Koychev, I. and Schwab, I. (2000). Adaptation to Drifting user's interests. *Proceedings of the Workshop on Machine Learning in New Information Age (MLNet 2000) at the 11th European Conference on Machine Learning (ECML 2000)*, Barcelona, Spain, pp. 39–45.
25. Koychev, I. and Lothian, R. (2005). Tracking drifting concepts by time window optimisation. *Proceedings of AI-2005, the 25th SGAI International Conference*

on Innovative Techniques and Applications of Artificial Intelligence. Bramer, Max; Coenen, Frans; Allen, Tony (Eds.), Springer, London, pp. 46–59.
26. Linden, G., Jacobi, J., and Benson, E. (2001). Collaborative recommendations using item-to-item similarity mappings, *US Patent 6266649B1*.
27. Masthoff, J. (2004). Group modelling: selecting a sequence of television items to suit a group of viewers. *User Modelling and User-Adapted Interaction* 14(1), pp. 37–85.
28. Micarelli, A. and Sciarrone, F. (2004). Anatomy and empirical evaluation of an adaptive web-based information filtering system. *User Modelling and User-Adapted Interaction* 14(2–3), pp. 159–200.
29. Padmanabhan, V.N. and Mogul, J.C. (1996). Using predictive prefetching to improve World Wide Web latency. *Computer Community Review* 26(3), pp. 22–36.
30. Papadogiorgaki, M., Papastathis, V., Nidelkou, E., Waddington, S., Bratu, B., Ribière, M., and Kompatsiaris, Y. (2007). Distributed user profile management and adaptation for personalised news content delivery, submitted to the special issue "Data Mining for Personalisation" of *User Modelling and User Adapted Interaction* (UMUAI) Journal.
31. Rajagopalan, B. and Deshmukh, A. (2005). Evaluation of online personalisation systems: a survey of evaluation schemes and a knowledge-based approach. *Journal of Electronic Commerce Research* 6(2), pp. 112–122.
32. Smith B.R., Linden, G.D., and Zada, N.K. (2005). Content personalisation based on actions performed during a current browsing session, *US Patent 6853983B2*.
33. Song, L., Ma, J., Li, L., and Chen, Z. (2006). Fuzzy similarity from conceptual relations. *IEEE Asia-Pacific Conference on Services Computing* (APSCC 2006), pp. 3–10.
34. Tsymbal, A. (2004). The problem of concept drift: definitions and related work, *Technical Report TCD-CS-2004-15*, Department of Computer Science, Trinity College Dublin, Ireland.
35. Vallet, D., Castells, P., Fernández, M., Mylonas, P., and Avrithis, Y. (2007). Personalised content retrieval in context using ontological knowledge. *IEEE Transactions on Circuits and Systems for Video Technology* 17(3), *Special Issue on the Convergence of Knowledge Engineering Semantics and Signal Processing in Audiovisual Information Retrieval*, pp. 336–346.
36. White, R.W., Jose, J.M., van Rijsbergen, C.J., and Ruthven, I. (2004). A simulated study of implicit feedback models. *Proceedings of the 26th European Conference on Information Retrieval* (ECIR 2004), Sunderland, UK, pp. 311–326.
37. Zamir, O.E., Korn, J.L., Fikes, A.B., and Lawrence, S.R. (2005). Personalisation of placed content ordering in search results. *US Patent Application 20050240580*.

Analysing the Behaviour of Students in Learning Management Systems with Respect to Learning Styles

Sabine Graf[1] and Kinshuk[2]

[1] Vienna University of Technology, Women's Postgraduate College for Internet Technologies, Vienna, Austria, sabine.graf@ieee.org
[2] Athabasca University, School of Computing and Information Systems, Athabasca, Canada, kinshuk@ieee.org

Summary. Learning management systems (LMS) are successfully used in e-education but they provide the same courses for all learners rather than considering the learners' individual needs. In recent years, more and more research is done on incorporating individual characteristics such as learning styles in technology enhanced learning. According to educational theories, learners with a strong preference for a specific learning style might have difficulties in learning if their learning style is not considered by the teaching environment. On the other hand, providing courses that fit to the individual learning styles makes learning easier for students. As a requirement for taking learning styles into consideration in LMS, the behaviour of students in online courses needs to be investigated. In this chapter, we analyse the behaviour of 43 students during an online course within an LMS with respect to their learning styles. The results show that learners with different preferences for learning styles act also differently in the course. From these results, information about the preferred way of learning and their favoured features in the LMS can be gained. On one hand, this information can be used to incorporate different features in a course in order to support different learning styles. On the other hand, the information can act as basis for providing adaptive courses. Moreover, we analysed the behaviour of students and their learning styles with respect to correlations. As a result, we found several significant correlations which can be used to investigate and develop an automatic approach for detecting learning styles based on the behaviour of learners in LMS.

1 Introduction

Learners have different ways of learning. Some learners prefer to work in groups, whereas others like to learn alone. Some prefer to learn concrete material such as data and facts; others prefer abstract content like theories and their underlying meaning. Besides, the students' perception of a course or a learning environment might be different. For some students, a course seems

S. Graf and Kinshuk: *Analysing the Behaviour of Students in Learning Management Systems*, Studies in Computational Intelligence (SCI) **93**, 53–73 (2008)
www.springerlink.com

to be easy, whereas others have problems with learning in the same course. According to Jonassen and Grabowski [1], the reason can be seen in the individual differences of learners which also include their individual learning styles. Felder, for example, pointed out that learners with a strong preference for a specific learning style might have difficulties in learning if their learning style is not supported by the teaching environment [2,3]. On the other hand, incorporating learning styles makes learning easier and leads to better achievement. Bajraktarevic, Hall, and Fullick [4], for example, confirmed this by a study showing that students attending an online course that matches their preferred learning style (either sequential or global) achieved significantly better results than those who got delivered a course that did not match their learning style.

In web-based learning systems, more and more attention is paid on incorporating learning styles and providing courses that fit to the students' individual learning style. Some examples of such adaptive systems are CS383 [5], IDEAL [6], INSPIRE [7], and TANGOW [8].

While supporting adaptivity is a big advantage of these systems, they also have severe limitations. For example, adaptive systems lack integration, supporting only few functions of web-enhanced education, and the content of courses is not available for reuse [9]. On the other hand, learning management systems (LMS) such as Moodle [10], Blackboard [11], and WebCT [12] provide a lot of simple features to administer and create courses. As such, they have become very successful in e-education, but they provide very little or, in most cases, no adaptivity [13].

As a requirement for incorporating learning styles in LMS, analysis needs to be carried out about the behaviour of learners with respect to their learning styles. In this chapter, we investigate the behaviour of learners in an online course within Moodle [10]. Our investigations are based on the learning style model by Felder and Silverman [2], which is described in more detail in Sect. 2. Based on this model, we identified several patterns of behaviour (Sect. 3), which on the one hand seem to be relevant with respect to the learning style model and on the other hand are commonly used features in LMS. This concept makes our results applicable for other LMS as well.

The performed study (Sect. 4) aims at two issues: Firstly, we investigate whether learners with different learning style preferences act differently in the online course. The results (Sect. 5) show the different preferences and needs of students with different learning styles. Since LMS currently provide the same course for each student, these results can act as the catalyst to make teachers and course developers aware of the needs of their students in order to incorporate these needs into the course development process by providing features for each learning style. Furthermore, the results can be used as recommendation for providing adaptivity based on learning styles in LMS.

Secondly, we investigate correlations between the learning style preferences and the behaviour of the students in the course. From these correlations, it is not only possible to draw conclusions from learning style preferences to the behaviour but also to obtain indications from the behaviour of students about their learning style preferences. These results provide information which can be used to investigate the identification of learning styles in LMS based on the actual behaviour of students during an online course.

The benefits of both analyses are discussed in detail in Sect. 6. In Sect. 7, we describe related work and point out differences to our work. Section 8 concludes the chapter.

2 Felder–Silverman Learning Style Model

While several learning style theories exist in the literature, e.g. the learning style models by Kolb [14] and Honey and Mumford [15], Felder-Silverman learning style model (FSLSM) [2] seems to be the most appropriate for use in computer-based educational systems [5, 16]. Most other learning style models classify learners in few groups, whereas FSLSM describes the learning style of a learner in more detail, distinguishing between preferences on four dimensions.

The first dimension distinguishes between an active and a reflective way of processing information. Active learners learn best by working actively with the learning material, e.g. working in groups, discussing the material, or applying it. In contrast, reflective learners prefer to think about and reflect on the material.

The second dimension covers sensing versus intuitive learning. Learners with preference for a sensing learning style like to learn facts and concrete learning material. They tend to be more patient with details and more careful about their work. Furthermore, sensing learners tend to be more practical than intuitive learners and like to relate the learned material to the real world. Intuitive learners prefer to learn abstract learning material, such as theories and their underlying meanings. They like to discover possibilities and relationships, and tend to be more innovative than sensing learners. This dimension differs from the active/reflective dimension in an important way: the sensing/intuitive dimension deals with the preferred source of information whereas the active/reflective dimension covers the process of transforming the perceived information into knowledge.

The third, visual/verbal dimension differentiates learners who remember best what they have seen, e.g. pictures, diagrams and flow-charts, and learners who get more out of textual representations, regardless of the fact whether they are written or spoken.

In the fourth dimension, the learners are characterized according to their understanding. Sequential learners learn in small incremental steps and therefore have a linear learning progress. They tend to follow logical stepwise paths in finding solutions. In contrast, global learners use a holistic thinking process and learn in large leaps. They tend to absorb learning material almost randomly without seeing connections, but after learning enough material they suddenly get the whole picture. Then they are able to solve complex problems and put things together in novel ways but find it difficult to explain how they did it. Since the whole picture is important for global learners, they tend to be more interested in overviews and a broad knowledge whereas sequential learners are more interested in details.

3 Investigated Patterns of Behaviour

The aim of our investigations is to analyse the behaviour of students in an online course with respect to their learning styles. These investigations are based on predefined patterns which on one hand are related to the FSLSM and on the other hand are based on commonly used features in LMS.

FSLSM is based on traditional learning rather than online learning and therefore describes the preferences of students in traditional learning environments. To apply FSLSM in online environments, some sort of mapping between the behaviour in traditional environments and in online environments is necessary. Therefore, we chose patterns in online environments that are related to the traditional behaviour and tested their significance with respect to learning styles.

Additionally, our findings should be applicable for LMS in general rather than only for Moodle. Since different LMS provide authors and course developers with the opportunity to integrate different features in an online course, we used only those features which are implemented in most LMS and which are also commonly used by authors and course developers.

The incorporated features include content objects, outlines, examples, self-assessment tests, exercises, and discussion forums. Furthermore, we investigated the navigation preferences of students in the course as well as the sequence in which they visited specific features. In the following paragraphs, a brief description of the features as well as the related patterns to each feature is provided.

Content objects are used to present the content of the course. These content objects can have different formats, depending on the LMS. For example, content can be presented as html-files or pdf-files. Related to the content objects, we considered the number of visits as well as the time learners spent on content objects. Additionally, we tracked the time learners spent on content objects including graphics.

We also included patterns regarding *outlines* of chapters since they are explicitly mentioned in FSLSM. Therefore, we again looked at the number of visits of outlines and the time learners spent on it.

Another feature is *examples* which aim at illustrating the theoretical content in a more concrete way. Again, the number of visits and the time learners spent on these objects are used as patterns.

Furthermore, *self-assessment tests* are included where students can check their acquired knowledge. Regarding these tests, we considered more detailed information such as the number of questions a learner answered, whether a learner performed all available tests at least once, the results a learner achieved, how often a learner revised his/her answers before submitting, how long a learner spent on the tests, and how long a learner checked his/her results. Furthermore, the questions contained in a test can be about facts or concepts, refer to an overview or to details, deal with interpreting or developing solutions, or can be based on graphics rather than on text. The results learners achieved on each kind of questions act as pattern as well.

Another element includes *exercises* which serve as practice area where students can try things out or answer questions about interpreting predefined solutions or developing new solutions. The number of visits and the time student spent on exercises is considered as pattern. Information about the number of revisions as well as students' performance on interpreting and developing solutions is gathered and combined with the data from self-assessment tests.

For communication issues, *discussion forum* is considered. As patterns, we incorporated the number of visits to the forum, how long learners stayed at the forum, and how many messages they posted.

Additionally, we incorporated the *navigation* between learning objects as well as the number of logins in the course. We considered how often learning objects were skipped in the course sequence, how often learners jumped back to the previous learning object, as well as how often and how long they stayed at the course overview page.

Furthermore, we incorporated patterns regarding the *sequence*, in which learners visited specific features. We considered which feature was visited first and last in a course chapter, distinguishing between content objects, outlines, self-assessment tests, exercises, examples, and marked assignments. Moreover, we used the order in which content objects, self-assessment tests, exercises, and examples were visited by comparing, for instance, in how many course chapters a student was visiting a content object before or after he/she was performing a self-assessment test. This is done for all combinations of the four features, resulting in six patterns.

4 Design of the Study

In this section, information about the design of the study is provided. Therefore, the course itself and its structure are described. Subsequently, we introduce the instrument which was used for identifying learning styles according to FSLSM. Although Moodle provides already quite comprehensive tracking mechanisms, some extensions were necessary in order to track all information that we aimed at investigating in our study. The last subsection presents the performed extensions in Moodle.

4.1 Description of the Course

The study is based on the data from a laboratory course about Web Engineering which was taught at Vienna University of Technology, Austria, in summer term 2006. The course was divided into two parts, XML and Java. Only for the XML part, all features which were mentioned in the previous section such as content object, examples, exercises and so on, were included in Moodle [10]. Therefore, our investigations deal with the XML part of the course only.

The XML part itself consisted of three chapters that included 182 content objects (39 include graphics) and 14 examples in total. Students could solve eight different exercises which allowed them to parse their entered source code and provided feedback. Self-assessment tests were provided for five topics, and included 123 questions overall.

Each chapter also included one marked assignment which had to be done in groups of two. Few days after the submission, each student had to present the solution individually and had to answer questions about it. At the end of the course, each student had to pass a written exam.

Although parts of the assignments were done in groups of two, the course was designed in a way that all students needed to learn everything and they were examined on all topics; hence the course was appropriate for investigation of individual learning.

4.2 Instrument for Identifying Learning Styles

In order to investigate the behaviour of students during the course with respect to their learning styles, these learning styles needed to be identified. Therefore, we used the Index of Learning Styles (ILS), a 44-item questionnaire developed by Felder and Soloman [3]. The ILS identifies learning styles according to FSLSM and is available online.

As mentioned earlier, each learner has a personal preference for each of the four dimensions of FSLSM. These preferences are expressed with values between +11 and −11 per dimension. This range comes from the 11 questions that are posed for each dimension. When answering a question, for instance, with an active preference, +1 is added to the value of the active/reflective dimension, whereas an answer for a reflective preference decreases the value by 1.

The ILS is an often used and well investigated instrument to identify the learning style. An overview of studies dealing with analysing the response data of ILS as well as with verifying the reliability and validity of the instrument is provided by Felder and Spurlin [17].

4.3 Extensions in Moodle for Tracking Students' Behaviour

In order to make our results applicable for most LMS, we incorporated only commonly used features of LMS. On the other hand, the investigations about the behaviour regarding these features asked for detailed information. Although Moodle provides much information about learners' behaviour in the courses, the need for some extensions emerged. These extensions focused on two issues. On the one hand, they dealt with getting additional descriptions about the learning material in order to distinguish it and be able to draw conclusions about students' behaviour related to specific kinds of learning material. On the other hand, extensions dealt with improving tracking mechanisms.

Moodle provides a number of different features to include in an online course. For our investigations and with respect to the above introduced patterns, only some of these features are of particular interest, namely the resources, the quiz, the assignments, and the forum. In Moodle, learning material regarding all proposed features can be created by using these four different types of learning objects. But for investigations regarding the learners' behaviour, the learning material has to be distinguished with respect to the proposed features and for some features, an additional description of the material is necessary. In most cases, this differentiation and additional descriptions are not supported by Moodle. Therefore, we added the possibility for authors to specify information about the created learning material by the use of meta-data. For resources, we ask authors to specify whether the learning objects can be considered as content objects, outlines, or examples. Furthermore, we provide authors the opportunity to specify whether the material includes graphics. For quizzes, we give the possibility to specify whether the quiz can be considered as exercise or as self-assessment test. Moreover, the authors are asked to specify the questions in more detail by indicating whether the questions are about facts or concepts, refer to an overview or to details, are based on graphics rather than on text, or deal with interpreting predefined solutions or developing new solutions. This detailed specification provides information about which kind of question a learner handles easily or with difficulty.

Since Moodle provides comprehensive tracking mechanisms, only one extension was necessary. This extension deals with the tracking of behaviour during students are performing quizzes. So far, only the final answers of the questions were stored. In order to get more detailed information about what learners are doing during a quiz, we extended the tracking function by storing each answer which is given by a learner, even if this answer is revised later. When the question asks for a textual answer rather than for choosing between predefined options, additionally the sequence of keys is tracked with attention to the delete and backspace key. This additional data provides information about how often students are revising their answers.

5 Results

We investigated two different issues within this study: Firstly, we analysed the given data in order to draw conclusions about whether students with different learning styles, or more precisely with different preferences for the questions of ILS, act differently in the online course. Secondly, we aimed at finding correlations between the answers to the questions and the behaviour of students during the course.

Forty-three students participated in our study. Since all students have either a visual or a balanced learning style and no student indicated a verbal style, further investigations are focused only on the active/reflective, sensing/intuitive, and sequential/global dimension.

5.1 Behaviour Vs. Learning Style Preferences

In order to identify significant differences of behaviour in the online course from different answers to questions of the ILS, we divided the students for each question, according to their answer (+1 or −1), into two groups. Then we tested these two groups respectively for significant difference for each pattern of behaviour described in Sect. 3.

Two tailed t-test was applied for patterns where data was normal distributed and two tailed Mann–Whitney U test (u-test) for patterns where data was not normal distributed. To check whether data was normal distributed, we used Kolmogorov–Smirnov test.

The results are presented in Table 1. Only significant values ($p < 0.05$) are shown. The table shows the patterns and respectively the ILS questions which lead to a significant result according to t-test or u-test. The T and U values as well as whether t-test or u-test was conducted, the significance level (p), and the direction of the relationship (d) is presented. Regarding the direction, 1 indicates that a high value concerning the pattern refers to the group answered the ILS question with +1 and vice versa.

Regarding the patterns dealing with visiting specific features first or last in a course chapter, we considered only patterns where the occurrence was higher than 5. Therefore, only assignments and examples were considered with respect to the first visited learning object, and content objects, examples, self-assessments, exercises, and assignments were considered regarding the last visited learning object.

In the following discussion, for all significant results the respective question is in semantic relation with the pattern unless mentioned otherwise.

Active/Reflective Dimension

According to the results of the active/reflective dimension, it can be seen that spending more time on examples and dealing more intensively with outlines (visiting and spending time) seems to be significant for reflective learning. These findings are in agreement with FSLSM, since reflective learners are described as learners who think and reflect more deeply about the provided learning material.

When looking at the sequence of visited learning objects, we found a significant preference for reflective learners to visit examples first and perform then exercises. In contrast, active learners preferred to perform exercises first

Table 1. Results of *t*-test/*u*-test

Group	Pattern	Question	*t*-test/ *u*-test	T or U	p	d
Active/reflective	Outline_visit	q29	t	−2.24	0.031	−1
	Outline_stay	q29	u	65.50	0.002	−1
	Example_stay	q33	u	143.50	0.045	−1
	Selfass_stay_results	q5	u	25	0.007	−1
	Ques_visit	q5	u	154.00	0.050	1
	Ques_facts	q5	t	3.21	0.005	1
	Ques_interpret	q9	t	−3.32	0.004	−1
	Forum_visit	q25	t	−2.92	0.006	−1
	Navigation_overview_stay	q13	t	2.17	0.036	1
	Navigation_overview_stay	q25	t	−3.02	0.005	−1
	Sequence_exercise/example	q21	u	136.50	0.049	1
	Sequence_selfass_last	q25	u	177.50	0.043	−1
	Sequence_selfass_last	q29	u	129.50	0.044	−1
	Sequence_assignment_last	q13	u	164.50	0.040	1
Sensing/intuitive	Content_visit	q26	t	2.69	0.012	1
	Outline_visit	q22	t	2.04	0.048	1
	Outline_stay	q34	u	123.00	0.036	−1
	Example_visit	q2	u	104.00	0.044	1
	Example_stay	q10	u	111.50	0.043	1
	Ques_overview	q42	t	−2.61	0.018	−1
	Ques_revisions	q10	t	2.47	0.021	1
	Forum_stay	q10	t	2.79	0.008	1
	Forum_stay	q22	t	2.63	0.012	1
	Forum_post	q22	u	117.00	0.001	1
	Navigation_back	q22	u	161.50	0.048	1
	Sequence_content/selfass	q10	u	104.00	0.011	1
	Sequence_exercise/selfass	q42	u	153.00	0.029	−1
	Sequence_example_first	q26	u	154.00	0.003	1
	Sequence_assignment_first	q10	u	99.50	0.014	1
	Sequence_example_last	q10	u	135.00	0.022	1
	Sequence_example_last	q38	u	170.00	0.035	1
Sequential/global	Outline_visit	q12	t	2.99	0.005	1
	Outline_stay	q44	u	114.50	0.005	1
	Selfass_visit_different	q36	u	101.00	0.028	1
	Selfass_stay_results	q20	u	33.00	0.024	1
	Ques_concepts	q44	t	−2.11	0.049	−1
	Ques_graphics	q32	t	2.86	0.010	1
	Ques_revisions	q28	t	3.04	0.007	1
	Forum_post	q20	u	149.00	0.014	1
	Navigation_skip	q20	u	176.00	0.038	−1
	Navigation_overview_visit	q44	t	−2.71	0.010	−1
	Sequence_content/exercise	q28	u	34.00	0.020	1
	Sequence_content/example	q4	u	76.50	0.038	−1
	Sequence_content_last	q12	u	171.00	0.021	1
	Sequence_assignment_last	q24	u	145.50	0.037	−1
	Sequence_assignment_last	q32	u	127.00	0.007	−1

and looked then at examples. According to FSLSM, active learners prefer to try things out and work actively, as they can do with exercises. By looking at examples, active learners see how others have solved a problem rather than being able to solve the problem by themselves. Therefore, they preferred exercises over examples and focused on exercises first. On the other hand, reflective learners gain more from examples, where they can reflect on an already given solution. Therefore, they looked at examples first and afterwards performed exercises.

Furthermore, it can be seen that reflective learners performed better on questions about interpreting predefined solutions (in terms of source code). This is again in line with the argumentation above. Moreover, they spent more time on looking at the results of their self-assessment tests. Again, this behaviour can be referred to the preference of reflecting.

In addition, results also show that reflective learners visited the forum significantly more often than active learners. This is because the forum in the course was mainly used for asking and clarifying questions regarding the assignments which were then answered by a tutor or a teacher. When the forum would be used for active discussions between students, maybe active learners would visit the forum more often.

Regarding active learning, it can be also seen that learners with an active preference performed significantly more self-assessment questions than reflective learners. This is in agreement with FSLSM as well, since active learners are characterized to prefer trying things out. It seems also to be significant that active learners performed better on questions dealing with facts. Further investigations about this finding need to be done since FSLSM does not include this behaviour in their description of an active/reflective learning style.

Considering the preferred first and last learning objects of active and reflective learners in a course chapter, it can be seen that reflective learners seem to perform self-assessment tests more often as last object in a course chapter than active learners. In contrast, active learners tended more often to submit their assignment and then went to the next course chapter. Since we also found out that active learners performed self-assessment test more often, these findings indicate that active learners used self-assessment tests as support for doing their assignments. In contrast, reflective learners were using these self-assessment tests also after the submission for preparing themselves for the presentation or the exam.

When looking at the pattern indicating how long students spent on the overview page, it can be seen that for one question, students answering with an active preference spent more time on it and for another question students with a reflective preference did. Hence, it seems that a preference for an active or reflective learning style does not provide significant information about this pattern.

Sensing/Intuitive Dimension

Sensing learners are described by Felder and Silverman as learners who prefer concrete material. This can be also seen by our findings, showing that sensing learners visited more often examples and spent more time there than intuitive learners. Moreover, the results show that sensing learners started a course chapter more often with an example than intuitive learners. Also, their last visited learning object in a course chapter was more often an example than for intuitive learners. This indicates that sensing learners were using examples for their preparation for the presentation of the assignments and the written exam.

Another characteristic of sensing learners according to FSLSM is that they tend to be more patient with details and careful about their work. Looking at the pattern about revising their answers in self-assessment tests and exercises, it can be seen that learners with a sensing preference significantly more often changed their answers. It can also be seen that sensing learners spent more time in the forum and posted more often than intuitive learners. So, it can be argued that due to their preference for details, they wanted to clarify the specifications by asking in forums and were also interested in the questions and answers of others. Again, when the forum would be used more for discussion, these results may change. As can be seen from the results, sensing learners also tended to visit content objects and outlines more often and also navigated back more often to the previous page. This behaviour may also results from their patience and accuracy.

Another characteristic of sensing learners is that they tend to be more practical oriented and more interested in the application of the learned material in the real world. According to our results, it can be seen that sensing learners tended to start a course chapter more often with looking at the assignment than intuitive learners did. This behaviour may be due to their interest in applications. On the one hand, the assignments present the tasks which have to be done for the course chapter, but on the other hand, assignments are programming tasks that also show how the learned material can be applied.

Intuitive learners are characterized by Felder and Silverman as learners who like challenges. From our results, this is indicated by the sequence of visited learning objects. It can be seen that intuitive learners had a higher preference than sensing learners for performing self-assessment tests first and afterwards looking at the content objects. Also, we found from the results, that intuitive learners tried to do exercises first and then performed self-assessment tests. In our setting, exercises can be considered as more challenging since they ask students for programming tasks, whereas self-assessment tests provide students with the opportunity to check their theoretical knowledge about the learning material and are less comprehensive.

Two more significant patterns could be found for intuitive learners. One is dealing with the time students spent on outlines, the other one is about the results achieved for questions about overview. The second one may be explained by the preference of details for sensing learners and that they therefore achieve worse than intuitive learners on questions about overview. However, further investigations are necessary for both relations with regard to FSLSM.

Sequential/Global Dimension

According to FSLSM, a main characteristic of sequential learners is that they learn in a linear way, going through the material step by step. Accordingly, our results show that learners with a sequential preference tended to cover all/more topics of self-assessment tests and that they dealt more often with outlines which indicates that they started at the beginning of each chapter rather than jumping in and starting somewhere in between. Moreover, we found that sequential learners significantly more often visited the content first and afterwards performed exercises, as it was recommended in the course structure. In contrast, global learners tend to prefer a more non-sequential path through the course material. This can be seen when looking at the results of skipping learning objects which show that global learners skipped learning objects more often.

From our results, it can also be seen that learners with a global preference visited more often the course overview page. This is in agreement with FSLSM, since global learners are described to prefer getting an overview of the topic/course. While for global learners the overview is very important, sequential learners are more inclined to the details. According to Felder and Spurlin [17], it has been proven that the sequential/global dimension correlates slightly with the sensing/intuitive dimension. This may be caused due to the overlapping of the preference for details. Accordingly, our results show that sequential learners posted more often in the forum, looked more detailed at the results of their tests, and made more revisions when answering questions. In contrast, global learners performed significantly better on questions about concepts than sequential learners. Moreover, it was found that global learners had higher preference for submitting the assignments and then going to the next course chapter, whereas for sequential learners we found a preference for content objects as the last visited material in a course chapter. Again, this gives indication that sequential learners tend to be more accurate and careful by preparing themselves for the presentations and the exam after submitting the assignments.

Sequential learners seem to perform also better on questions about graphics. This might be because they remember better the details of the graphics. However, further investigations on this issue needs to be done.

Another result that needs further investigations is dealing with the preferred sequence of visiting examples and content. According to our results, sequential learners visited more often examples before content objects which is not in agreement with the recommended order of the course structure but might be explained by the correlation to the sensing learning style, where examples play an important role to facilitate learning for sensing learners.

5.2 Correlations Between Behaviour and Learning Style Preferences

The previous analysis pointed out relations where learners who answered questions of ILS differently also acted differently in the online course. In the next analysis, we investigate the correlation between both, answers of ILS questions and the behaviour of the learners in the course based on the specified patterns. Thus, the resulting relations additionally allow drawing conclusions from the behaviour of the learners to their preferences of learning styles.

Since the values of the patterns are on a continuous scale and the possible answers to the questions of ILS can only be either +1 or −1, point-biserial correlation was performed using SPSS. Table 2 presents the results of the point-biserial correlation analysis. Again, only significant results are shown ($p < 0.05$). Furthermore, we included only patterns where more than five students had a value greater than zero, as done in the previous analysis.

From the results, it can be seen that most of the significant relations found by t-test and u-test were also found by the point-biserial correlation. Therefore, in the following subsections, we discuss only the additional relations as well as relations which were found by t-test or u-test but were not confirmed by correlation analysis.

Table 2. Results of correlation analysis

	Pattern	Question	Rpb	p	d
Active/Reflective	Content_stay_graphics	q21	0.34	0.037	1
	Outline_visit	q29	−0.33	0.031	−1
	Outline_stay	q21	−0.34	0.026	−1
	Outline_stay	q29	−0.43	0.004	−1
	Example_visit	q33	−0.31	0.042	−1
	Selfass_visit_different	q5	0.35	0.022	1
	Selfass_stay_results	q1	−0.49	0.016	−1
	Ques_visit	q5	0.43	0.004	1
	Ques_facts	q5	0.59	0.005	1
	Ques_interpret	q9	−0.64	0.004	−1
	Ques_develop	q5	−0.64	0.036	−1
	Forum_visit	q25	−0.41	0.006	−1
	Navigation_overview_stay	q13	0.32	0.036	1
	Navigation_overview_stay	q25	−0.43	0.004	−1
	Sequence_exercise/example	q21	0.34	0.025	1
	Sequence_assignment_last	q13	0.33	0.030	1

(continued)

Table 2. (continued)

	Pattern	Question	Rpb	p	d
Sensing/Intuitive	Content_visit	q26	0.39	0.009	1
	Outline_visit	q22	0.30	0.048	1
	Example_stay	q10	0.35	0.023	1
	Example_stay	q42	−0.43	0.004	−1
	Ques_detail	q10	0.43	0.050	1
	Ques_overview	q42	−0.52	0.018	−1
	Ques_develop	q34	0.66	0.028	1
	Ques_revisions	q10	0.46	0.021	1
	Exercise_visit	q10	0.38	0.011	1
	Exercise_stay	q10	0.39	0.010	1
	Forum_stay	q10	0.40	0.008	1
	Forum_stay	q22	0.38	0.012	1
	Forum_post	q22	0.48	0.001	1
	Sequence_content/selfass	q10	0.43	0.004	1
	Sequence_content/selfass	q22	0.33	0.032	1
	Sequence_exercies/selfass	q42	−0.32	0.038	−1
	Sequence_example_first	q26	0.45	0.002	1
	Sequence_assignment_first	q10	0.38	0.013	1
	Sequence_exercise_last	q10	0.35	0.021	1
	Sequence_example_last	q10	0.37	0.015	1
	Sequence_example_last	q38	0.31	0.045	1
Sequential/Global	Outline_visit	q12	0.42	0.005	1
	Outline_stay	q44	0.34	0.024	1
	Selfass_stay	q12	−0.41	0.038	−1
	Selfass_stay	q16	−0.40	0.042	−1
	Selfass_stay	q20	−0.39	0.046	−1
	Selfass_visit_different	q36	0.34	0.024	1
	Selfass_stay_results	q28	0.52	0.010	1
	Ques_concepts	q44	−0.45	0.049	−1
	Ques_graphics	q32	0.56	0.010	1
	Ques_develop	q20	−0.78	0.004	−1
	Exercise_stay	q40	0.33	0.032	1
	Forum_post	q20	0.35	0.021	1
	Forum_post	q32	−0.33	0.031	−1
	Navigation_skip	q40	0.33	0.032	1
	Navigation_overview_visit	q44	−0.39	0.010	−1
	Sequence_content/selfass	q28	0.31	0.045	1
	Sequence_content/exercise	q28	0.39	0.010	1
	Sequence_exercise_last	q12	0.30	0.047	1
	Sequence_exercise_last	q28	0.41	0.007	1
	Sequence_content_last	q12	0.34	0.028	1
	Sequence_assignment_last	q24	−0.33	0.033	−1
	Sequence_assignment_last	q32	−0.38	0.013	−1

Active/Reflective Dimension

Regarding the active/reflective dimension, additionally a relation can be seen between active learners and their preference for performing most or all self-assessment tests. This result is in agreement with FSLSM. The preference of reflective learners to finish a chapter with a self-assessment test more often than active learners could not be confirmed according to correlation analysis.

Furthermore, a relation can be seen between active learners and their interest in graphics. This interest in graphics may be explained by the fact that active learners tend to be less interested in reading and reflecting about text but instead look more details at graphics. Nevertheless, further investigations seem to be necessary since this behaviour is not explicitly described according to FSLSM.

While the time spent on examples could not be confirmed as an indication for a reflective preference, the number of visits was found as significant pattern. Regarding the performance on questions dealing with interpretation and development of source code, both seem to correlate with a reflective preference according to the results of the correlation analysis.

Sensing/Intuitive Dimension

While for learners with a sensing preference the number of visits of examples seems to be not significant according to the calculated correlation, exercises plays an important role. The number as well as the time spent on exercises correlates significantly with a sensing learning preference. Furthermore, it can be seen that learners with a sensing preference performed more often exercises as their last learning object in a chapter. This indicates that they used exercises to prepare themselves for the oral presentations and the written exam. The preferences of sensing learners for exercises are in agreement with FSLSM.

Regarding the time spent on examples, a significant correlation is found for a sensing as well as for an intuitive learning preference which necessitate further investigations.

An additional relation between a sensing learning preference and a better performance in questions about details and code development was found. Both are in agreement with FSLSM.

The impact of navigating to previous learning objects could not be confirmed by the results of the correlation analysis. Also the pattern indicating that intuitive learners spent more time on outlines was not found as significant according to correlation analysis. Anyway, this pattern is not explicitly supported by FSLSM.

Sequential/Global Dimension

Regarding the sequential/global dimension, results show that a correlation was found indicating that learners with a global preference spent more time on self-assessment tests and performed better when developing source code. This is in line with FSLSM since the self-assessment tests are based on the learning material and therefore can be answered more easily when learning the material step by step, which tends to be the preferred way of learning for sequential learners. In contrast, for developing source code, more overview knowledge about the concepts is necessary, which tend to favour global learners.

According to the results of the correlation analysis, two further patterns were found, indicating the step-by-step navigation of sequential learners. First, it can be seen that sequential learners more often visited content objects before they performed self-assessment tests, as it is recommended by the course structure. In contrast to the *u*-test in the previous section, we did not found a significant correlation indicating that sequential learners looked at examples before they went to the content objects. Anyway, this pattern needs further analysis since it is not explicitly supported by FSLSM.

According to the *u*-test, sequential learners tended to look more often at content objects before they went to the next chapter. Regarding correlation analysis, an additional preference for exercises as last learning object of a chapter was found. However, both patterns indicate that sequential learners tend to be more accurate and prepare themselves for the presentation and the exam. On the other hand, the preference for assignments as last learning object of a chapter for global learners was confirmed by the correlation analysis.

Another correlation was found between the time students spent on exercises and a sequential learning preference. This relation needs further investigations with respect to FSLSM. Regarding the number of postings, once a positive and once a negative correlation was found. A similar disagreement was found for skipping learning material since *u*-test and correlation analysis indicate different directions of the relationship. Therefore, further investigations are necessary for both of these cases. Furthermore, the relation for revising answers in self-assessment tests and exercises could not be confirmed by correlation analysis.

6 Benefits

Table 3 summarizes the results of our study. These results show that learners with different preferences for learning styles act differently in the online course. They used different features such as examples, exercises, and so on with different frequency, performed differently on specific kinds of questions, navigated differently through the course, and visited particular features in a different sequence. The results can also be interpreted in a way that each feature is needed to support a specific learning style and therefore plays an important role in the course. According to Felder and Silverman [2], learners might have difficulties in learning if their learning style is not supported by the teaching environment. As a solution, they recommended to provide courses with many different features which support different learning styles rather than providing courses that suit only one learning style.

Table 3. Summary of the results

	Active/reflective		Sensing/intuitive		Sequential/global	
Patterns	t/u-test	Corr.	t/u-test	Corr.	t/u-test	Corr.
Content_visit			✓	✓		
Content_stay_graphics		✓				
Outline_visit	✓	✓	✓	✓	✓	✓
Outline_stay	✓	✓	✓		✓	✓
Example_visit		✓	✓			
Example_stay	✓		✓			
Exercise_visit				✓		
Exercise_stay				✓		✓
Selfass_stay						✓
Selfass_visit_different		✓			✓	✓
Selfass_stay_results	✓	✓			✓	✓
Ques_visit	✓	✓				
Ques_revisions			✓	✓	✓	
Ques_facts	✓	✓				
Ques_concepts					✓	✓
Ques_detail				✓		
Ques_overview			✓	✓		
Ques_graphics					✓	✓
Ques_develop		✓		✓		✓
Ques_interpret	✓	✓				
Forum_visit	✓	✓				
Forum_stay			✓	✓		
Forum_post			✓	✓	✓	
Navigation_skip					✓	✓
Navigation_back			✓			
Navigation_overview_visit					✓	✓
Navigation_overview_stay						
Sequence_content/selfass			✓	✓		✓
Sequence_content/example					✓	
Sequence_content/exercise					✓	✓
Sequence_exercies/selfass			✓	✓		
Sequence_exercise/example	✓	✓				
Sequence_example_first			✓	✓		
Sequence_assignment_first			✓	✓		
Sequence_content_last					✓	✓
Sequence_example_last			✓	✓		
Sequence_selfass_last	✓					
Sequence_exercise_last				✓		✓
Sequence_assignment_last	✓	✓			✓	✓

The results of this study can act as catalyst to make teachers and course developers aware of the different needs of their students and also the different ways of learning from the course material. The results point out the preferences of learners with different learning styles. Furthermore, it can be seen that all investigated features are used differently at least regarding one learning style dimension. This shows the importance of each feature and highlights the requirement for providing different features to support each learning style.

At the current stage, learning management systems provide the same course for each learner. Learners then have the possibility to use the provided learning material in different ways and as can be seen from our results, they also do so. Besides providing a high amount of learning material that includes some features for each learning style, courses can also be adapted to the individual learning styles of learners. For providing proper adaptivity, it is important to know the preferences of learners with respect to their learning styles. Since FSLSM is developed for learning in traditional learning environments rather than for learning in technology enhanced learning environments, the behaviour of students in such environments has to be investigated and incorporated in the design of adaptation features. Accordingly, the results of this study can be used as basis for the generation process of adaptation features, especially in learning management systems.

Another requirement for providing adaptivity is to know the learning styles of the learners. While the most often used approach in adaptive systems is to let students fill out a questionnaire which aims at detecting their learning styles, recent research is done on developing an automatic approach which identifies the learning styles based on the behaviour of students during an online course. Such an automatic student modelling approach can be more accurate by avoiding answers that are not in agreement with students' actual behaviour or preferences, given intentionally or non-intentionally. Moreover, students just need to use the learning system and do not have any additional effort.

The second part of this study aims at finding correlations between the behaviour of students in an online course and their learning styles. Such a correlation allows, on one hand, inferences from the learning styles to the behaviour of students. On the other hand, it allows drawing conclusions from the behaviour of the students to their learning styles. This information can be used to investigate and develop an automatic approach for detecting learning styles based on the behaviour of students in LMS.

In the following section, we introduce related work on automatic approaches for identifying learning styles.

7 Related Work

Some recent work already exist dealing with systems that analyse the behaviour of students in an online course and use this information to generate or update the student model with respect to learning styles.

For example, García et al. [18] proposed an automatic approach for the system SAVER. The approach considers the active/reflective, sensing/intuitive, and the sequential/global dimensions of the Felder–Silverman learning style model. The visual/verbal dimension is not incorporated since no relevant learning material is presented in the course. Overall, 11 patterns are included for the three dimensions. These 11 patterns are derived from literature. Based on the data from these patterns, Bayesian networks [19] are used to calculate the preferences of the learning style dimensions for each learner. An evaluation of the approach with 10 users shows a high degree of precision when comparing the calculated learning styles of the users with their results from the ILS questionnaire.

Another approach for automatic detection of learning styles was investigated by Cha et al. [20]. Again, they observed the behaviour of learners during an online course in an intelligent learning environment based on specific patterns. An experiment with 70 learners was conducted in order to test the effectiveness of Decision Trees [21] and Hidden Markov Models [22] for detecting learning styles according to FSLSM. For both techniques, several patterns were incorporated for each learning style dimension. Results of the experiment show a high degree of precision for two dimensions, and moderate degree of precision for the other two dimensions.

All these approaches used patterns that were derived from literature. However, most learning style models and also FSLSM are based on traditional learning rather than on online learning. In our investigations, we look whether students with different learning styles really behave differently in learning management systems and whether a correlation exists between their behaviour and their learning styles. This information can be used to improve the selection of patterns from literature and therefore enhance the detection process of learning styles.

Furthermore, the above described approaches are developed for specific systems and therefore used only those features and patterns which are incorporated in the system. When investigating the behaviour of learners in LMS and aiming at inferring learning styles from this behaviour, we have to consider that authors have a lot of options to create courses by using or not using particular features. Therefore, it is important to have many different patterns from which information can be gained. Moreover, neither the LMS itself nor the structure of most courses is developed in consideration of learning styles. Thus, investigations about the actual behaviour of students in LMS are even more important for developing an approach that automatically detects students' learning styles.

8 Conclusion and Future Work

In this chapter, we presented investigations about the behaviour of students in an online course within a learning management system with respect to the students' learning styles. By analysing the behaviour of students based on predefined patterns, we found significant results for several patterns, indicating that students with different learning style preferences also behave differently in the online course. These results can act as recommendation when aiming at providing courses in LMS that fit to the learning styles of students.

Furthermore, the behaviour of students and their learning style preferences were analyzed with regard to correlations. Again, several significant correlations were found. These results additionally allow drawing conclusions from the behaviour of students to their learning styles and can be used to develop an automatic approach to detect learning style in learning management systems.

The results of our study provide information to facilitate the consideration of learning styles in learning management systems. Future work will deal on one hand with using the gathered information as basis for providing adaptivity in LMS. On the other hand, the results regarding correlations between learning styles and behaviour will be used to investigate and develop an approach for automatic identification of learning styles from the behaviour of students in LMS.

Acknowledgements

This research has been partly funded by the Austrian Federal Ministry for Education, Science, and Culture, and the European Social Fund (ESF) under grant 31.963/46-VII/9/2002.

References

1. Jonassen DH, Grabowski BL (1993) Handbook of Individual Differences, Learning, and Instruction. Lawrence Erlbaum Associates, Hillsdale, NJ
2. Felder RM, Silverman LK (1988) Learning and teaching styles in engineering education. Engineering Education 78(7):674–681. Preceded by a preface in 2002: http://www.ncsu.edu/felderpublic/Papers/LS-1988.pdf
3. Felder RM, Soloman BA (1997) Index of Learning Styles questionnaire. Retrieved 30 April, 2007, from http://www.engr.ncsu.edu/learningstyles/ilsweb.html
4. Bajraktarevic N, Hall W, Fullick P (2003) Incorporating learning styles in hypermedia environment: empirical evaluation. In: Proceedings of the Workshop on Adaptive Hypermedia and Adaptive Web-Based Systems, Nottingham, UK, pp. 41–52
5. Carver CA, Howard RA, Lane WD (1999) Addressing different learning styles through course hypermedia. IEEE Transactions on Education 42(1):33–38
6. Shang Y, Shi H, Chen S-S (2001) An intelligent distributed environment for active learning. ACM Journal of Educational Resources in Computing 1(2):1–17
7. Papanikolaou KA, Grigoriadou M (2003) An instructional framework supporting personalized learning on the web. In: Proceedings of the International Conference on Advanced Learning Technologies. IEEE Computer Society, Los Alamitos, CA, pp. 120–124

8. Paredes P, Rodríguez P (2004) A mixed approach to modelling learning styles in adaptive educational hypermedia. Advanced Technology for Learning 1(4):210–215
9. Brusilovsky P (2004) Knowledge tree: a distributed architecture for adaptive e-learning. In: Proceedings of the International Conference on World Wide Web. ACM Press, New York, USA, pp. 104–113
10. Moodle (2007). Retrieved 30 April, 2007, from http://www.moodle.org
11. Blackboard (2007). Retrieved 30 April, 2007, from http://www.blackboard.com
12. WebCT (2007). Retrieved 30 April, 2007, from http://www.webct.com/
13. Graf S, List B (2005) An evaluation of open source e-learning platforms stressing adaptation issues. In: Proceedings of the Fifth International Conference on Advanced Learning Technologies. IEEE Press, New York, pp. 163–165
14. Kolb DA (1984) Experiential learning: experience as the source of learning and development. Prentice-Hall, Englewood Cliffs, NJ
15. Honey P, Mumford A (1982) The Manual of Learning Styles. Peter Honey, Maidenhead
16. Kuljis J, Liu F (2005) A comparison of learning style theories on the suitability for e-learning. In: Hamza MH (ed.), Proceedings of the IASTED Conference on Web Technologies, Applications, and Services. ACTA Press, pp. 191–197
17. Felder RM, Spurlin J (2005) Applications, reliability and validity of the index of learning styles. International Journal on Engineering Education 21(1):103–112
18. García P, Amandi A, Schiaffino S, Campo M (in press) Evaluating Bayesian networks' precision for detecting students' learning styles. Computers & Education
19. Jensen FV (1996) An Introduction to Bayesian Networks. Springer, Berlin Heidelberg New York
20. Cha HJ, Kim YS, Park SH, Yoon TB, Jung YM, Lee J-H (2006) Learning style diagnosis based on user interface behavior for the customization of learning interfaces in an intelligent tutoring system. In: Ikeda M, Ashley KD, Chan T-W (eds.) Proceedings of the Eigth International Conference on Intelligent Tutoring Systems, LNCS, vol. 4053. Springer, Berlin Heidelberg New York, pp. 513–524
21. Dunham MH (2002) Data Mining: Introductory and Advanced Topics. Prentice Hall, Upper Saddle River, NJ
22. Rabiner LR (1989) A tutorial on hidden Markov models and selected applications in speech recognition. Proceedings of the IEEE 77(2):257–286

Improving Image Retrieval Using Semantic Resources

Adrian Popescu, Gregory Grefenstette, and Pierre-Alain Moellic

CEA LIST-LIC2M, France adrian.popescu@cea.fr, gregory.grefenstette@cea.fr, pierre-alain.moellic@cea.fr

Summary. Many people use the Internet to find pictures of things. When extraneous images appear in response to simple queries on a search engine, the user has a hard time understanding why his seemingly clear request was not properly satisfied. If the computer could only understand what he wanted better, then maybe the results would be more precise. The introduction of an ontology, though hidden from the user, into current image retrieval engines may provide more accurate image responses to his query. The improvement of the results translates into the possibility of offering structured results, to disambiguate queries and to provide more interactivity options to the user, transforming the current string of character based retrieval into a concept based process. Each one of these aspects is presented and examples are used to support our proposals. We equally discuss the notion of picturability and justify our choice to work exclusively with entities that can be directly represented in a picture. Coordinating the use of a lexical ontology (an OWL representation of WordNet) with image processing techniques, we have developed a system that, given an initial query, automatically returns images associated with the query using automatic reformulation (each concepts is represented by its deepest hyponyms from the ontology). We show that picking randomly from this new set of pictures provides an improved representation for the initial, more general query. We also treat the visual aspects of the images for these deepest hyponyms (the leaves of WordNet). The depictions associated to leaf categories are clustered into coherent sets using low-level image features like color and texture. Some limitations (e.g. the quality and coverage of the semantic structure, the impossibility to answer complex queries) of the ontology based retrieval are equally discussed.

1 Introduction

Images represent an important part of the information searched for on the Internet. Major search engines like Yahoo! or Google report having indexed over one billion images. Existing Internet image retrieval frameworks are fast and robust but they present a number of problematic aspects. First, even when simple and clear queries are formulated, the obtained results are often not representative for the search term. Second, there is no semantic structure

A. Popescu et al.: *Improving Image Retrieval Using Semantic Resources*, Studies in Computational Intelligence (SCI) **93**, 75–96 (2008)
www.springerlink.com

in the responses offered by popular search engines such as Google, Yahoo, Ask, or Picsearch. Moreover, current image retrieval systems are keyword based and make little or no use of image processing techniques. Their image related advanced options are limited to image features like: file format, image size or type of picture (color or black and white) which are not definitory for the content of the picture and there is no content-related treatment. Even collaborative image repositories such as Flickr only use textual user-added tags to structure search results. Google and Yahoo! propose combinations of keywords that include simple logical operators like AND, OR and NOT. In addition to those options, Picsearch proposes a rudimentary ontology, grouping some 120 concepts into 6 higher order categories: *animals*, *classic cars*, *flowers*, *landmarks* and *legend*. In some cases, Yahoo offers specializations of the initial query, thus introducing some context for the initial concept. For the query *dog*, Yahoo! Image Search also suggests searching for *dog breeds*, *dog names*, *dog the bounty hunter*.

The most advanced semantic structure proposed by a search engine is Ask's AskX service[1]. Aside answers for the current query, narrower, expanded and related queries are proposed. When one asks for images of *dog*, queries like *puppy*, *free dogs* as narrower search; *cat* or *lion* as expanded search and *rabbit*, or *boxer* are presented as related. It is unclear how these relations are determined because they are, in part, counterintuitive. For example, the presentation of *cat* and *lion* as extended queries for *dog* is not sustained by semantic relations between these categories. It is equally unclear why *rabbit* is considered a related category of *dog*, the same as *boxer* which is clearly a subconcept of the initial category.

While simple and computationally efficient, the current approaches to image search depend heavily on the quality of the text found near images, or in their filename. This text unfortunately can sometimes have no direct relation to the image content. New approaches should exploit the textual information around Web pictures so as to maximize the relation between the selected keywords and the picture's content.

We exploit ontological knowledge contained in WordNet [14] to improve image retrieval results. WordNet is a lexical resource intended to structure commonsense knowledge into a semantic hierarchy. Initially built for psychological experimentation, and exploiting common dictionaries in its construction, its semantics was hoped to correspond to the way that people see the world. By using this hierarchy, we hope that a commonly accepted model for organizing natural categories can be reflected in the responses proposed by the image retrieval system.

In this chapter we describe an approach to image-to-language association. We propose a system that employs a simple lexical ontology, derived from WordNet, and image processing techniques to automatically associate picture classes to concepts. In the process we build a large scale image catalogue, an

[1] http://www.askx.com.

ontology containing both lexical and image data, considering only picturable entities extracted from the lexical hierarchy. The use of WordNet is two-fold: first, it provides the list of terms that are used to query the Web for images that constitute the raw data in our system. Second, after its transformation into an OWL [23] ontology, it constitutes a taxonomical base for our system. We exploit this taxonomy in the following way: since specialized categories (leaves in the lexical hierarchy) are usually less ambiguous than higher order concepts in WordNet, given a query term, we use the hyponymy relation to browse the taxonomy down to the concepts found at the leaf nodes. We then use these specialized concepts to query the Web rather than the original query term. For each picture set associated to a leaf concept in the hierarchy, we employ image processing techniques to index and cluster the raw data collected from the Web into visually coherent sets of images.

The remainder of this chapter is organized as follows: in Sect. 2 we describe related work and their relation with our approach. In Sect. 3 we discuss some ontological issues related to our work. Section 4 is dedicated to a presentation of the techniques we employed to build the structured image catalog. We discuss some problems we encountered in Sect. 5. Before concluding, we present and discuss preliminary evaluations of the performances of our system compared to those of a current search engine, Yahoo!.

2 Related Work

As proven by the increasing number of works in the image retrieval field, exploring the text–image association is an interesting and challenging task. In [1], the authors employ a tree of concepts and probabilistic methods in order to associate words from the tree to images. Their tree of concepts has more general concepts in higher positions and specific words in lower ones but there is no inheritance relation that governs the structure of the tree. The image processing includes picture segmentation and characterization in terms of color and texture. The results of their method for image clustering are assessed using three different settings: using only associated words, only visual descriptors and combining the two methods. The authors state the last type of clustering gives the best results. The entire work is performed on pictures from the Corel database and we are not aware of the application of a similar method for larger image sets.

In [17], the authors describe M-OntoMat-Annotizer, a complete framework for multimedia documents annotation and analysis is presented. The system in [17] is structured as an ontology that is intended to represent and link low level image descriptors (color, texture, shape), spatial relations in the image and linguistic concepts extracted from domain ontologies. The system requires the use of a training set where an association between regions in the picture and an object names has been manually performed. Given this learning set, the presented results are very good.

In [16], the author describes OntoVis, a domain model that addresses interior crime scenes. The system includes detailed models for a limited number of objects (20 in the current version). OntoVis notably includes a detailed description of partonomic relations and the 3D modeling of the included objects. Unfortunately, no quantitative evaluations of OntoVis are reported in this paper. Given the level of detail of the modeling and the associated efforts, it would be hard to extend the approach to broader domains.

An interesting utilization of ontologies for image retrieval is reported in [3]. The authors employ domain ontologies (from the cultural heritage domain) to improve the quality of image retrieval results for a limited set of items (200 pictures are considered in the experiment). An important contribution introduced in [3] is the joint use of semantic relations and content based image retrieval to improve the quality of the results.

One important difference between our approach and those of [1,3,16,17] is that we consider images that cover all picturable entities represented in a commonsense knowledge resource while they work on relatively narrow domains. Moreover, what separates the present work from those in [1, 3, 16] is we use raw data from a highly unstructured resource, the Web. We do not aim at finely modeling any particular domain. Consequently, the level of detail of ontological knowledge in our approach is smaller than that in [16, 17].

WordNet was already used in a number of applications in the image retrieval field and we present here some of the most relevant for our current work. In a recent work [25], Wang and collegues present an ontology based approach to image retrieval. A domain related ontology (containing about 600 classes representing animals) is employed for retrieving Web images. Their experiments on 20 canine species show that with the use of ontological knowledge (both at the conceptual and low-level image features), the retrieval precision is consequently improved when compared to Google results. The approach in [25] is close to ours in that it uses an ontology but a significant difference arises in what is the coverage of the conceptual hierarchy. Their system is domain dependant while we propose the use of a far more general ontology.

In [24], the authors propose the constitution of an image thesaurus using images on the Web. They extract weighted key terms from the text around the image and try to match these keywords to regions in the images. Both low-level descriptors (color and texture information are used) and high level, linguistic concepts are integrated in the system and, consequently, image and keyword queries are supported. Taxonomic relations in WordNet are used to expand queries for given concepts and to filter word senses. Wang's approach is closely related to ours. Both approaches are aimed at constructing image catalogues using raw data collected from the Web. One key difference between Wang's system and ours is our exclusive use of leaves in the hierarchy to collect data from the Web while the authors of [24] use keywords on several levels. Another important difference in Wang's system is that, for polysemic terms (concepts appearing in more than one synset), they retain the first sense only, losing one important advantage offered by the WordNet structure: sense separation. We preserve sense separation using automatic query expansion in order to differentiate word senses for ambiguous terms. An important distinction between our approach and the majority of those described above is that our technique does not imply a learning phase, a time consuming step that becomes critical when working with large data collections.

An interesting approach is described in [26] where the authors present a picture search application that supports both conceptual and content based browsing. A notable contribution of [26] is the definition of dynamic semantic hierarchies which allow the proposition of close categories to be browsed. The interest of these hierarchies relies in the fact that they reduce the number of presented categories, retaining only the most pertinent ones. This selection becomes critical when large scale hierarchies are constructed and the number of potential navigation options the system can present to the user is too high. The system described in [26] employs relevance feedback cycles to improve results but this technique is time costing and often rejected in large public applications. Similarly to [24], the sense separation for ambiguous words is lost because uniquely the first meaning of a word is retained.

A key point in our approach to image retrieval is automatic query reformulation. Related work is to be found in [12], where the authors present LiveImage, a meta-search engine drawing on Google. Their approach consists in refining user queries employing relevant modifiers. If images for *building* are demanded, LiveImage proposes a structured presentation of results for: *building*, *office building*, *tall building*, *school building* and *modern building*. We propose a different approach to reformulation: the use of subconcepts of a category to render images. A second noteworthy contribution of [12] is the realisation of an interactivity study comparing LiveImage and Google. This experiment shows that the users prefer the structured presentation over the unstructured list proposed by Google and they find that the automatic reformulation of queries accelerates the retrieval process.

3 Ontological Issues

In this section, we describe some ontology related aspects that are relevant for our work. We justify the choice of WordNet as taxonomical base for the current application and propose a way to separate picturable concepts from the others. We equally discuss current methods for automatic ontology creation.

3.1 Hierarchies

The "IsA" relation is fundamental to the way people organize entities in the world. We currently dispose of some comprehensive hand built systems based on this relation (i.e. WordNet in lexicography, Cyc [8] in formal ontologies). We are aware that it is probably illusory [4] to attempt to construct a hierarchy that performs best in all situations. The choice of one particular hierarchy is directed by the envisioned application. Since we wish to respect commonsense knowledge in our system, it is desirable to use a resource that accounts for the way people organize entities in the world. For example, WordNet contains *lionet* or *lion cub* as subcategories of *lion* while it is highly improbable to find these hierarchical structure in a scientific ontology. Another interesting characteristic of WordNet is that is supports multiple inheritance. For example *dog* inherits both from *canine* and *domestic animal* and if a query with these last two categories would be expressed, dog will be in both answers sets.

WordNet, which is organized as a tangled hierarchy and covers most common English terms seems a good choice for large public image retrieval applications. The root concept is entity. There are two relations that are fundamental in WordNet, "IsA" and synonymy. The first structures the hierarchy in depth (a part of the path from *golden retriever* to entity is *golden retriever-retriever-dog-animal-entity*), while the second gives rules for constituting its basic units, the synsets. A synset includes one or more terms that describe the same entity (e.g. *dog* is synonym to *domestic dog* and *Canis familiaris*). If it appears, ambiguity can be resolved by attaching a sense number to all defined terms. *Dog* has seven senses in WordNet and the corresponding classes in the ontology will range from *dog_1* to *dog_7*. Thus, each modelled entity is uniquely described by a WordNet synset and it is possible to provide separate image sets for each sense of a word.

3.2 Picturable Objects

One way to separate high order categories is to distinguish between *nominals*, *natural concepts* and *artifacts* [10]. The first do not correspond to *physical entities* in the world and there are no coherent pictural representations of such categories. It would be very hard to imagine a set of pertinent pictures for *truth*, *association*, or *nothing*. We illustrate our supposition by presenting, in Fig. 1, the first results obtained with Google for the three abstract concepts cited above. The images presented here do not have a coherent content as the queried concepts cannot be directly represented in pictures and this is the reason why we do not associate pictures to *nominal concepts* in WordNet. *Biologic concepts* and *artifacts* however are picturable categories and we can envision the construction of image classes that properly represent the associated linguistic concepts.

There are differences between the categorical distinctions in [10] and those in WordNet but these differences do not affect the distinction of entities in picturable or not. We decided to associate picture clusters only to concepts ranged under *physical entity* in the WordNet hierarchy. This is an initial choice and it is an open discussion if we should further restrain the categories to which we associate image classes. We think that, for example, subconcepts of *physical process* like *iteration*, *looping* or *irreversible process* probably do not

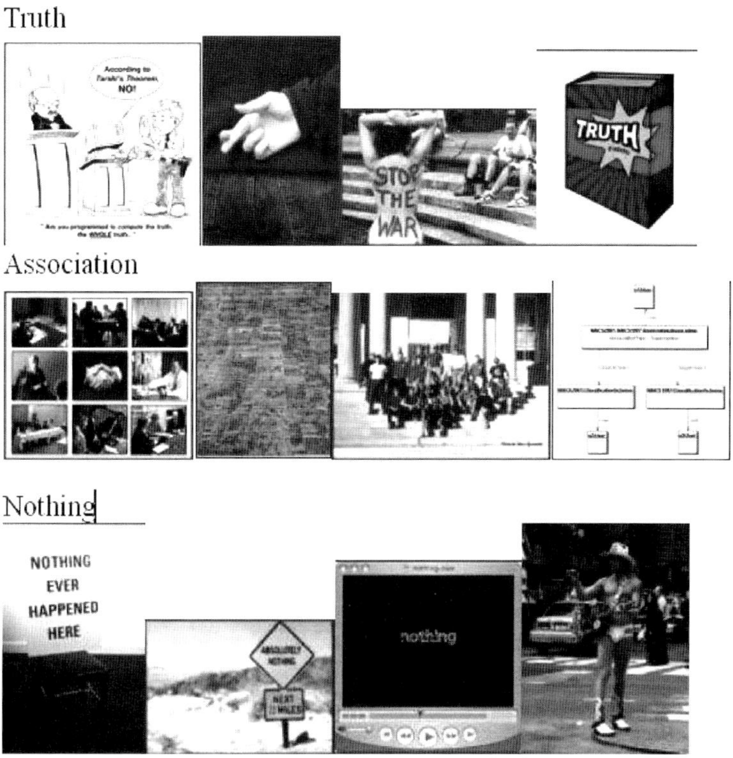

Fig. 1. Images representing *truth*, *association* and *nothing* (Google image)

have a coherent visual representation though they are found as hyponyms of the concept *physical entity*.

3.3 Automatic Ontology Building

Manual construction of taxonomies is a time consuming process, especially when we deal with large quantities of data. There exists an important current in ontology engineering that addresses problems related to the automatic construction of ontologies [2]. The standard procedure is to process specific domain textual documents and develop hierarchies for the relevant domain concepts. When one wants to cover broad domains, this approach is, for the moment, impractical. One other possible solution is to reuse existing resources. We adopted this last idea and automatically transformed the WordNet nouns hierarchy into an Ontology Web Language (OWL) ontology [23]. Our version is similar to that of [22] but differences arise given that we translate for an identified goal: the use of WordNet in image retrieval tasks. The two versions are not contradictory and, given that the translation described in [22] emerged from an official Semantic Web task force, further work might include alignment of our translation to the official one.

4 Visual Ontology Construction

In this section, we describe some aspects related to WordNet and its utility in image retrieval tasks. This resource has a double role in our image retrieval framework:

- During picture gathering – the use of less ambiguous specialized concepts, instead of more general terms, reduces noise
- During picture retrieving – the use of hierarchic relations provides a method for structuring results

Finally, we present the image clustering module we used in the image retrieval framework described in this chapter.

4.1 The WordNet Nouns Hierarchy

Our translation of the WordNet nouns hierarchy into an OWL format did not include instance-of synsets, resulting in 73,733 OWL classes, rather than 81,246 total synsets. There are about 60,000 leaves in the entire hierarchy. As we mentioned, we decided that leaf concepts under *physical entity* (the great majority of leaves) would be used to represent corresponding higher level concepts. Studies on concept representation [20] show that specialized concepts offer a good visual coherence, which led us to consider the most specialized nodes (leaf synsets) as possibly providing a better visual representation of general concepts than a direct representation of these last (the images for a general category is the union of image sets for its subconcepts).

Since WordNet is a representation of commonsense knowledge, the ontology is far from complete from a domain specialist's point of view. For example, the *placentals* hierarchy in WordNet contains only 1,112 synsets whereas another knowledge repository, Wikipedia[2] includes nearly 600 entries for *dog breeds* alone. Nevertheless, WordNet provides sufficient coverage of common terms for our image search application. To precise the level of detail of the knowledge contained in WordNet, we discuss the case *bear* and its hyponyms (Fig. 2).

The class *bear* includes immediate subtypes such as *ice bear* or *brown bear*, *bruin* or *bear cub*, *American* and *Asiatic black bear*, and *sloth bear*. *Brown bear* in turn is subdivided into *Syrian bear*, *grizzly* and *Alaskan brown bear*. We use the most specialized concepts (leaf nodes like *grizzly* or *ice bear* and their equivalents in the respective synsets) to collect pictures from a commercial

[2] http://wikipedia.org.

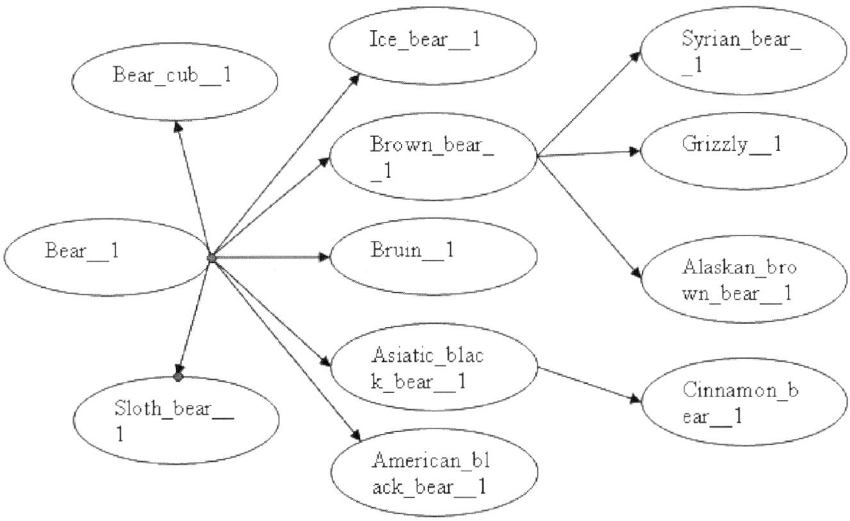

Fig. 2. WordNet hierachy for *bear*

picture search engine, Yahoo![3]. The image answers for *bear*, an intermediary node, are obtained by merging the results for all its leaf subtypes.

4.2 Image Database Constitution

The first step in our approach to constituting a structured picture repository is to gather images using Yahoo!. We describe here the procedure we developed in order to form an image database.

Ambiguity is one of the main sources of noise in the sets of images currently rendered by search engines when queried with a given term. In [11], it is shown that for WordNet2.0, 23,772 unique terms out of 81,795 are ambiguous. Moreover, ambiguous terms are by far more frequently used than monosemous ones. A good example of ambiguity is the term *angora* which stands for several concepts in the world. A part of the first page of picture results displayed by Google is presented in Fig. 3. It is to be noted that three senses of *angora* appear: *rabbit*, *cat* and *goat*. With the use of WordNet, a proposition of separate image sets for these different meanings of the same term is straightforward. For ambiguous terms (i.e. terms appearing in more than one WordNet synset), the query is expanded by adding the immediate hypernym of the concept. The utility of query expansion image retrieval is discussed in [9]. A good example of disambiguation, that of *drill*, is provided in the next section.

Terms in a language differ in the degree of generality of the concept they cover. For example, in WordNet, a *pit-bull* is a *pit-bull*, a *dog*, a *placental*,

[3] http://yahoo.com.

Fig. 3. Images for *angora* obtained with Google

an *animal*, a *physical entity* and, finally, an *entity*. Specialized concepts like *pit-bull* point to well determined entities in the world while a general term like entity can be (and is) used to point towards arbitrary things. We gathered images using queries with the leaf synsets under *physical entity* in WordNet. When querying the Web for the synset *letter opener*, we formed separate queries with all the members of the synset: *letter opener*, *paper knife* and *paperknife*. In WordNet, all these terms refer to the same type of objects in the world and should form an unified image class for them.

4.3 Image Retrieval Phase

Once the image database including pictures for categories under *physical entity* constituted and the images clustered into visually similar sets, it is possible to answer user queries about entities in the ontology. We present here some areas where the use of hierarchies of concepts is beneficial for image retrieval.

Use of Hypernymy

Existing picture search engines present results as a straight ranked list. The integration of the hypernymy relation in WordNet allows us to aggregate specialized concepts and to present conceptually structured responses. For example, in the image retrieval framework described here, the answers for *dog* or *car* are composed of all the leaf terms under these concepts.

Visualization problems appear when a concept has many leaves but this problem can be resolved for example by favouring frequent terms. One way to calculate term frequency is to exploit the information about the number of images for a concept offered by current search engines. If one queries for *dog*, there are over 100 classes that can be displayed and it is necessary to order them. With this measure, the first subclasses are: *puppy*, *pug*, *Newfoundland*, *basset*, *beagle*, *cairn*, *Airedale*, *Doberman*, *German shepherd* and

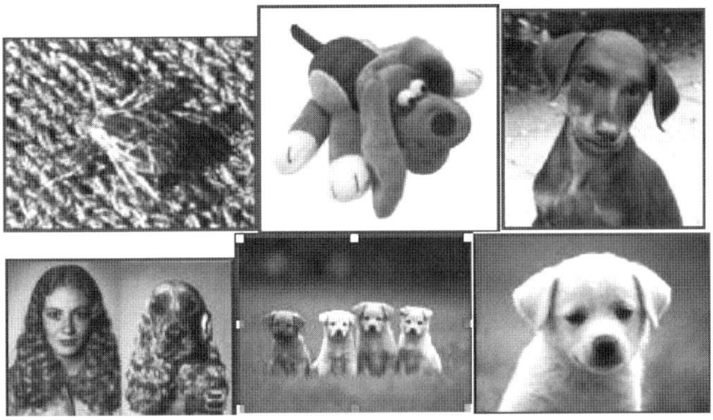

Fig. 4. Images for *dog* obtained with Yahoo!

basenji. Among the classes that are less frequent we cite *keeshond*, *Sealyham terrier*, *Old English shepherd dog* or *briard*. A query for *car* (a concept with 32 subtypes in the hierarchy) would be represented in priority by leaf concepts like *bus*, *electric car*, *coupe*, *jeep*, *roadster* or *convertible* while less representative subtypes are *gypsy cab*, *stock car*, *pace car* or *horseless carriage*. The order of concepts obtained in the two cases generally respects the knowledge a user has about *dogs* and *cars*.

Since WordNet was built to capture common knowledge, it is likely that the structure of the category is close to the way people represent that category and the results to be considered adequate by the users. The utility of a structure for the results is supported by the findings in [12], where it is shown that an organized presentation of image results is preferred over a plain list.

In Figs. 4 and 5, we compare an excerpt from the representation for the query *dog* from an existing application (results furnished by Yahoo!), and from our system which structures the results as subconcepts from WordNet.

The aggregation of specialized concepts into higher level ones allows the presentation of rich sets of answers while providing an economic way of stocking data. The image database contains only images for leaf concepts in the hierarchy but this is sufficient answer queries addressing any term in the ontology.

Images on the Web are frequently annotated with words that are not related to the image content. The probability of having such spurious annotations is higher for general concepts than for specialized ones. The use of leaf concepts for image gathering determines an improvement of the precision for the image sets associated to concepts (see Sect. 6 for a comparative test).

Sense Disambiguation

Current search engines do not discriminate among the several meanings a word can have. If a term points towards two or more separate entities in the

Fig. 5. Images for *dog* using leaf subconcepts and image clustering

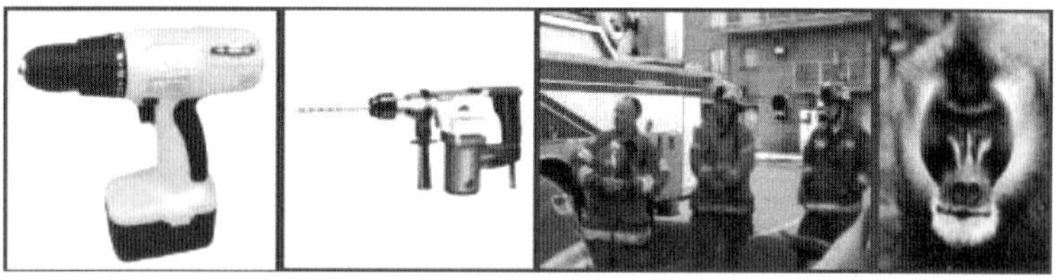

Fig. 6. Images for *drill* obtained with Yahoo!

world, the answer includes images related to all these terms. The average polysemy of nouns in WordNet[4] is 1.23 meanings (when a word is polysemic, it appears in an average of 2.77 synsets).

If a user queries for an ambiguous concept and a semantic resource like WordNet is employed, the image retrieval system can present separate sets of images for each meaning of the term. Reconsider the example of *drill.* Yahoo! returns both senses of the term as a *monkey*, as a *tool* and unrelated these two senses (Fig. 6). But knowing that this word is ambiguous allows the system automatically propose either *drill baboon* as one result set, for which we give an example in Fig. 7.

[4] http://wordnet.princeton.edu.

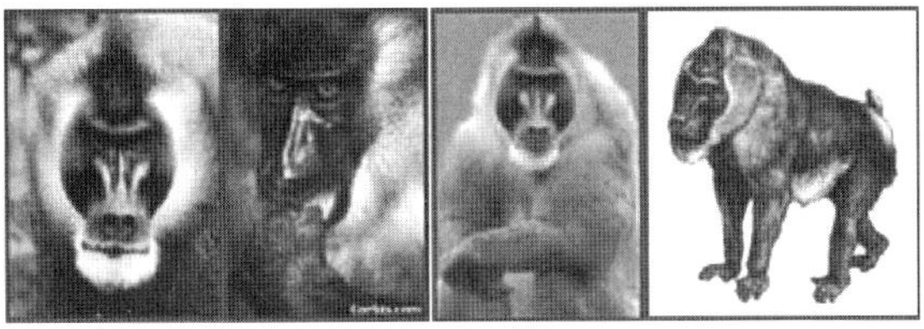

Fig. 7. Images for *drill* (and) *baboon* obtained with Yahoo!

Augmented Interactivity

A third advantage of using a semantic hierarchy such as WordNet is the possibility of offering an augmented interaction for the application. While the "keep it simple" directive is important for search engines, increased interactivity options may be useful if they account for user preferences. In [12], the authors present an user study which confirms that an increased interactivity for image retrieval applications is favorably assessed by the users. We are currently performing a study to assess user opinions concerning improved interactivity for search engines and their preferred options for browsing through image answers. Preliminary results confirm the findings in [12]. Some guidance can prove useful especially for cases when the user is browsing the results without a clear idea of the category or specific image he is looking for. A box labelled "related queries" on the answers page may help them to hone their query.

One of the major existing search engines, Ask, recently introduced the Askx platform, which employs a thesaurus to propose related classes in regard to a given query. For example, when a user queries for images of *dog*, the system equally proposes the following classes: *puppy*, *free dogs* as narrower search; *cat*, *horse*, *monkey*, *lion* as expanded search and *rabbit*, *wolf*, *boxer* as related names. These propositions are not always intuitive. First, is unclear why a query with *cat* is an expansion of a query with *dog* while a query with *rabbit* ranges in the related classes category. Second, *boxer* is subcategory of *dog* and it should be proposed in the narrower search class. Third, the proposed classes are only partially related to *dog*. For example, *monkey* or *lion* are weakly related to the seed class. Using an ontology can be an alternative to the use of a thesaurus to propose related classes. If we go back to the example of *dog* and use the knowledge in an ontology to display close concepts, we obtain: *poodle*, *corgi*, *spitz* or *hunting dog* as narrower classes; *wolf*, *fox*, *hyena* and *domestic cat* as related concepts and *canine*, *domestic animal*, *animal*, *organism* as expanded queries. When compared to the results proposed by Askx, the related classes in an ontology are better ordered and much more relevant in respect to the original query.

Since there is a danger of rejection of a computer application when one proposes too much content in a web page, the augmentation of interactivity should not exceed the cognitive capacities of the user. Many sets of images can be proposed for browsing and it is important to choose those that are the most relevant for a particular search. Some image retrieval practices (e.g. encyclopaedic, professional search or browsing for fun) are found to be frequent [7] and it is possible to propose adapted views of the results. These views (that would differ both in content and in the form of presentation) can be encoded in the ontology, with the instantiation of some class properties. We present here an anecdotal example of what an adaptation of the answers to the type of query might mean. If one asks for images of *dogs* in a encyclopaedic perspective, the system can focus on displaying results for *dog* races, eventually grouped following the probability for those races to be known by the user. If the same query appears in a browsing for fun perspective, the system might elicit first *funny dogs* like *pugs* or famous *dog cartoons* like *Scooby Doo* or *Pluto*. It is equally possible to propose user profiles that account for his preferences when using a search engine.

4.4 Image Clustering

Described elsewhere [27], we have developed a clustering tool that takes a textual query, fetches Internet images and furnishes clustered pictures as results. We exploited this clustering program to create our large pictorial image dictionary in the following manner. First, we created a dictionary entry for each leaf synset under *physical entity* in WordNet. The dictionary entry is indexed by the same name as in the OWL hierarchy we created. For example, the entry *grizzly_1*, corresponds to the synset *grizzly*, *grizzly bear*, *silvertip*, *silver-tip*, *Ursus horribilis*, *Ursus arctos horribilis*. Second, to collect images for this synset, search engine queries for each term in the synset were generated and launched. The union of all gathered images will be connected to the ontology class *grizzly_1*. But before attaching the images, we apply our clustering tool using a border/interior pixel classification algorithm [5] designed to index images from broad domains as we consider here. This tool clusters the indexed images using a k-SNN (k-Shared Nearest Neighbors) algorithm [21] in which the similarity between two images is related to the number of neighbors they share. The clusters are formed around a group of core images that possess the best connectivity with respect to other images, called aggregated images. This algorithm fits to our purposes as it is flexible: it does not impose a fixed number of picture clusters, or a fixed number of images in each class and, equally important, not all elements in the raw data set have to belong to a class.

The main role of the image clustering algorithm is to group images into visually coherent sets. In Fig. 8, some raw image data for *grizzly*, obtained by querying the Yahoo! search engine is shown. In Fig. 9, we present two image clusters obtained for the same concept after visually clustering the first 1,000 images returned by Yahoo!.

Fig. 8. Raw Yahoo! image data for *grizzly*

(A)

(B)

Fig. 9. Two image clusters for *grizzly* – obtained using visual characteristics of retrieved images

Comparing images from Figs. 8 and 9a and b, we observe that latter are thematically structured while the raw images of Fig. 8 do not present the same kind of coherence. Generally, the clustering step performs well and groups together visually similar images. The clustered images are finally attached to the dictionary entry, and then attached to the OWL hierarchy.

5 Limits

There are two types of limits we discuss here: image processing related ones and hierarchy related. We present them in the following subsections.

5.1 Image Processing Limitations

One important limitation of our system comes from the fact that existing image indexers are not capable to perform reasonably outside very narrow domains. When dealing with pictures having a highly variable content (as it is the case for Web images), it is impossible to improve precision in the image clusters when passing from the raw data to clustered data. This is to say that, currently, the image processing step does not introduce an improvement of precision for leaf concepts in the hierarchy.

Another problem is related to the fact that not all leaves are equally represented on the Web. Generally, we obtain hundreds of images for each leaf class in the ontology, but the variation between concepts is large. There are concepts that are represented by very few images or, at worst, not at all represented on the Web, but these are rare cases.

5.2 WordNet Coverage

Given the frequency of appearance of celebrity names as queries in commercial search engines [7], the inclusion of this kind of information in our ontology would be useful. The semi-automatic extension of WordNet constitutes an important research subject [15] and it might be possible to use such techniques in order to extend our OWL translation of WordNet. Since we are considering a service that is user-oriented, it is important to extend the ontology in accordance to the preferences of the public. We detain partial information concerning the queries people formulate when using image search engines [7] and the practices related to the use of image search engines are a subject that needs to be further researched.

There also exist meanings of terms that are not present in WordNet. There are two main cases that appear: idiosyncratic or metaphorical meanings or novel meanings that are not integrated in the hierarchy. The former are sources for noise in image retrieval as we build an application that is destined to a large public (supposing a social acceptance of term meanings). Languages are moving corpuses [4] and, even for initiatives like WordNet where updated versions appear constantly, dictionaries only present static and partial views of language use. There are word meanings that become socially known before being introduced in dictionaries. In these situations, it is impossible to present associated image classes.

5.3 Complex Queries

The approach we have adopted integrating WordNet in an image retrieval framework is applicable exclusively to terms designing single entities. It provides no solution for combinations of terms. This limitation does not restrain drastically the utility of the framework we present here because extensive studies of search engines [7] show that the users type short queries when searching for Internet images. In [7], the mean length of a query is 3.74 unique strings per query. It should be noted that this length includes image related terms (e.g.: "picture of", "photo") which were used to spot the picture related queries. If these last are subtracted, the mean length of image query is smaller than 2.74. Note that there is a difference between class names in WordNet and unique strings in [7]. In the former case, we count composed names like golden retriever or domestic cat as a term while they contain two unique strings.

The system might respond to more complex queries if other semantic resources were introduced in the framework. For example it is possible to use ConceptNet [13], a semantic network that includes commonsense knowledge about English terms. With its integration in the system, we might obtain information about the most current contexts for a concept. For example, we can provide answers for queries like: object and its habitual locations or object and related concepts. It might also be possible to generate queries involving all possible combinations of two or three entities. A small part of composed queries can be answered even in the current configuration. For example, a query like *dog* (and) *animal*, formed of two terms that can be found on the same vertical in line in the hierarchy and produce the same results as *dog*.

5.4 Quality of the Semantic Structure

The knowledge contained in WordNet is of good quality but, as it is the case with any man made hierarchy [4], there are flaws and inconsistencies. Many papers raise questions about the ontological status of this lexical hierarchy [6] but, at least in the Semantic Web community, people agree about its value as semantic resource. Its official translation to OWL [23], coauthored by people that have initially raised doubts about the ontological value of WordNet [6], demonstrates the perceived utility of the hierarchy. As one example of the types of inconsistencies we find in WordNet, consider the case of *iteration*, defined as "(computer science) a single execution of a set of instructions that are to be repeated"[5]. This concept is found under the *physical object* branch in the WordNet hierarchy, and it is hard to imagine a visually coherent image classes for *iteration* as it does not correspond to an entity in the world.

[5] http://wordnet.princeton.edu.

6 Evaluation

We evaluated our ontology based image retrieval method against a classical search engine, Yahoo!. Classical evaluation of image retrieval systems includes measures for precision and recall. In this paper, we present uniquely an evaluation of the precision measure. The rationale for this choice is simple: it is impossible to measure recall for Web images because we are not aware what is the total number of pictures representing a concept.

6.1 Experimental Settings

We proceeded as follows: a naive user was presented with result images corresponding to 15 search terms and asked, for each image, if it corresponded to the given search term. Two criteria guided the selection of the concepts to be evaluated:

- Coverage of different types of *physical entities*
- The chosen terms should be commonly known

Terms standing for both *natural objects* and man-made *artefacts* [10] were chosen. For the *natural objects*, both *biologic* (*animals* and *plants*) and *non-biologic entities* were selected. The familiarity with the terms was ensured by using basic concept names [20], the names we use in priority to name objects around us. For ambiguous terms like *rock*, only one meaning was retained as correct for evaluation.

For each search term, 100 pictures were assessed, 50 for each of the two evaluated image retrieval methods. Each image was separately prompted on the screen, with the corresponding search term and two buttons were used to decide if it was representative or not.

6.2 Results

In Table 1, we present the results of the evaluation. The values in the table correspond to the number of positive assessments in each of the random 50 image sets. The last line contains a mean of the precision for the two approaches. The mean precision in the image sets approaches 80% when using concept merging, while the same measure scarcely exceeds 60% for Yahoo!. Remarkable results are obtained for concepts like *car*, *cloud*, *dog*, or *eagle* where precision approaches 100%. Other concepts for which important improvements are obtained when employing an ontology in the retrieval process are: *apple*, *hammer*, *rock* or *toy*. Yahoo! performs better for two concepts, *knife* and *pants*, but the differences are not very important. For other concepts, like *dolphin*, *flower*, or *truck*, *slight* ameliorations of the results quality are obtained when employing the ontology.

The results for artefacts are less convincing than those for *natural concepts*. Still, there are encouraging results for *toy*, *car*, or *hammer*. This situation

Table 1. Comparison of image results using a general term and using its merged hyponyms

	Correct answers [out of 50]	
	Yahoo! query	Concept merging
Apple	10	30
Butterfly	35	35
Car	37	48
Cloud	22	47
Dog	35	50
Dolphin	33	35
Eagle	26	48
Flower	45	48
Hammer	11	23
House	39	39
Knife	40	36
Pants	41	40
Rock	32	42
Toy	19	38
Truck	34	37
Mean	**30.6**	**39.7**

might be explained by the fact that *natural terms* are well described as taxonomic constructions, while the same is not always true for *artefacts*. More study is needed in order to propose ontological structures for *artefacts* that will allow an increase of the precision when using subconcepts for image gathering.

With the use of subtypes to represent higher order concepts, we obtain more representative images of these concepts. Moreover, using the ontology and the clusters associated to leaf concepts, we have the additional advantage of being able to present the results in a structured and visually coherent manner, using the ontology labels as well as the clusters.

7 Perspectives

The results we presented in this paper encourage us to continue the efforts to define a semantic framework for image search engines. Our current work concentrates on:

- Enlarging the multilingual search framework presented in [19], which exploits the work in [18]. An extension of the this multilingual ontology in is possible, both in terms of coverage and of the number of included languages. Common sets of image answers, indexed in any of these languages, can be presented to the user in this case.
- Enriching parts of the ontology. For the moment, the ontology we use is minimal, as it reduces to a taxonomy, a definition of synonymy relations

and textual definition for the included concepts. Following an approach somewhat similar to [15] domain information that could assist image retrieval is to be included in the ontology. For example, for *animals* we know that they have certain development stages that make them visually different and this could be a structuring axis for the hierarchy of *animals*. There are equally frequent locations or colour characteristics we can introduce in the ontology and that could result in a richer structure for the answers provided to the user. These kinds of extensions of the ontology can be performed semi-automatically and they can help increase the interactivity of image retrieval systems and the quality of the results.

- Assessing user behaviour in image retrieval applications in order to formally describe practices related to this type of systems and to design future applications so as to account for common usages. This adaptation to practices can result in the creation of more interactive and understandable image retrieval applications. Ontologies can play a key role in such systems, as they can be employed to present results in a structured fashion.
- Proposing ways to identify and instantiate culturally encoded representations for abstract concepts in a language. There are concepts, like *human emotions*, *imaginary characters*, *justice* or *thinking* that can be indirectly represented using coherent image sets. These images do not represent the concepts directly, but a second order recognition can be elicited in the user.
- Employing ConceptNet [13], in conjunction with WordNet, to propose contextualized images for the concepts in the hierarchy. This joint use of the two resources would allow our system to respond to some queries that are formed of a conjunction of two terms. The inclusion of ConceptNet in the architecture of our system would enable it to provide answers to questions like: initial concept and its frequent locations, initial term and frequently associated concepts.

8 Conclusions

We have presented techniques for automatically associating images to terminal and non terminal classes in a large scale ontology derived from WordNet. Hypothesizing that a joint use of this lexical semantic resource and of image processing techniques can improve image retrieval, we have shown that we can ameliorate precision in a small number of image sets associated with general concepts in the ontology. We transformed the WordNet nouns hierarchy into an OWL ontology and used the leaves under physical entity to gather data from a classical image search engine and then indexed the raw data using image processing techniques to create visual signatures. These indexed images were then clustered to provide visually coherent image classes associated to leaf concepts in the hierarchy. As to the pertinence of the using leaf node labels compared to more general terms, we compared our results to those obtained by querying an image search engine for 15 familiar concepts and showed that the use of our technique produces improved results for these general terms. We equally discussed some other advantages of using a conceptual hierarchy in image retrieval, namely resolution of ambiguity, increased interactivity and proposition of structured results. The last three aspects are not yet fully tested but our current research is focused on their evaluation and the first results are very encouraging.

References

1. Barnard K and Forsyth D (2001) Learning the Semantic of Words and Pictures. In: Proc. of ICCV 2001, Vancouver, Canada: 408–415
2. Cimiano P, Handschuh S and Staab S (2004) Towards the Self-Annotating Web. In: Proc. of WWW 2004, Manhattan, NY: 462–471
3. Doulaverakis C, Nidekou E, Gounaris A and Kompatsiaris Y (2006) A Hybrid Ontology and Content-Based Search Engine for Multimedia Retrieval. In: Proc. of the 10th East - European Conference on Advances in Databases and Information Systems, ADBIS 2006, Thessalonki, Hellas
4. Eco U (1997) The Search for the Perfect Language. Blackwell Publishers
5. Ertoz L, Steibach M and Kumar V (2003) Finding Topics in Collections of Documents. A Shared Nearest Neighbor Approach. In: Wu W, Xiong H and Shekar S (eds) Clustering and Information Retrieval, Kluwer
6. Gangemi A, Navigli R and Velardi P (2003) The OntoWordNet Project: Extension and Axiomatisation of Conceptual Relations in WordNet. In: Proc. of CoopIS/DOA/ODBASE, Catania, Sicily, Italy: 689–706
7. Goodrum A and Spink A (2001) Image Searching on the Excite Web Search Engine. International Journal of Information Processing and Management 37, 2: 295–311
8. Guha R V and Lenat D B (1990) Cyc: A Midterm Report, AI Magazine 11, 3:32–59
9. Hollink L (2006) Semantic Annotation for Retrieval of Visual Resources. Vrije Universiteit Amsterdam
10. Keil F C (1992) Concepts, Kinds, and Conceptual Development. Bradford Books
11. Kuo C H, Huang Y T, Lan Y H and Chou T C (2004) Building Semantic Indexing for Image Retrieval Systems. In: Proc. of International Computer Symposium, Taipein, Taiwan: 208–213
12. Liao S P, Cheng P J, Chen R C and Chien L F (2005) LiveImage: Organizing Web Images by Relevant Concept. In: Proc. of the Workshop on the Science of the Artificial Hualien, Taiwan: 210–220
13. Liu H and Singh P ConceptNet (2004) A Practical Commonsense Reasoning Toolkit. BT Technology Journal, Kluwer Academic 22, 4: 211–226
14. Miller G A (1990) Nouns in WordNet: A Lexical Inheritance System. International Journal of Lexicography 3, 4: 245–264
15. Missikoff M, Navigli R and Velardi P (2002) Integrated Approach to Web Ontology Learning and Engineering. IEEE Computer, 35(11): 60–63
16. Pastra K (2006) Image - Language Association: Are We Looking at the Right Features?. In: Proc. of the Workshop on Language Resources for Content-based Image Retrieval, LREC 2006, Genoa, Italy: 40–44

17. Petridis K, Bloehdorn S, Saathoff C, Simou N, Dasiopoulou S, Tzouvaras V, Handschuh S, Avrithis Y, Kompatsiaris Y and Staab S (2006) Knowledge Representation and Semantic Annotation of Multimedia Content. IEEE Proceedings on Vision, Image and Signal Processing, 153/32: 55–262
18. Pianta E, Bentivogli L and Girardi C (2002) MultiWordNet: Developing an Aligned Multilingual Database. In: Proc. of the 1st International Conference on Global WordNet, Mysore, India: 293–302
19. Popescu A, Grefenstette G and Moellic P A (2007) Image Retrieval Using a Multilingual Ontology. accepted for RIAO2007, Pittsburgh, USA
20. Rosch E, Mervis C B, Gray W D, Johnson D M and Boyes-Braem P (1976) Basic Objects in Natural Categories. Cognitive Psychology, 8: 382–439
21. Stehling R O, Nascimento M A and Falcao A X (2002) Compact and Efficient Image Retrieval Approach Based on Border/Interior Pixel Classification. In: Proc. of CKIM 2002, Mc Lean, USA: 102–109
22. van Assem M, Gangemi A and Schreiber G (2006) RDF/OWL Representation of WordNet. http://www.w3.org/TR/2006/WD-wordnet-rdf-20060619
23. W3C (2004) OWL Web Ontology Language Overview. www.w3.org/TR/owl-features/
24. Wang X J, Ma W Y and Li X (2004) Data-driven Approach for Bridging the Cognitive Gap in Image Retrieval. In: Proc. of ICME 2004, Taipei, Taiwan: 2231–2234
25. Wang H, Liu S and Chia L T (2006) Does Ontology Help in Image Retrieval? - A comparison between Keyword, Text Ontology and Multi-Modality Ontology Approaches. In: Proc. of ACM Multimedia, Santa Barbara, CA: 109–112
26. Yang J, Liu W, Zhang H and Zhuang Y (2001) Thesaurus-Aided Approach for Image Browsing and Retrieval. In: IEEE Conference on Multimedia and Expo, Tokyo, Japan
27. Zinger S, Millet C, Mathieu B, Grefenstette G, Hede P and Moellic P A (2006) Clustering and Semantically Filtering Web Images to Create a Large Scale Image Ontology. In: Proc. of IS&T/SPIE 18th Symposium Electronic Imaging, San Jose, California, CA

Implementing an Internet Image Search Service Based on the AJAX Web Programming Techniques

I. Anagnostopoulos, C. Anagnostopoulos, I. Psoroulas, and I. Maglogiannis

Summary. Due to the extremely rapid growth of the web, search engines cannot spider all the new pages at the same time or with the same priority. Besides, the search engines index their catalogues with different algorithms, having as a result, different response time in updating their directories. Consequently, the user may lose some useful information resources if he chooses to use only one search service. Thus, the user has to resubmit his queries to more search engines, adding in this way, to a process that is already time-consuming [11]. Using a meta-search engine, additional information is provided, without the need to know the query language for the search services that contribute to the search. However, in this kind of retrieval a large amount of inappropriate and useless information is often returned to the user, especially when requiring images.

1 Introduction

This chapter proposes a meta-search engine, which combines fuzzy logic rules for human skin recognition joint with a probabilistic neural network for face detection. By these methods the proposed meta-search engine increases significantly the accuracy of the searches. In addition, the proposed information system continuously 'learns' according to the user's search preferences, maintaining in parallel record with his searches. The involved search engines are AlltheWeb, AltaVista, Excite, DMOZ (Open Directory Project), DirectHit, HotBot, Lycos, Northern Light and Yahoo. AJAX stands as a new web development model for several web applications (J.J. Garret, Ajax: A New Approach to Web Applications, http://www.adaptivepath.com/publications/essays/archives/000385.php). Even if, it does not employs new programming technologies, the whole approach is quite promising for future implementations [15]. AJAX is the abbreviation of Asynchronous Javascript and XML and thus its generic model consists of seven Internet programming technologies such as XML, JavaScript, HTTP and XHTML.

I. Anagnostopoulos et al.: *Implementing an Internet Image Search Service Based on the AJAX Web Programming Techniques*, Studies in Computational Intelligence (SCI) **93**, 97–116 (2008)
www.springerlink.com

There many web search services that use AJAX technology like the Google mapping (http://maps.google.com/), SNAP (http://www.snap.com/) and many others. AJAX technology is highly associated with many web services, which implement several widely-used and web programming-related technologies such as XML, DOM and HTTP. The following section presents the related work in the field of AJAX and web services as well as a generic model, which integrates both of them.

1.1 Internet Programming with Asynchronous Javascript and XML

As mentioned in the previous section AJAX is an emergent model in the client–server web programming, using up to six web programming technologies (JavaScript, HTML, CSS, DOM, XML, and XSLT) over the HTTP. In classic client–server web applications, a web page is provided by a web browser, displaying mainly static information. Dynamic nature requires the submission of new requests, which lead to an additional load in the network's traffic. However, in an AJAX-based web application, the downloaded web document includes code that allows the page to behave in a 'smart' way (Nexaweb, white paper, www.openajax.org/member/wiki/images/8/89/NexawebAjaxCharacteristics.pdf).

In other words, the additional code is loaded in the background, providing thus transparent functions while the web document is presented to the user. The transparent functions involve events (key strokes, mouse click, mouse-over, etc.) performing actions responding to these events, without submitting additional requests to the server. The left part of Fig. 1 depicts the way a web server provides web pages in the common client–server model, while in the right illustrates how a significant traffic load is reduced by transferring the server logic and its functions over the client-side in the engine that provides the AJAX-based services. In other words, when the server serves data in the AJAX engine instead of continuously serving web pages to the user's interface, more flexibility is offered to the client-side, avoiding extra client–server request and responses. Thus, user interactions with the server are asynchronously, while the information is provided without having to deliver server responses in the client browser.

2 System Architecture

In this section the major subsystems as well as the overall architecture of the proposed meta-search machine is presented. As far as the web interface and the respective technologies are concerned, the meta-search engine was created using JDK 1.4, while C++ was used for the applied fuzzy rules and the neural network implementation. Figure 2 presents the overall architecture of the proposed system and how its major parts interact, in order to collect, store evaluate and present the meta-results to the user. The 'PhotoSearch' function

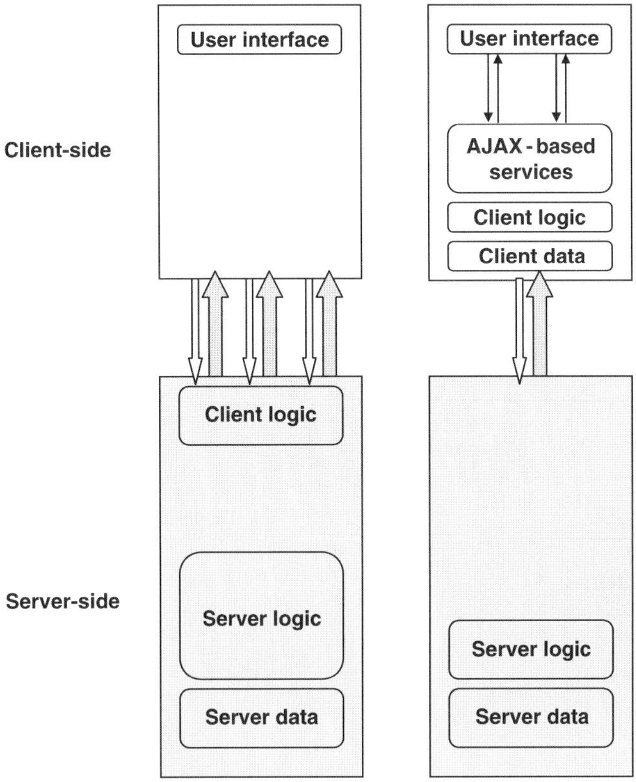

Fig. 1. Client/Server web programming vs. AJAX-based web programming

employs a probabilistic neural network fed by fuzzy Logic rules, aiming to offer a more precise search through image information analysis and retrieval. This function as well as the role of the fuzzy logic rules and the neural network is described in the followings. Figure 3 presents all the web programming layers in respect to the AJAX based technology used for implementing our Internet search service.

2.1 The User-Interface (UI)

Through the User Interface (UI), which resides in the upper web programming layer, the user expands the search into multiple search services. As mentioned before, the query is simultaneously submitted to different search engines. An important aspect of the UI includes a choice of which search engines to involve in the search. The UI of the proposed meta-search engine is presented in Fig. 4.

There are four major tabs namely 'Multi Download', 'Statistics', 'Meta-Results' and 'PhotoSearch'. Activating the 'Multi Download' tab, the user can optionally provide his username, in case he wants to personalise his queries.

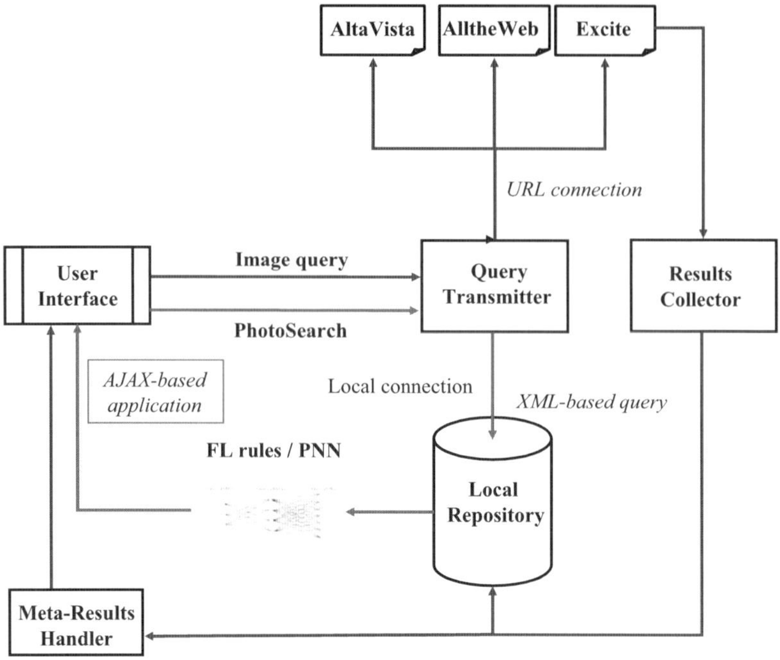

Fig. 2. The major system components

Studies presented that search engine users, usually restrict their queries to a specific subset of topics on the web [9]. Thus, when a user wants to search for a similar topic, he can work offline and search over the already returned meta-results. The user enters his query according to the proposed unified format. Finally, he decides upon the number of the involved search engines and defines their order of priority. This order of priority is crucial, since it initiates the meta-Results Handler. Afterwards, the user can either submit his query to the selected search services (on-line submission), or search in the local repository over previous downloaded results (off-line AJAX-based submission).

This feature significantly improves the response time and the efficiency of the service over the large amount of the disseminated information. Also, by default, the UI presents to the user his last query, which is submitted to each search engine respectively. Furthermore, Fig. 4 presents a screenshot of the UI in the case that the user decides to submit his image queries. The current version of the proposed meta-search engine uses three search engines (AltaVista, Excite, AlltheWeb) while the rest are inactive, since they do not support queries for images. Thus, in case the user selects one or more of the other search engines, the search status of the system is automatically set to the inactive mode. As a result, the query is not transmitted to those search engines that do not support image queries, while the respective priority order is ignored. Figure 4, presents a specific user query for images, where the system

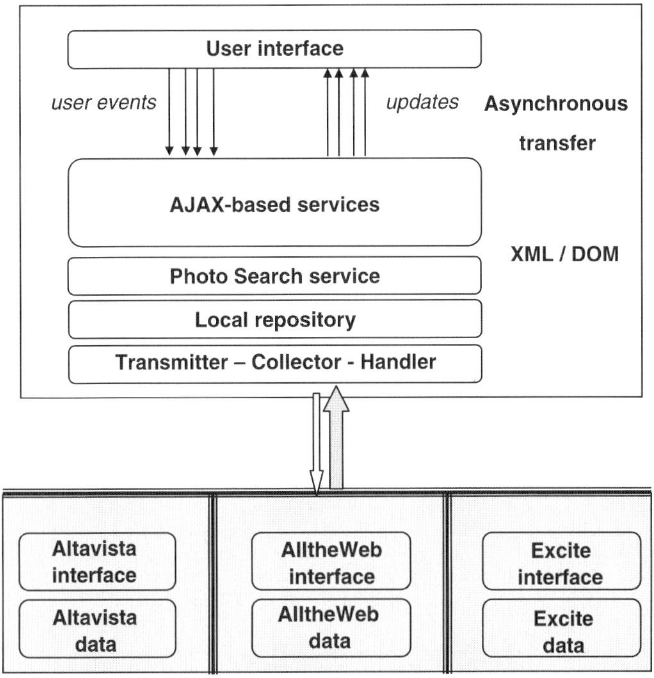

Fig. 3. The AJAX-based system architecture

collects and compare results from AltaVista, Excite and AllTheWeb in respect to this relative priority order. The UI of the proposed meta-search engine was designed for research purposes and according to the project's Technical Annex. Thus, the prototype which was delivered to the European Commission had the functionalities discussed in this chapter. Since the prototype meta-search engine was not developed for commercial use, the minimization of the information loss using visual features outclassed a user-friendly design orientation.

2.2 The Query-Transmitter

The transmission phase is initiated upon the submission of a query to the system. This phase begins with the simultaneous query submission to the selected web search services. The transmitter can be considered as an equivalent of the broker in the resource discovery system [10]. It translates the user query to the suitable script conforming to the interfaces of the selected search engines, creates a URL object and through an 'openStream()' method, sends the results to the Results-Collector. Since a meta-search engine depends on results extracted from other search engines, it does not need to spider or index sources. Instead, it only communicates with the search services. The transmitter needs to understand and query each service in its own interface, integrating

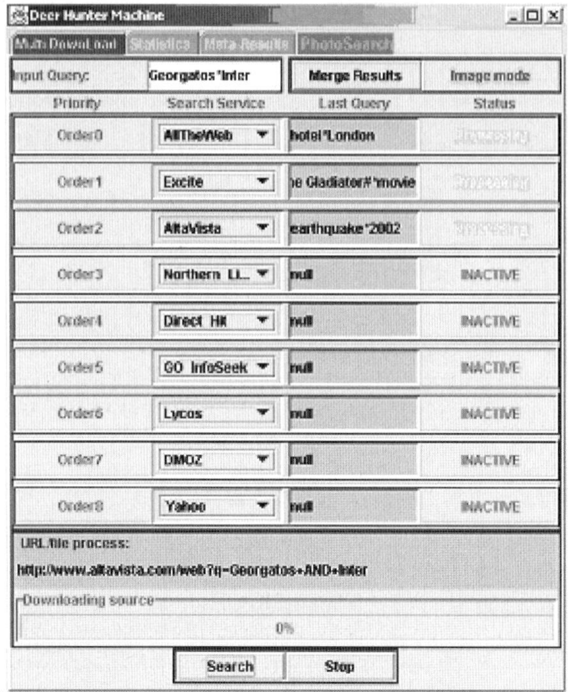

Fig. 4. Image query interface

varied search interfaces into a Generic format. The user configures the query-transmitter from the UI, by providing the query and defining the number of search engines to query in parallel. Furthermore, before query submission and before making any use of network resources, the proposed meta-search interface performs a check of resource accessibility such as network bandwidth and connection availability.

2.3 The Results-Collector

The next step is to collect the results returned from every submitted query. The query transmitter sends to the collector specific information fields, which correspond to the returned results of the used search services. The collector, which resides in the local server 'reads' the appropriate fields, such as the URLs and the respective description information areas, and stores them in the local repository. This procedure is performed in parallel for each selected search engine and is terminated when all the returned streams are indexed. It should be noted that indexing the necessary information fields is not an easy task, since each search service publishes its results in a different way. For example, the Excite search engine returns a stream, in which an additional URL object must be created for the collection of the URLs. This is due to the fact that the respective search engine is powered by Overture, transmitting an argument, which cannot be immediately resolved from the collector, affecting the collection procedure in terms of time. Finally, the results of the varied search engines are converted and stored in a unified format for further process.

2.4 The Local Repository and XML Handler

As mentioned in the previous paragraphs, the useful information collected from the related query streams is stored in the local repository as well as in XML-based files. For each search service, a unique table space is assigned in order to allow the user to view the contents of the local repository. Additionally, when a search query is submitted and results are obtained, the system keeps a record of each query instance, in order to provide statistics. These statistics can highlight several search engines features such as user query per used search services, as well as network statistics such as the download rate and the returned results parsing time. Furthermore, the local repository is kept totally independent of the previous subsystems (Query-Transmitter and Results-Collector), meaning that it can be processed off-line.

The user can observe how the ranking of the meta-results varies according to different search engine priority. For instance, working offline, the user can define a new order of priority and check whether for a particular request, a better ranking position for a specific result can be obtained. Finally, it should be mentioned that the local repository is responsible for handling all the requests for viewing, searching and modifying the meta-results and is also dedicated to handling requests according to user profiles.

2.5 The Meta-Results Handler

This is the last subsystem, dedicated to tackling the meta-results merging problem. The system uses a rank-based isolated merging method, since it uses readily available information from search servers, without requiring any other server functionality [3]. In other words, the proposed method employs server-assigned ordinal ranks in order to generate the merged list of the meta-results. Having solved the query-translation problem, the Results-Collector 'extracts' the required information from all the submitted services and updates the respective table spaces in the local repository with the meta-results and the user profile information. Then the duplicate results are removed from the table space with the $(p+1)$ priority in respect to the table space with the p priority, where p takes values from 1 up to the total number of the used search services. In this chapter the problem addressed is the problem of having a search engine ranking $S = \langle R, r\rangle$, consisted of a set of results and an ordering r. Given N ranking from N different search engines, the anticipated outcome is the generation of a single ranking $S_m = \langle R_m, r_m\rangle$, such that $R_m = R_1 \cup \ldots \cup R_N$ and r_m is the derived meta results ranking. Table 1 depicts the correlation

Table 1. Correlation between priority (p) and retrieved rank position (r)

		Search engine priority rank (p)					
		1	2	...	p	...	N
Result rank (r)	1	$p_1\ r_1$	$p_2\ r_1$	...	$p_p\ r_1$	...	$p_N\ r_1$
	2	$p_1\ r_2$	$p_2\ r_2$	...	$p_p\ r_2$	...	$p_N\ r_2$
	...	...	...	...	...	...	...
	r	$p_1\ r_r$	$p_2\ r_r$	...	$p_p\ r_r$	...	$p_N\ r_r$
	...	...	...	...	...	...	...
	K	$p_1\ r_k$	$p_2\ r_k$	...	$p_p\ r_k$	...	$p_N\ r_k$

between the user-defined search engine priority (p) and the returned ranking position (r).

In other words, the merging algorithm compares whether the information source retrieved in the rth rank position of search engine with priority p, exists until the $(r-1)$th rank position of the other selected search engines. The duplicate fields in the above sequence are eliminated while the procedure ends with the assignment of the last meta-result. The number of the meta-results is the total of returned results from all the involved search engines, after removing duplicated fields. The system allows the user to adjust the number of the returned results from each search service used. This number has a large impact on the total number and the presentation time of the meta-results, since it affects the response time of both the Results-Collector and the meta-Results Handler.

The collected results are considered as elements $R(p_x r_y)$ where x is the priority rank set by the user and y is the rank position of the result ordered by the selected search engine. If, for example, the user set AltaVista as his first search service preference, then the pair $p_1 r_r$ corresponds to the collected result, which AltaVista ranks at the rth position. The algorithm concurrently checks the pairs of the results, eliminates duplicated fields and assigns the meta-results, as it is shown in the following pseudocode:

```
Rm=0,  rm=0
for (y=1;  y ≤ k;  y++) {
           for (x=1;  x ≤ N,  x++) {
            if (x=N AND y≠k)
// ** check results between the last element of row y and the
first element of row y+1, y≠k
      Q(x)=R {pNry} ∩ R {p1ry+1}
              else
              // **check results on row y
                 Q(x)=R {pxry} ∩ R {px+1ry}
            if (Q(x)=0) {
            // **assign the meta-result Mi
```

```
                M_i=R {p_x r_y}
                R_m++
              }
            }
}
```

$$r_m = \bigcup_{i=1}^{R_m} M_i \quad \text{// proposed meta-results ranking}$$

where x is priority of the used search service, y is rank position of each collected result, N is the maximum number of the used search services, K is the maximum number of the results returned from a used search service, $R\{p_x r_y\}$ is the result in the rth position of the search service with pth priority, and M_i is the next meta-result, provided by the proposed meta-search engine.

3 Retrieving Images with Human Appearance on the Web

This section presents an example of using the proposed meta-search engine to reveal all possible results for an image query. As it was mentioned previously, AltaVista, Excite and AlltheWeb are engaged in the image search.

3.1 A Case Study

The case study presented here involves the boolean image query 'Georgatos AND Inter', where the user seeks photos of a football player who plays in the Italian team Inter. The system simultaneously translates the query and submits it to the three selected search services, respectively. Table 2 holds all the returned results in terms of returned images that contain the player and other images that are irrelevant to images with human appearance. After merging the results and removing the duplicates, the meta-search engine returned 79 images, 52 of which were photos that actually fulfil the submitted query (images with human appearance).

Table 2. Returned meta-results (case-study)

Search service	Photos with humans	Other images
AlltheWeb	19	15
AltaVista	37	4
Excite	7	14
Duplicate results	11	6
Total returned results (excluding duplicates)	52	27

It is obvious that a significant amount of information is actually irrelevant to the respective query. However, the innovation in this chapter stands in the fact that the user can further investigate among all the returned results in order to isolate photos with human presence. The existence of a human face, which is triggered by the potential human skin chromatic detection, is decisive since it implies human appearance and distinguishes a photo of a human from an image in general. Thus, the 'PhotoSearch' AJAX-based function initiates a two-step off-line mechanism, which is briefly explained in the following. A fuzzy logic mechanism checks for potential appearance of human skin, according to specific chromatic rules and in case of a positive identification, a neural network further verifies the existence of a human through face detection.

Service 1: Fuzzy Logic Rules

Every candidate image is analysed according to fuzzy logic rules for chromatic analysis in order to distinguish the web pages with possible human presence. This is possible due to the fact that the colour of human skin is distinctive from the colour of many other objects and, therefore, the statistical measurements of this attribute are of great importance for the problem addressed [12,18]. By evaluating skin tone statistics, it is expected that the face colour tones will be distributed over a discriminate space in the RGB colour plane, respectively. Thus, the first function of the proposed system is to locate potential skin areas in the image, using RGB colour and chrominance (CbCr) information [5,13].

Combination of two or more colour spaces have already been proposed in recent works as a single space is definitely not adequate for modelling human skin colour of different races under various illumination conditions. In the proposed system, the skin-masking algorithm is partially used along with RGB cluster groups that represent skin colour extracted from experimental tests in a large database of images with human presence [17]. The above measurements and the skin-masking algorithm formed the basis for the definition of the fuzzy logic rules. The aforementioned if-then rule statements are used to formulate the conditional statements that comprise the fuzzy logic-based skin colour detector. RGB or HSV are the most widely used model for colour modelling in computers and form the basis for other colour space transformation. However, in the RGB domain, each component of the picture (red, green and blue) has a different brightness. Therefore, in order to overcome this drawback, the proposed meta-search engine uses the YCbCr model since it was proved to be more representative for the choice of regions that suit human skin for face detection [6,16,20]. In addition, after a comparison among various colour spaces it features three advantages. It is the most skin tone independent model, it is widely used in video compression standards and the influence of luminosity can be removed during image processing [1,2,7,8,14]. Every image is mapped in two colour space models, namely the RGB and YCbCr. The following conversions are used to segment the RGB image into Y, Cb and Cr components:

$$Y = 0.257 * R + 0.504 * G + 0.098 * B + 16$$
$$Cb = 0.148 * R - 0.291 * G + 0.439 * B + 128$$
$$Cr = 0.439 * R - 0.368 * G - 0.071 * B + 128$$

However, despite the fact that RGB or Hue-Saturation-Value (HSV) are the most common colour spaces discussed for detecting skin region, an issue arises when different race colour is considered. The latest developments that have been presented in the literature propose the YCbCr model as the most appropriate one to provide good coverage of all human races [1, 8, 14]. These works describe that the apparent difference among skin tones is due to the brightness or the intensity. It is highlighted that the intensity value has little influence on the distribution of skin colour in the CbCr plane and that the skin colours form a small and very compact cluster in the CbCr plane. This conclusion indicates that CbCr plane is the most important one for the definition of skin colour distribution and that the Y value can be neglected. More specifically, skin colours classification was performed directly in the chrominance plane CbCr without taking the intensity value Y into account as initially suggested in [19]. Thus, by adding fuzzy logic rules, the performance of the proposed skin detection algorithm is significantly improved.

Through the application of fuzzy Logic rules, the proposed system decides whether a specified window in the inspected image contains a potential skin region. However, a skin region does not represent always a face, and therefore the candidate area should be further checked in order to discern whether it represent a face or not.

According to these methods, the initial step is to receive the inputs and determine the degree in which they belong to each of the appropriate fuzzy sets, through membership functions. Once the inputs have been fuzzified, the fuzzy logical operations have to be implemented. For this application the OR operator was used. The weights in every rule were set equal to one and the aggregation method for the rules was the maximum value. Finally, the defuzzification method is the middle of maximum (the average of the maximum value) of the output set. Since the processing time depends on the size of the tested image, before the described procedure the candidate image is transformed in a unified size of 250×250 pixels. In a Pentium IV at 3.0 MHz with 1,024 MB RAM, the required time for skin area detection was measured at nearly 0.3 s. Figure 5 illustrates the detection of possible human skin areas in a downloaded image (input) and its transformation to a grey scale image (output) according the proposed fuzzy logic rules.

Service 2: The ANN for Image Classification

Having collected images with possible skin areas, the next step involves the correct identification of images with human faces. This requires further image

Fig. 5. Human skin detection

processing steps in order to properly feed the image classifier. The image-processing operations consist of four distinct parts.

Firstly, potential skin areas are clustered to form the Region of Interest (RoI), roughly describing its shape, on the basis of the FL output. Each image is transformed in grey scale and in the specific size of 100×100 pixels. Then two morphological operations, which help to eliminate some of the noise in the tested image, are involved. In particular, simple erosion with a 10×10 matrix of ones is performed followed by dilation. Further on, the created image is parsed through a skeletonisation technique, removing simultaneously all the areas that are considered as 'holes'.

Having defined the RoI in the previous part, in the second step, the algorithm is applied to the initial tested image, merging objects that belong to one defect and performing a simple dilation once again, with a structural element, which is a 5×5 matrix of ones. With this technique, segmented pixels in the same neighbourhood, are merged in one region. All the image parts that are included in the defined RoIs, are then transformed to grey scale. In the following part, all the segmented images are resized to a specific size of 225×225 pixels. Finally, the 225×225 pixel images are divided into non-overlapping sub-images of size 15×15 and the mean value for each is calculated, followed by histogram equalization, which expands the range of intensities in the window. During this procedure, a lower resolution image in respect to the RoI is created, forming in parallel a descriptor vector that consists of 225 grey scale values from 0 to 255.

The proposed ANN is trained to identify which of the skin regions detected from the service 1 represent facial photos. The training set of the ANN consists of a large group of images sized 15×15, representing face regions or other skin areas. The idea of this approach was motivated by the observation that human faces present a high degree of resemblance when they are sampled in low-resolution [4]. This is quite natural, since all faces have darker areas, which represent the eyes and the mouth. It is undoubtedly easier for an ANN to recognize the presence or absence of a face, judging from a low quality image.

The ANN is a two layer Probabilistic Neural Network with biases and Radial Basis Neurons in the first layer and Competitive Neurons in the second one. Training a neural network for the task of face detection is quite challenging due to the difficulty in characterizing prototypical 'non-face' images. Unlike in face recognition, where the classes to be discriminated are different faces, in face detection, the two classes to be discriminated are 'face area' and 'non-face area'.

A sample of 172 frontal view face images was used as the training set for the class 'Face', as well as a large sample of 395 images corresponding to other correct skin areas, such as hands and legs, or to erroneously detected objects. The training set consists of 567 sub-images sized 15×15 in a vector form, as these were extracted from 124 colour images according to the proposed image processing steps. In other words, the neural network 'learned' to identify 170 from the 172 sub-images corresponding to human faces, as well as 391 from the 395 sub-images corresponding to other skin areas and objects. The time needed for the completion of one training epoch in a Pentium IV at 1.5 MHz with 512 MB RAM, was 22 s. The topology of the proposed neural network is 225-567-2.

Performance of the Off-line Services

Prior to testing the proposed meta-search engine, the performance of the add-in service for human face detection was examined using 412 colour images of various extensions, types and sizes that contained 723 human faces. The fuzzy logic rules segmented 1,261 skin areas. However, 42 skin areas that contained a human faces were not selected among them and therefore the precision of the system was measured at 94.19% ($P_{FL} = 681/723 = 0.9419$). Following the fuzzy logic system, the ANN received the 1,261 skin areas and decided that 610 of them represent faces. Thus, the precision of the ANN was 89.57% ($P_{ANN} = 610/681 = 0.8957$). Finally, the overall system performance was measured at 84.37%, since 610 from a total of 723 faces were successfully identified. All the results are presented in Table 3.

3.2 The Meta-Search Engine Performance

In the specific submitted query regarding the case study, a large amount of the returned images are irrelevant to photos of the soccer player. For instance, some of them represented stadium images, t-shirts of the team, or logos. Having removed the duplicated fields, the returned meta-results, after excluding the duplicated ones, were 79, as presented in the Appendix 1. By initiating the two services (FL rules and PNN), the user narrows his search in retrieving photos only. After analysing the tested images with the FL rules, 68 images were sent for further identification. Finally, the PNN returned 33 images in respect to the submitted query as presented in the Appendix 2, detecting correctly the human faces among the candidate group (68 images), successfully excluding those images that included regions of interest with potential skin areas but

Table 3. Photo-search performance

Submitted image queries	421	
Total returned meta-results	Photos	Other
20,871	12,576	8,295
	FL Rules	
Human skin identification	12,032	3,179
	ANN	
Face recognition	11,310	246
	'PhotoSearch' returned meta-results 11,556	
	Excluded meta-results 9,315	
	Excluded Relevant	Excluded Irrelevant
FL rules	544	5,116
ANN	722	2,933
Total	1,266	8,049
	Reduction in Irrelevant information	
8,049/8,295		97.03%
	Information Loss	
1,266/12,576		10.06%

not a human face. The time required for each tested image was approximately 0.3 s, involving the fuzzy logic rules and the artificial neural network processing. Consequently, the user got a significantly precise amount of information concerning photos from the Boolean image query 'Georgatos AND Inter'.

Table 3 holds some results over 421 mixed Boolean image queries that concern persons as they were collected by the three search services, which support image retrieval. Therefore, over 20,871 images, from which 12,576 were actually images with human presence, fuzzy logic rules identified skin regions in 12,032 among them. It was evaluated that the failure in this step occurred due to the fact that the chrominance fuzzy logic rules do not cover sufficiently all the variation in terms of the human skin colour. Other reasons come from the area of image processing and involve limitations in illumination conditions or image texture and inconsistency.

The ANN further processed the above set of images and decided that 11,310 are actually photos that contain human faces, while 722 images were erroneously excluded from this class. On the other hand and in accordance with the images that really presented human faces, the procedure added erroneously 246 images, which are unrelated to the photo query submissions. Using the AJAX-based 'SearchPhoto' meta-search procedure a significant amount of irrelevant information was excluded (8,049 images over 8,295). However, the system fails and erroneously rejects 1,266 photos, defining by this an

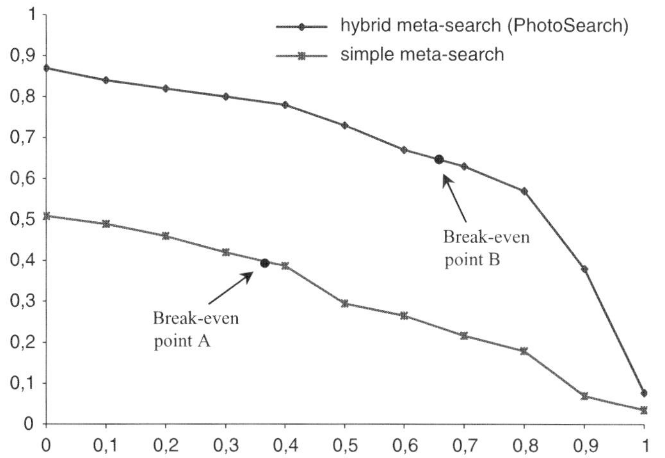

Fig. 6. Precision performance

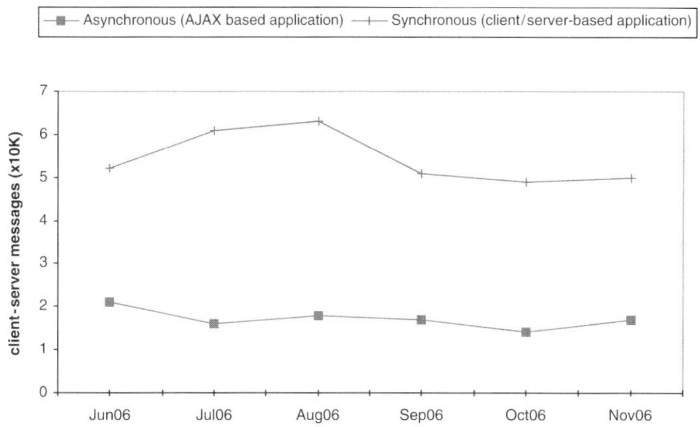

Fig. 7. Traffic analysis (AJAX-based vs. classic client/server model)

information loss close to 10%. Figure 6 outlines the Precision-Recall diagrams in case of using the proposed image meta-search mode comparing with the simple search mode. In order to measure Precision and Recall, results that contain at least one human face were considered as relevant while the other as irrelevant. As it is shown, the break-even point is significantly increased from 39% (point A) to 68% (point B) when the user selects to use the 'PhotoSearch' mode.

3.3 Asynchronous–Synchronous Analysis Performance

Figure 7 depicts the amount of client–server messages (request/responses) regarding a pilot study conducted over a five-month period. In this pilot study we measured the traffic overhead of the round-trip information exchanged between the client and the server in case where the proposed search engine was running over a AJAX engine and over a classic client/server model. A significant reduction (in average values) verifies the positive influence of using the AJAX technology when performing meta-search on the web. Comparing the classic client/server request/response model with the AJAX architecture two different concepts are arising. In the server-centric model the large amount of round-trip messages make the model not able to deliver the same performance and availability as the client-centric model of AJAX technology. The main benefit in client-centric models is that significant application logic is deployed on the client side and most of user interactions can be processed locally without incurring a round trip to the server (Nexaweb, white paper, www.openajax.org/member/wiki/images/8/89/NexawebAjaxCharacteristics.pdf). In addition, the whole system is robust in terms of several network connectivity failures and traffic overload over the time where the user interacts with the system.

4 Conclusions: Future Considerations

This chapter proposes an AJAX-based image meta-search engine, aiming to provide more precise and accurate search results for queries that concern images with human appearance. For the purposes of this work an additional services were implemented in the client-side, involving fuzzy Logic rules for identifying possible human presence and an artificial neural network for face detection. The user can receive more precise amount of information in respect to the submitted query, record his search preferences and narrow his search when seeking photos in the web. Despite the limitations confronted in terms of different image types, sizes and illumination conditions, the procedure is quite effective, since it excludes a large amount of irrelevant information in image search. Another potential application of the proposed system is the direct installation of the proposed system to the image database/repository of a search engine in order to reduce the processing time, which in this proposal is performed locally in the client-side.

The AJAX architecture is a way of developing richer and more interactive web applications at the client-side. Such kind of client-side applications are called as RIA (Rich Internet Applications). However, AJAX stands as a scripting based RIA approach with many advantages and disadvantages as well. Scripting-based approaches have the advantages of easy programming (scripting), they are compatible with most existing hypertext infrastructure, while they offer built-in support in most browsers. On the other hand, Object-Oriented Programming-based (OOP-based) approaches have the advantages of object oriented programming and thus are suitable for team development, while they offer scalability and reliability. Future work involves further investigation over the proposed meta-search engine's performance when the prosperity in the client-side is provided with an OOP-based approach, such as using the .NET framework.

Appendix 1 (Initially Retrieved Images)

AllTheWeb

Excite

AltaVista

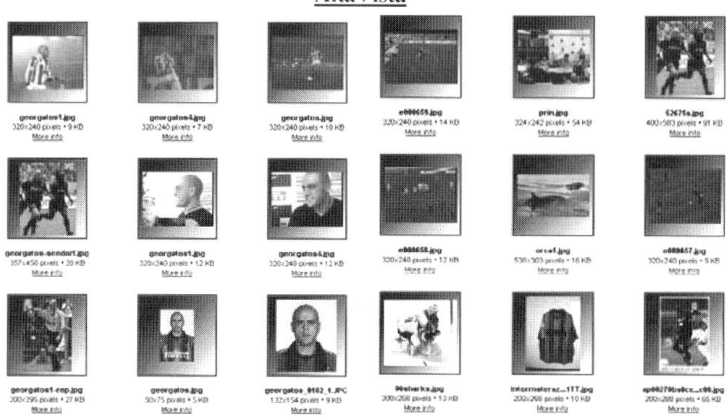

(continued)

AltaVista (continued)

Appendix 2 (Provided Photos from the Proposed Meta-Search Engine)

References

1. Alatan AA, Onural L, Wollborn M, Mech R, Tuncel E, Sikora T (1998) Image sequence analysis for emerging interactive multimedia services – the European COST 211 framework. IEEE Transactions on Circuits and Systems for Video Technology. 8(7):802–813
2. Chiou-Shann F, Shun-Wen C, Essig K (2000) Hierarchical color image region segmentation for content-based image retrieval system. IEEE Transactions on Image Processing. 9(1):156–162
3. Craswell N, Hawking D, Thistlewaite PB (1999) Merging results from isolated search engines. In: Proceedings of the 10th Australasian Database Conference, pp. 189–200
4. Dai Y, Nakano Y (1998) Recognition of facial images with low resolution using a Hopfield memory model. Pattern Recognition. 31(2):159–167
5. Garcia C, Tziritas G (1999) Face detection using quantized skin color regions merging and wavelet packet analysis. IEEE Transactions on Multimedia. 1(3):264–277
6. Habili N, Cheng-Chew L, Moini A (2002) Automatic human skin segmentation based on color information in the YCbCr color space. In: Proceedings of the Information, Decision and Control Conference Final Program and Abstracts, pp. 377–382
7. Habili N, Cheng-Chew L, Moini A (2004) Segmentation of the face and hands in sign language video sequences using color and motion cues. IEEE Transactions on Circuits and Systems for Video Technology. 14(8):1086–1097
8. Hsu RL, Abdel-Mottaleb M, Jain AK (2002) Face detection in color images. IEEE Transactions in Pattern Analysis and Machine Intelligence. 24(5):696–706
9. Jang MG, Myaeng, SH, Park SH (1999) Using mutual information to resolve query translation ambiguities and query term weighting. In: Proceedings of the 37th Annual Meeting of the Association for computational Linguistics. Washington DC, USA, pp. 223–229
10. Lawrence S, Giles CL (1999) Searching the Web: general and scientific information access. IEEE Communications Magazine. 37(1):116–122
11. Liu F, Yu C, Meng W (2004) Personalized Web search for improving retrieval effectiveness. IEEE Transactions on Knowledge and Data Engineering. 16(1): 28–40
12. Mirmehdi M, Petrou M (2000) Segmentation of color textures. IEEE Transactions on Machine Intelligence and Pattern Analysis. 22(2):142–159
13. Nikolaidis A, Pitas I (2000) Robust watermarking of facial images based on salient geometric pattern matching. IEEE Transactions on Multimedia. 2(3):172–184
14. Phung SL, Bouzerdoum A, Chai D (2002) A novel skin color model in YCbCr color space and its application to human face detection. In: Proceedings of International Conference on Image Processing (ICIP2002), Vol. 1, pp. 289–292
15. Sayar A, Pierce M, Fox G (2006) Integrating AJAX approach into GIS visualization Web services. In: Proceedings of International Conference on Internet and Web Applications and Services/Advanced International Conference on Telecommunications (AICT–ICIW 2006), p. 169
16. Sigal L, Sclaroff S, Athitsos V (2004) Skin color-based video segmentation under time-varying illumination. IEEE Transactions on Pattern Analysis and Machine Intelligence. 26(7):862–877

17. Umbaugh SE (1998) Computer Vision and Image Processing. Prentice Hall PTR, Upper Saddle River, NJ, USA
18. Voth D (2003) Face recognition technology. IEEE Intelligent Systems. 18(3):4–7
19. Wang H, Chang SF (1997) A highly efficient system for automatic face region detection in MPEG video. IEEE Transactions on Circuits and Systems for Video Technology. 7(4):615–628
20. Zhi-fang L, Zhi-sheng Y, Jain AK, Yun-qiong W (2003) Face detection and facial feature extraction in color image. In: Proceedings of the 5th International Conference on Computational Intelligence and Multimedia Applications ICCIMA 2003, pp. 126–130

A Rule-Based Adaption Model for Ontology-Based Personalization

Thanh Tran, Philipp Cimiano, and Anupriya Ankolekar

Institute AIFB, University of Karlsruhe, Germany
dtr@aifb.uni-karlsruhe.de,pci@aifb.uni-karlsruhe.de,
aan@aifb.uni-karlsruhe.de

Summary. Various adaptive hypermedia systems have been proposed to alleviate information overload on the Web by personalising the delivery of information and resources to the user. These systems have however been afflicted with difficulties in the acquisition of an accurate user model, a limited degree of customization offered to the user as well as general lack of user control on and transparency of the systems' adaptive behavior. In this chapter, we argue that the use of rules on top on ontologies can enable adaptive functionality that is transparent and controllable for users. To this end, we present ODAS, a domain ontology for adaptive hypermedia systems, and a model for the specification of ODAS-based adaptation rules. We demonstrate the use of this model by showing how it can be instantiated within a knowledge portal to arrive at rules that exploit ODAS semantics to perform meaningful personalization.

1 Introduction

Without assistance, the vast information space of the Web leads to confusion for the average user, manifested by "comprehension and orientation problems" and a general "loss in information space" [19]. A common approach to reduce such confusion, used by many existing Web portals, such as Yahoo[1], Ebay[2] and Amazon[3], is to filter out irrelevant information for the user and only provide personalized content. These portals are often both **adaptable**, i.e. allow for manual configuration by the user, as well as **adaptive**, i.e. providing proactive personalization to address the ad-hoc needs of the user. Such adaptive behaviour is typically realised by either collaborative filtering [15], which identifies content found relevant by similar users, or content-based filtering [16], which exploits similar content to identify relevant resources for a certain user.

[1] http://www.yahoo.com.
[2] http://www.ebay.com.
[3] http://www.amazon.com.

T. Tran et al.: *A Rule-Based Adaption Model for Ontology-Based Personalization*, Studies in Computational Intelligence (SCI) **93**, 117–135 (2008)
www.springerlink.com

In order to provide better personalization functionality, more dimensions such as the task, the system and the environment need to be considered. The increase in contextual information and the use of advanced machine learning techniques leading to better user models can be expected to provide improved personalization functionality. However, apart from the inherent difficulties in collecting model information, these approaches are criticized as being black-boxes that cannot be inspected.

A major drawback of both filtering approaches is that the models they develop of users and content are black-box-like and not amenable to user inspection or modification. If, however, we could represent the adaptation logic in the form of rules, users would have much greater control over the system. They would be able to inspect, possibly understand, and even modify the underlying adaption model. In this chapter, we present a model for the specification of such adaptation rules. In addition, we present several examples to illustrate how this rule-based model can be instantiated to specify different styles of adaptive behavior.

In this paper, we argue that an ontology-based approach to personalization can successfully address these drawbacks. We discuss an ontology capable of capturing information about the user, the task, the system, the environment, and various aspects of the content (structure and presentation), i.e. all dimensions that have been proven to be useful in existing approaches. This maximizes the amount of contextual information that can be used to accomplish sophisticated adaptation. Moreover, current adaptive hypermedia systems rely on their own formalism and vocabulary for data representation. By the use of a standardized ontology, we argue that systems can share and reuse model information to solve the inherent lack of data that hinders sophisticated personalization.

In the remainder of the paper, we first discuss in greater detail the benefits of an ontology-based personalization approach in Sect. 2. Then, in Sect. 2.3, we briefly present a domain ontology, ODAS, to introduce some concepts that we will use in the specifcation of adaptation rules. The main bulk of this paper is contained in Sect. 3, which introduces and illustrates by way of examples, a rule-based model for ontology-based adaptation. Finally, we review other ontology-based approaches in Sect. 5, before pointing towards future work in Sect. 6.

2 Overview of Adaptation Approaches

In this section, we present a brief survey of the main adaptation approaches and their drawbacks, arguing that ontology-based approaches plus the use of rules can overcome most of these drawbacks.

2.1 State-of-the-Art

Many authors agree on three different levels of adaptation. Whereas **adaptive content** focuses on selecting the appropriate content in different multimedia resources, **adaptive presentation** is concerned with the visualization of the content, the interactive elements and the appropriate media. Finally, **adaptive structure** deals with the number, appearance, order and target of links [2].

In current commercial applications, filtering-based approaches are state-of-the-art. Content-based filtering applies different similarity measures on a representational model of the content. So, if the user prefers a product, similar product pages will be suggested [16]. As the system always recommends more of what the user has already indicated an interest for, there is a potential problem of **overspecialization**, reducing the possibility of serendipitous finds.

Collaborative filtering assumes similar users have similar interests, so recommendations are based on user similarity [15]. This assumption reduces the role of individual preferences. In addition, it is not possible to recommend a **new item** not yet rated by users. When there is only **sparse data**, i.e. when the number of users is small relative to the amount of content, typically only a small percentage of the resources will be accessed by the user. The spare coverage of ratings then results in an accordingly sparse set of recommendable items [20].

Noting that both approaches are complementary, some authors suggest combining them [1]. Subjective ratings of user groups counterbalance the lack of subjective data about resources. New items can be recommended on the basis of their content. Also, sparsity is solved when recommendations do not depend merely on user ratings. Yet, incorporating even more sources can further improve personalization. It has been shown that information related to the task [3], the environment and the system [13] can be relevant for adaptation. Furthermore, by means of an explicit presentation model, an adapted presentation of the content can even be generated on the fly (compare [23]).

2.2 On the Use of Ontologies and Rules for Adaptation

While these advances have lead to sophisticated adaptive hypermedia systems, there are still two drawbacks that we think can be addressed by the use of ontologies and rules.

First of all, these systems are seen as **black boxes**, computerized oracles which give advice but do not make their model explicit such that it can be inspected [11]. This is due to the fact that the underlying algorithms compute latent factors and heuristics that cannot be translated to rationales for the adaptive behavior. Consequently, the reasons behind the recommendations cannot be explained to the user. This is different when using rules. When adaptive behavior is captured by rules, inference engines processing them produce recommendations that are more accessible to the user. Underlying inferences can be analyzed, provided with a proof and the rules can be made available for the user to be inspected and modified. This could allow for feedback loops, user-control and thereby enhance user trust towards the system. These benefits will be illustrated in section 3.

Second, current adaptive hypermedia systems face difficulty in **obtaining model information**. Sophisticated user models are hard to develop in systems where the duration and the type of interactions do not provide much relevant information. Yet users may expose lots of useful information at applications they frequently use. Note that the quality of adaptation depends on the amount of information that can be drawn on. Consequently, adaptive hypermedia systems benefit from the ability to exchange and reuse information. This ability is particularly important in the context of open-corpus adaptation [9]. Such a corpus may be the Web and the ultimate objective is to reduce the users' cognitive overload by providing a personalized view on the resources available on the Web.

Currently, the exchange of information is made possible mostly by using XML-based protocols. An XML schema defines the syntax, i.e. the structure and labels of the data elements that are to be exchanged across systems. In this respect, it may be sufficient to exchange standard metadata and model information that are known to the involved parties, e.g. the title and author information of a PDF document. They have implicitly agreed on the semantics of these data elements and a priori know how to process them. However, exchanging data in the domain of adaptive hypermedia systems, and especially in an open-corpus, requires more flexibility. In fact, adaptive behavior may build on explicit, semantic descriptions of the current task of the user, the resources he/she is currently reading, etc. in form of an ontology. Thus, the semantics of the information transferred and reused across systems must be formalized so that it can be exploited in the adaptation process. For instance, the semantics of the content syndicated from external sites must be processable to the user's system in order for them to be personalized in a dynamic and straightforward manner.

Ontologies can enable such semantic interoperability. Adding further capabilities to the stack already available with XML, ontologies can be used to specify the semantics of data elements shared across systems. With an XML schema, the semantics is only available to the people that have specified it. Using ontologies, the semantics can be determined automatically by the systems at runtime [22]. Therefore, the use of ontologies can address the lack of data in current adaptive hypermedia systems. They can share and reuse data, and on the basis of the semantics formally specified in the accompanied ontologies, they can make sense of the data to exploit it for adaptation. In the same manner, ontologies can facilitate adaptation in an open corpus where the semantics of the available resources is made explicit to the systems.

2.3 ODAS - A Domain Ontology for Adaptive Hypermedia Systems

A crucial point in the development of adaptive hypermedia systems is the identification of information sources other than related to the user and content. Each of these sources establishes one aspect of the adaptation context that can help to improve the sophistication and effectiveness of adaptive behavior.

Figure 1 shows a portion of the subclass hierarchy of ODAS, an ontology for the domain of adaptive hypermedia systems we have developed. We will now illustrate how the different aspects of the context can be represented by the use of this ontology and why they are useful for adaptation. The different adaption dimensions are highlighted by rectangles in Fig. 1 and, henceforth, will be referred to as models.

Central to the representation of the adaptation context is the notion of `Process`. `Application Interaction`, for instance, tells the system that a particular `User` is currently interacting with a `Content` resource (**resource model**) of the `Application` (**system model** part of the **environment model**) to accomplish a task. Indirectly, this task may have been modelled in the system as a `Composite Process`. Precisely, the workflows required to accomplish the tasks supported by the system can be represented in the system as instances of a `Computer-aided Process` (**task model**). Such a process-orientated representation of a task comprises many `Application Interactions`. When the current interaction is identified as part of such a workflow, the modelled `output` of the respective `Process` can be implicitly assumed to be the user's goal. Ideally, the adaptation should then be performed in a way that supports the user in achieving this goal. Section 3.3 shows rules that make use of the relations among sub-activities of this task model, e.g. `is pre` and `is post`, to perform task-based adaptation.

`Application Interactions` thus contain information about the `Content` currently processed by the `User`. Different content types are distinguished

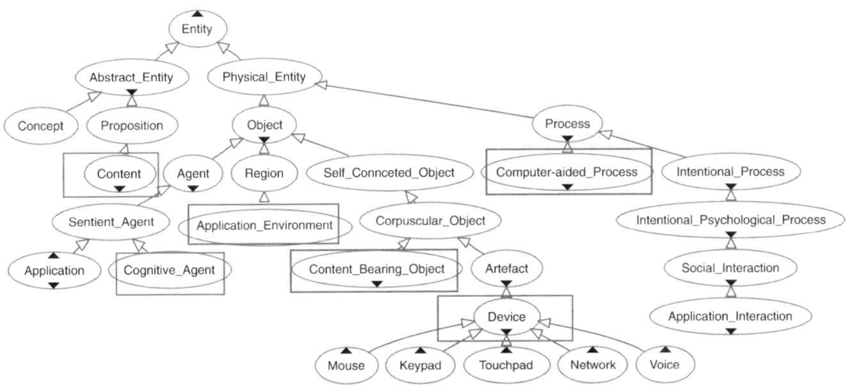

Fig. 1. ODAS concept hierarchy

by the subjects they describe. While `Content about Entity` has exactly one ontology entity as subject, `Content about Relation` has several such subjects. The subject is in fact the "content semantics" that ODAS aims to capture. Section 3.3 contains several content-based adaptation rules that exploit this notion of semantics to trigger recommendations representing different types of `Content` resources related to the one the user is currently interacting with.

Interesting is also the `Executable Content` concept, which differs from other types in that it is embodied in a `UI Element` and represents a `Service`. This is useful because in addition to information resources, services can also be considered in adaptation. In fact, any services can be associated with an individual of `Executable Content` and be adapted to the `User` just like other content types. Furthermore, the concept of `Content Bearing Object` (`CBO`) allows for a distinction of the actual materialization from the abstract `Content` embodied in it. Different variations of layout- and presentation of the same resource can be captured by an instance of `CBO` and the resulting **presentation model** supports the concept of adaptive presentation.

Further concepts that deliver contextual information are `User` and `Environment`. User properties such as `has credential`, `has read`, `knows`, `has interest` or `is able to` can be used to reflect constraints that need to be considered (**user model**). Also, characteristics of the `Environment` can play a similar role in adaptation (**environment model**). Restrictions given in the `Environmental Components` such as `bandwidth` of the `Network` and `size` and `resolution` of the `Display Device` can be incorporated to only deliver resources which comply with the specified constraints.

3 A Rule-Based Adaptation Model

While the ontology represents the different adaptivity dimensions in terms of user, resource, task, environment and system model, this section demonstrates that the logic underlying the system behavior can also be explicitly captured on the basis of a rule-based adaptation model. This section begins with an illustration of the main ideas on the basis of a personalized portal. Then, we present a formalization of the abstract adaptation model as well as a concrete instantiation based on DL-safe rules. Finally, the different part of the rules and different possible instantiations of the adaptation model will be discussed.

3.1 Applying Rules-Based Adaptation to a Portal

Figure 2 shows a personalized portal, a prototype implementation that shall demonstrate the main ideas of the approach proposed here. The adaptive part of the portal is an extension of the Liferay architecture framework that encompasses four functional modules. Apart from basic navigation and search,

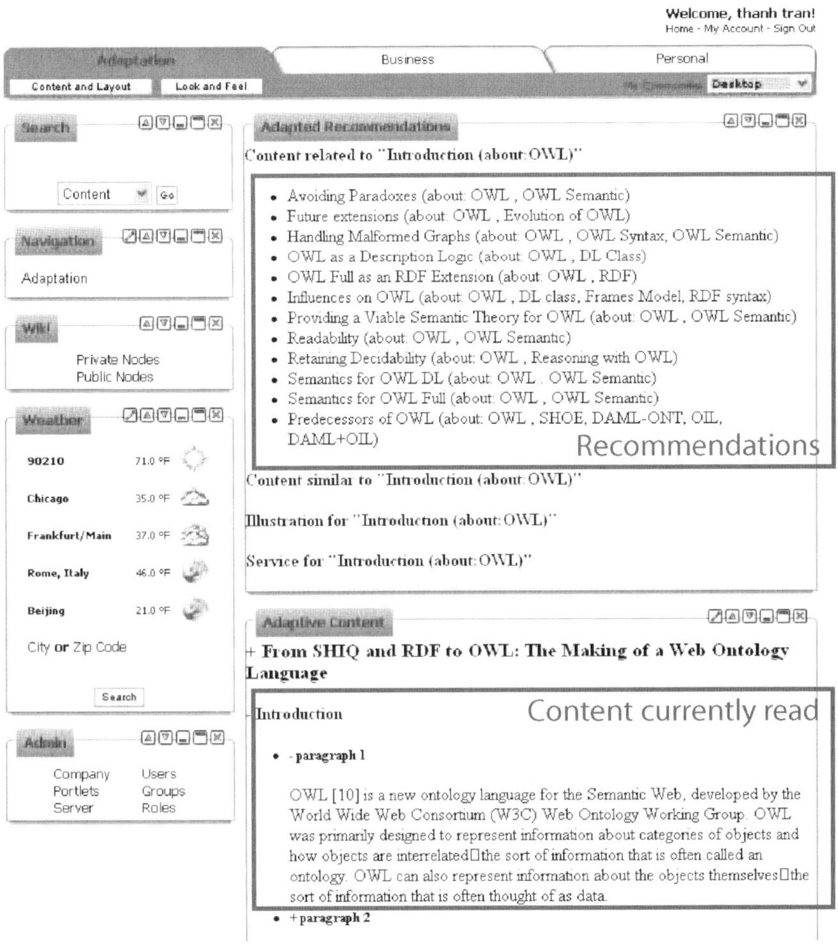

Fig. 2. An ODAS-based personalized portal

the system is able to track user interactions, generate recommendations in the form of links and apply different layouts to the content chosen by the user.

The presentation module in Fig. 2 shows a user reading "Introduction" – a section of a `Composite Content` unit, i.e. a scientific paper, with the title "From SHIQ and RDF to OWL...". This establishes a context that the system can exploit for adaptation. The result of adaptation is a list of recommendations, i.e. context-relevant resources that are presented by the recommendation module as hyperlinks. As shown, resources assessed by the system as relevant to the context also have `OWL` as subject – and additional ontology entities such as `OWL semantics` and `RDF` as shown in brackets. Though only content-related information is exploited by the system in its current implementation, in principle the approach also supports adaptation dimensions related to structure and presentation. In fact, the actual context model is more complex and involves many of the models discussed previously.

Firstly, such recommendations based on the context semantics are possible by the annotation of each paragraph with the entities that it deals with. In order to account for this so called resource model, object properties of `Content` such as `subject` are used. Different layout variations of the contents are represented via the `CBO` concept and serve as presentation models. In the future, the context employed by the system will be extended with further dimensions. System- and environment-related information such as bandwidth, resolution and other properties of the client device will be taken into account. Technically, precise metrics are assessable on a desktop application, but with the portal system it may suffice (and is more feasible) to obtain the client device's properties as nominal values only. Also, task related information will be incorporated. An expert can model a flow of activities involving dependent content units (and services). For instance, booking a flight (represented as `Computer-aided Process`) can be modelled as consisting of a sequence of atomic interactions (represented as `Executing`, subclass of `Application Interaction`) such as destination selection, car booking and hotel booking. Using this information, the system can recommend the `Executable Content` units associated with these services to help the user accomplishing the task of flight booking.

Most crucially, the interactions between the user and the system are recorded. For making recommendations, the system needs to know what the user is currently reading, what service is being invoked, for how long, with what result etc. This establishes the actual context and allows the system to update the user model. In fact, relevant properties of the user which change the course of the interaction are propagated to the knowledge base (KB). For instance, suppose the user chooses one of the recommendations shown in Fig. 2. Then, if there is a timeout greater than the time amount estimated to be the `Content`'s minimum `processing time`, the system assumes that the user `has read` the content and will update the user model accordingly. And when the user has read the same content for a number of times, the system assumes the he/she then `knows` it.

Thus, there are many contextual dimensions that are relevant for adaptation and ODAS is used to represent them. The entire context being captured can then be regarded as a list of conditions. Given a relevant list, the system will generate the most appropriate recommendations. In other words, when a situation is recognized by the system, it will adapt the resources accordingly. Modelling this behavior in the form of rules is indeed the core of the approach presented here. The logic for adaptation is specified declaratively in the form of rules representing some generic if-then patterns. Such patterns, i.e. a logic program, is at a higher conceptual level than if-then statements in a purely imperative program.

In particular, some generic adaptation contexts are declaratively specified as a set of conditions part of adaptation rules that, when satisfied, lead to the inference that some resource is recommended. Consequently, the procedural code needed for the recommendation module is straightforward. It comprises of only a few retrieval operations. Most of the work is done by the inference engine. For the portal, we use KAON2 as knowledge backend [14]. It supports the persistent storage of the many employed models, i.e. ODAS concepts and instances. A modelling service continuously inserts new instances of `Application Interaction` and thus triggers the processing of DL-safe adaptation rules. Such rules are constrained in a way that allows for integration with Description Logics – OWL in particular – while keeping reasoning decideable (see [18]). They can be processed efficiently with the KAON2 inference engine. This engine checks if the inserted interactions match the conditions stated in the rules and fires them accordingly.

3.2 A Formal Definition of the Adaptation Model

We will begin with a formalization of our proposed adaptation model. Let CC be the set of all context conditions, AC the set of adaptation conditions, UC the set of user conditions, EC the set of environment conditions and R the set of recommendations; CC, AC, UC, EC and R are pairwise disjoint; let $V = (EC \cup UC \cup AC \cup CC) \cup R$ be the set of all conditions and recommendations. An adaptation model then is defined as a finite set of adaptation rules of the following form, where r_i^+ are positive head atoms, ec_i^+, uc_i^+, ac_i^+, cc_i^+ are positive body atoms and $\neg ec_1^-$, $\neg uc_1^-$, $\neg ac_1^-$ and $\neg cc_1^-$ are negative body atoms:

$$\begin{aligned}(r_1^+ \oplus ... \oplus r_n^+) \leftarrow\ & [ec_1^+ \oplus ... \oplus ec_m^+ \wedge \neg ec_1^- \oplus ... \oplus \neg ec_l^-] \wedge \\ & [uc_1^+ \oplus ... \oplus uc_k^+ \wedge \neg uc_1^- \oplus ... \oplus \neg uc_x^-] \wedge \\ & [ac_1^+ \oplus ... \oplus ac_x^+ \wedge \neg ac_1^- \oplus ... \oplus \neg ac_z^-] \wedge \\ & [cc_1^+ \oplus ... \oplus cc_v^+ \wedge \neg cc_1^- \oplus ... \oplus \neg cc_w^-]\end{aligned} \tag{1}$$

where $\oplus \in \{\vee, \wedge\}$.

These atoms stand for the conditions and recommendations captured in the rules, i.e. they are elements of V, more precisely, $ec_i \in EC$, $uc_i \in UC$, $ac_i \in AC$, $cc_i \in CC$ and $r_i \in R$. Hence, the rule head is a conjunction of recommendations expressed in the form of logical formulas. The rule body is a conjunction of elements in EC, UC, AC and CC, each consisting of an arbitrary combination of disjunctions and conjunctions of positive and negative conditions also expressed as logical formulas.

Note that this adaptation model is formalized in a way independent of a logical formalism. Thus, the semantics depends on the concrete formalism that is used to formulate a particular adaptation model. We now continue to illustrate the application of this abstract model on the basis of an instantiation expressed as DL-safe rules.

The DL-safe rule fragment constitutes a decidable combination of OWL-DL and function-free Horn rules. Such a rule has the form $H \leftarrow B_1, \ldots, B_i$

where H and B_i are atoms; H is the rule head, and the set of all B_i constitute the rule body. An atom has the form $A(t_1, \ldots, t_n)$ where A is a predicate and t_i are terms; a term is either a constant or a variable. A rule r is called DL-safe if each variable in r occurs in a non-DL atom in the rule body where a DL-atom is $A(s) \in N_C$ or $R(s,t) \in N_{R_a} \cup N_{R_c}$ and $N_c \cup N_{R_a} \cup N_{R_c} \subseteq N_p$, the set of predicate symbols of a description logic knowledge base. A DL-safe rule-based adaptation model is a finite set of such rules, i.e. a logic program P containing only of DL-safe rules. The semantics of a DL-safe rule is defined by the equivalence to the clause $H \vee \neg B_1 \vee \ldots \vee \neg B_n$ [18].

Note that as opposed to the abstract model, the expressivity of this DL-safe instantiation is more limited, e.g. rules are not allowed to contain negation or disjunctions in the rule body. Nevertheless, the DL-safe rule fragment has practical relevance. Using this formalism, statements expressed in OWL-DL can be incorporated into adaptation rules and be more or less efficiently processed by inference engines. The following adaption rule (2) illustrates how knowledge represented with ODAS can be exploited for the purpose of formulating adaptation rules. We will now discuss this and other possible instantiations of the adaptation model using ODAS.

$$\begin{aligned} needs(u,z) \leftarrow\ & [Reading(p), CognitiveAgent(u), agent(p,u), \\ & ContentAboutEntity(c), resource(p,c),] \\ & [ContentAboutRelation(y), CBO(z), Entity(e), \\ & hasSubject(c,e), hasSubject(z,e), contain(z,y),] \\ & [Credential(h), requiresCredential(z,h), \\ & hasCredential(u,h), interestIn(u,z),] \\ & [Voice(v), instrument(p,v), Audio(z)] \end{aligned} \tag{2}$$

3.3 ODAS-Based Instantiations of the Adaptation Model

As shown in rule (2), ODAS concepts and properties are used to express conditions on recommendations. Context-related conditions are captured by the notion of `Process` and its relations to other ontology entities. This concept serves as the "entry point" to access various models, i.e. `Content`, `Task`, `User` and `Environment`. Rule (2) shows that the `Content` concept can be used to express conditions of the adaptation part. In particular, the adaptation logic specified there is based on subconcepts of `Content`, i.e. `Content about Relation` and `Content about Entity`. While this is referred to as content-based adaptation, we will also discuss other possible instantiations of this part that lead to different adaptive behaviors. Eventually, these different styles of adaptation yield a set of resources related to the one the user is currently interacting with. In the last two parts, properties related to the `User` and the `Environment` act as constraints that, when applied, have a minimizing effect on this adapted set of resources. This ensures that the resources are indeed suitable to the user and the environment. Note that this is due to the fact that rules fire only when all conditions in all parts are satisfied. We will now elaborate on the instantiations of these parts of the rule in more details.

Content-Based Adaptation – Semantically-Related Contents

In ODAS, the (semantic) content of a resource is described by the property `has subject`. In essence, it tells which ontology entities a content unit deals with. Based on this representation, resources can be considered as related if they have the same entity or related entities as subjects. In this regard, entities can be considered related if they are directly or indirectly connected – through some other entities – via some properties specified in the ontology. Exploiting the ODAS conceptualization, the adaptation-related part of the rule may consist of the following conditions:

$$\begin{aligned} \leftarrow [\ldots]\ & Entity(z), Content(x), hasSubject(x,z), \\ & Content(y), hasSubject(y,z)\ [\ldots] \end{aligned} \tag{3}$$

This would lead to the recommendations of all content units y which are related to the content x currently processed by the user via the same entity z they have as subject – or to be precise, any entities that can be substituted for the variable z. The following are two other instantiations of the adaptation part of the rule. In a similar manner, they ensure that recommendations encompass related content units:

$$\begin{aligned} \leftarrow [\ldots]\ & Entity(u), Entity(v), isRelatedto(u,v), Content(x), \\ & hasSubject(x,u), Content(y), hasSubject(y,v)[\ldots] \end{aligned} \tag{4}$$

$$\begin{aligned} \leftarrow [\ldots]\ & Entity(u), Entity(v), Process(p), isInvolvedIn(u,p), \\ & isInvolvedIn(v,p), Content(x), hasSubject(x,u), \\ & Content(y), hasSubject(y,v)[\ldots] \end{aligned} \tag{5}$$

As opposed to (4), contents recommended by the examples (5) and (6) are related to the current content not because they describe the same but a related entity. In (5), entities that can be bound to u and v are involved in a particular relationship. Precisely, they are connected by `is related to`, or by any sub-properties thereof, respectively. Note that in ODAS, this is the top property, i.e. the super-property of any object properties defined. When the integration of domain ontologies into ODAS respect this semantics, i.e. domain properties are also mapped as sub-properties of `is related to`, then any pairwise related entities can be substituted for u and v. In (6), entities bound to u and v are indirectly related because they participate in the same `Process`. Since `agent`, `target` etc. are sub-properties of `is involved in`, there may be many entity types that are related in this way.

In fact, these instantiations of the adaptation model produce the results shown in the portal in Fig. 2. As the user is reading "Introduction" (`OWL`), which is a `Content about Entity` describing `OWL` (entities in brackets stand for the subjects), "Avoiding Paradoxes" (`OWL`, `OWL Semantics`) is one of the recommendations generated by the portal as a result of processing the instantiation in (4), "Predecessors of OWL" (`OWL`, `SHOE`, `DAML-ONT`, `OIL`, `DAML+OIL`) as a result of adaption rule (5) and "Future extensions" (`OWL`, `Development of OWL`) as a result of the adaption rule (6).

It is worth noting that we assume that this adaptation model mimics in some way the style of authors who start with a section describing a key entity, e.g. OWL, in a general, introductory way. In subsequent sections, the author goes into details by focusing on some of its relationships to other entities that are relevant from his/her point of view, e.g. relationships among `OWL`, `SHOE`, `OIL` etc. The counterpart to this narrative style is to start with an overview of a complex phenomenon involving many entities and proceed with subsections, each focusing on one of these entities. Using ODAS `Content about Relation` and a related `Content about Entity`, this may be captured in the adaptation part as follows:

$$\begin{aligned} \leftarrow [\cdots]\ & Entity(u), ContentAboutRelation(x), hasPrimarySubject(x,u), \\ & ContentAboutEntity(y), hasSubject(y,u)[\cdots] \end{aligned} \quad (6)$$

The concept `Content about Relation` is used to describe (and can be exploited by the reasoner to classify) instances having more than one entity as subjects. The aim is to reflect the semantics of content units that deal with a complex phenomenon – rather than one single entity like instances of `Content about Entity` do. Such a phenomenon may be a relationship, i.e. one-, two-, n-ary relations between entities that may reflect a causality, a correlation, an interrelation or a connection in general. Due to the limitation in the expressiveness of the language underlying ODAS, i.e. OWL DL, this semantics cannot be precisely axiomatized. Thus, `Content about Entity` and `Content about Relation` help to distinguish content units focussing on one single entity from those that describe "something" involving several entities. So, when the user reads a resource that deals with many entities including u, processing a rule containing the adaptation part as shown in (7) leads to recommendations encompassing all those resources that have u as the one and only subject.

Another type of adaptation, which exploits the content semantics as captured by sub-concepts of `Content`, is to go from resources containing pure text to related resources with figures, images etc. that can serve as illustration, i.e. from `Unliteral Content` to related `Literal Content`. While reading, it may also be helpful to browse through other resources that deal with the current one, i.e. from `Content` to `Meta-Content`. Examples of the type `Meta-Content`, which is axiomatized as `Content` that has `Content` as subject, are the summary or the reference section of an article. Besides, when the user reads a `General Content`, which deals with a concept, the system may also make recommendations for `Specific Content` which deals with an instance of this concept. Thus, recommendations of this type can be seen as examples that support the comprehension of resources with a rather abstract content.

Content-Based Adaptation – Narratively-Related Contents

While the adaption rules presented in the last section exploit the connections implied by a semantic description of the resources' content, this section describes conditions that make use of the given narrative structure. In fact, the order of atomic parts contained by a `Composite Content` resource could reflect a particular relationship (dependency, causality etc.) between the described entities - which might be only in the mind of the author and not directly encoded via properties of the ontology. Using properties modelling the structure of content resources such as `has part`, `is pre` and `is post`, the narrative sequence given by the author can be reproduced. The following conditions lead to resources annotated as the subsequent content of the one currently read:

$$\leftarrow [\cdots]\ AtomicContent(x), AtomicContent(y), isPostContent(y, x)[\cdots] \quad (7)$$

Note that the specific subject of the content is not decisive. Here, all resources y that have been placed by the author after x are considered by the system as relevant to the current context.

Task-Based Adaptation

Similarly to adaptation that is based on the structure of the content, the type of adaptation discussed in the following makes use of the sequence of processes given by the supported workflow. That is, we assume that the tasks supported by the application are specified as a `Computer-aided Process` as discussed in Sect. 2.3. Based on this notion, the adaptation part can be instantiated in a way that leads to recommendations suitable for the current task. This could be as accomplished as follows:

$$\begin{aligned} \leftarrow [\ldots]\ & ComputerAidedProcess(p), ApplicationInteraction(p1), \\ & isPartOf(p1, p), User(u), isInvolvedIn(u, p1), \\ & ApplicationInteraction(p2), isPostProcess(p2, p1), Service(s), \\ & instrument(p2, s), ExecutableContent(y), isRepreentationOf(y, s)[\ldots] \end{aligned} \quad (8)$$

So, when the `User is involved in` an interaction that is part of a `Computer-aided Process`, then the system recommends subsequent processes as given by the `is post process` property until the user accomplishes the task, i.e. until he/she obtains the output of the corresponding `Computer-aided Process`. More precisely, the system would recommend an `Executable Content`, which is a representation of the respective `Service`. This `Service` acts as the `instrument` of the subsequent process. Note that for this purpose, the modelling service of the portal tries to identify the current process, i.e. an instance of `Application Interaction`, or an instance of a sub-concept of `Application Interaction` such as `Executing`, `Reading` etc.

Also, the corresponding `Computer-Aided Process` has to be computed. This is a non-trivial task if the system supports many workflows because there are certain types of `Atomic Application Interaction`, e.g. `Browsing`, `Timeout` and `Reading`, that are part of almost all workflows.

Currently, the service attempts to match the type of the current interaction with the type of all the atomic parts of the supported workflows. For instance, there is a `Learning Description Logic`-workflow, which contains a sequence of `Reading` activities that involve articles dealing with description logic formalisms such as ALC, SHIQ, SHOIN(D) and OWL. As there are not many workflows currently available in the prototypical implementation of the system, there are no ambiguities. In the future, we aim to reduce ambiguities by considering further information in the matching such as the entity types involve in the respective `Application Interaction`.

The style of adaptation elaborated here exploits only structural information of `Process`. It might also make sense to incorporate more information such as other properties and entity types involved in `Process`. The recommendations would then encompass semantically-related services, and could be achieved in a way similar to what has been done for semantically-related content.

Applying Constraints to the Adapted Set

When conditions of the adaptation part are satisfied, the system could generate a large number of recommendations. However, other parts of the adaptation model reduce them to a set of relevant resources. Relevance here is assessed with respect to user and environment conditions. That is, resources are relevant if they meet users' requirements and can be appropriately presented to the user, given the environmental conditions.

Again, conditions that can achieve this result can be specified using ODAS. For instance, a `Content` can be seen as appropriate for a user in case he `has credential` for it, does not `know` it, `has` not `read` it and `has interest for` it (i.e. for entities that this resource deals with). The instantiation of the user-constraint part of the rule may be as follows:

$$\begin{aligned} \leftarrow [\ldots]\ & Content(y), Credential(c), User(u), \\ & requiresCredential(y, c), hasCredential(u, c)[\ldots] \end{aligned} \tag{9}$$

Note that as negation is not supported in DL-safe rules, negative constraints such as not `know` and not `has read` have to be formulated in a separate rule that results in instances of the form `needsFalsified(user, content)`. Correspondingly, the recommendation service neglects recommendations that have been "falsified".

The last part of rule (2) shows an instantiation of the environment-constraint part. These conditions make sure that the recommended resources are appropriate for being displayed within the user's environment. For example, in case the interaction with the user is taking place via `Voice`, the `CBOs` suggested to the user must also be an instance of `Auditive`. When the `resolution` and `size` of the client `Display Device` is taken into consideration, only a preview version of the resources may be recommended. Other user characteristics as well as environmental constraints may be applied similarly.

4 Implementation and Discussion

As a proof of concept, we have implemented a prototype personalized portal using the above described rule-based adaption model. We have already presented this prototype implementation in Sect. 3.1. Preliminary evaluations have shown that recommendations reasonably match the resources that would have been manually chosen by the user. However, evaluating ontology-based applications is far from straightforward. In fact, the widely adopted precision/recall measures only measure part of the relevant aspects of a system.

A relevant aspect which is neglected by precision/recall evaluations is the effort spent in providing the necessary knowledge to the system. In our approach, knowledge has been created manually, resulting in domain ontologies and annotations of resources based on them with high quality. This leads to high quality recommendations but implies a high investment of time and effort to get the system run. In fact, we think our approach (and any ontology-based application) can achieve high effectiveness, but it will be necessary to examine more closer the trade-off between modelling effort spent and effectiveness achieved. In fact, evaluations need always to be sensitive to the amount of background knowledge previously provided to the system. In order to lower the effort, we will in the future experiment with different systems for automatic knowledge acquisition [4] and annotation [8].

A further aspect is related to the fact whether the user is satisfied with the current adaption model of the system. In the current implementation, we experiment with many different adaptation styles, i.e. different configurations of rules. However, as only limited information can be acquired from the user and the environment, the constraints on relevant resources are rather generic and therefore lead to some overgeneration, and thus to a high number of recommended resources. Of these, possibly not all will be really relevant, such that precision will be actually negatively affected. In fact, in the implemented prototype, most of the resources are recommended due to the fact that their content is related to the resources currently inspected by the user. However, only a few of these recommendations have been assessed by the user as really relevant. That is, a high recall has been achieved at the expense of precision. We have tweaked this recall-precision tradeoff by using a subset of the discussed adaptation rules. However, a more principled solution is definitely required here. We think that a sophisticated ranking mechanism can improve precision. We aim to improve this by a prioritization of rules and in particular a mechanism to process weights attached to conditions of the rules in order to compute the "relevance" of the recommendations. Concerning the subjective "correctness" of the rules, we will work on a presentation mechanism that facilitates the comprehension of rules. This will make it easier for the user to adapt them to what he/she deems as correct. Since after all, relevance is subjective, we think this is a way to improve the effectiveness of the system.

Finally, an important aspect is runtime performance. Currently, the inference engine requires several minutes to process and update recommendations, despite using one of the fastest reasoners available (see evaluation of KAON2 in [17]) and a decideable rule formalism (see discussion on DL-safe rules [18]). In fact, the low performance is also due to the many richly axiomatized concepts used in the rules. In fact, state-of-the-art OWL reasoners including KAON2 do not perform well with the combined use of disjunctions and equality as they open up drastically the space of possible solutions that have to be explored for making inferences. However, eliminating these primitives would result in a loss of semantics and, hence, in recommendations of lower effectiveness and comprehensiveness (recommendations based on less sophisticated models). Thus, the tradeoff between the degree of comprehensiveness of personalization and performance needs to be examined. Improvement can be achieved here by finding the appropriate degree of axiomatization.

Therefore, we will elaborate on these ideas to accomplish improvements. Also, we will work on more comprehensive evaluation of useability that besides precision and recall also considers factors such intuitiveness (controllability), performance as well as the resources (time and effort) needed to run the system.

5 Related Work

All personalization approaches discussed in Sect. 2.1 are related to the approach presented in this chapter. However, we have already argued in Sect. 2.2 that an approach which builds on ontologies and rules can address several problems common to previous work. Therefore, this section considers only ontology-based approaches.

In this respect, our approach is distinct with respect to the degree of comprehensiveness and generality. Other approaches (see [5–7, 10, 21] are mostly geared towards specific problems of adaptation. Frasincar et al. [7], for instance, also employ a domain ontology to model system resources, i.e. ontology individuals for slices and properties for links between them. In their approach, the domain ontology has a different character as it specifically models the museum domain whereas our ontology deals with the domain of adaptive system in general.

In some cases, an explicit conceptualization of resources, e.g. via the concept `Document` is also employed [5, 6, 10]. However, the comprehensiveness of ODAS is more appropriate for a more fine-grained adaptation. It supports the representation of many models we deem as valuable for sophisticated adaptation (see discussion in Sect. 2.1). For instance, via `Content` and `CBO`, there is a clear distinction between content-, structure- (`Content`) and presentation-related (`CBO`) aspects of the resource. Therefore, ontology-based adaptive presentation is possible.

In many approaches, the employed ontology is rather a taxonomy of keywords, e.g., the ACM topic hierarchy[4] [5]. In our approach, resources are represented as entities that describe individuals of a domain ontology. While keywords have no formal meaning per se, the semantics of entities provided by the domain ontology can be exploited by a reasoner to classify resources as subtypes of `Content`, which can be used to recommend semantically-related resources.

Most importantly, the adaptation model introduced here differs substantially from the approaches discussed so far. Adaptation is currently supported by querying the ontologies with corresponding adaptation rules. The rules are essentially conjunctive queries [12]. However, the formal model we propose is general, domain- and formalism-independent. Personalization in any domains can be achieved by adapting the concrete grounding based on DL-safe rules. Also, we have shown different styles of adaptation that can be accomplished by different variations of this grounding.

6 Conclusion

We have proposed an adaptation model to be used in an ontology-based approach to personalization. This model has been formalized in an abstract, formalism-independent way. Also, a concrete grounding based on DL-safe rules has been provided. A comprehensive ontology capable of capturing many aspects relevant to adaptation has been exploited to express conditions in the DL-safe grounding. Many variations of this grounding, i.e. adaptation rules, have been discussed to illustrate the use of the model. We have shown that many different styles of adaptation can be achieved this way. In particular, the concept of adaptive content, adaptive structure and adaptive presentation can be supported by means of a declarative model of the adaptation rationales. The rules that make up this model can be modified (also by the user) for specific adaptation requirements. Also, we have argued that this approach can solve many drawbacks in current systems. We have further also discussed the bottlenecks and weaknesses of our approach.

[4] `http://www.acm.org/class/`.

References

1. M. Balabanovic and Y. Shoham. Content-based, collaborative recommendation. *Communications of the ACM*, 40(3):66–72, 1997.
2. P. Brusilovsky. Methods and techniques of adaptive hypermedia. *User Modeling and User-Adapted Interaction*, 6(2–3):87–129, 1996.
3. P. Brusilovsky and D. Cooper. Domain, task, and user models for an adaptive hypermedia performance support system. pages 23–30. ACM, 2002.
4. P. Cimiano and J. Völker. Text2onto - a framework for ontology learning and data-driven change discovery. In *Proceedings of the 10th International Conference on Applications of Natural Language to Information Systems (NLDB)*, pages 227–238, 2005.
5. P. Dolog, R. Gavriloaie, W. Nejdl, and J. Brase. Integrating adaptive hypermedia techniques and open RDF-based environments. In *Proceedings of 12th International World Wide Web Conference*, 2003.
6. P. Dolog, N. Henze, W. Nejdl, and M. Sintek. The Personal Reader: Personalizing and Enriching Learning Resources using Semantic Web Technologies. In *Proceedings of the Adaptive Hypermedia 2004*, pages 85–94. Springer, 2004.
7. F. Frasincar, G. Houben, and R. Vdovjak. Specification framework for engineering adaptive web applications. In *Proceedings of the 11th International World Wide Web (WWW) Conference*, 2002.
8. S. Handschuh and S. Staab, editors. *Annotation in the Semantic Web*. IOS Press, 2003.
9. N. Henze and W. Nejdl. Adaptation in open corpus hypermedia. *Special Issue on Adaptive and Intelligent Web-Based Systems*, 12:325–350, 2001.
10. N. Henze and W. Nejdl. Knowledge modeling for open adaptive hypermedia. In P. De Bra, P. Brusilovsky, and R. Conejo, editors, *AH*, volume 2347 of *Lecture Notes in Computer Science*, pages 174–183. Springer, 2002.
11. J. Herlocker, J. Konstan, and J. Riedl. Explaining collaborative filtering recommendations. In *Proceedings of the ACM conference on Computer supported cooperative work*, pages 241–250. ACM, 2000.
12. I. Horrocks and S. Tessaris. Querying the semantic web: a formal approach. In *Proceedings of the First International Semantic Web Conference (ISWC)*, volume 2342, pages 177–191. Springer, 2002.
13. G. Houben, P. Barna, F. Frasincar, and R. Vdovjak. Hera: Development of Semantic Web Information Systems. *Proceedings of the Third International Conference on Web Engineering*, pages 529–538, 2003.
14. U. Hustadt, B. Motik, and U. Sattler. Reducing shiq descrption logic to disjunctive datalog programs. In *Proceedings of the 9th International Conference on Knowledge Representation and Reasoning (KR)*, pages 152–162, 2004.
15. J. Konstan, B. Miller, D. Maltz, J. Herlocker, L. Gordon, and J. Riedl. GroupLens: applying collaborative filtering to Usenet news. *Communications of the ACM*, 40(3):77–87, 1997.
16. K. Lang. NewsWeeder: learning to filter netnews. In *Proceedings of the 12th International Conference on Machine Learning (ICML)*, pages 331–339. Morgan Kaufmann, 1995.
17. B. Motik and U. Sattler. A comparison of reasoning techniques for querying large description logic ABoxes. In M. Hermann and A. Voronkov, editors, *Proceedings of the 13th International Conference on Logic for Programming, Artificial*

Intelligence, and Reasoning (LPAR), volume 4246 of *Lecture Notes in Computer Science*, pages 227–241. Springer, 2006.
18. B. Motik, U. Sattler, and R. Studer. Query Answering for OWL-DL with Rules. In *Proceedings of the International Semantic Web Conference (ISWC)*, pages 549–563. Springer, 2004.
19. J. Nielsen and U. Lyngbaek. Two field studies of hypermedia usability. *Hypertext: State of the Art*, pages 64–72, 1989.
20. L. Terveen and W. Hill. Beyond Recommender Systems: Helping People Help Each Other. *HCI In The New Millenium, J. Carroll ed. Addison-Wesley*, 2001.
21. T. Tran, P. Cimiano, and A. Ankolekar. Rules for an ontology-based approach to adaptation. In *Proceedings of the 1st International Workshop on Semantic Media Adaptation and Personalization*, 2006.
22. T. Tran, H. Lewen, and P. Haase. Semantic software engineering: the role and usage of ontologies in information systems. *Proceedings of the 5th IEEE International Conference on Computer Science - Research, Innovation and Vision for the Future*, 2007.
23. J. van Ossenbruggen, J. Geurts, F. Cornelissen, L. Rutledge, and L. Hardman. Towards second and third generation web-based multimedia. In *Proceedings of the 10th International World Wide Web (WWW) Conference*, pages 479–488, 2001.

Management and Processing of Personalized Annotations in Image Retrieval Systems

Johanna Vompras and Stefan Conrad

Heinrich Heine University, Institute of Computer Science, Databases and Information Systems, Duesseldorf, Germany vompras@cs.uni-duesseldorf.de, conrad@cs.uni-duesseldorf.de

Summary. Due to the importance of semantic meaning in image retrieval, manual or semi-automated annotation still remains indispensable in both professional and personal retrieval applications. Annotations are used to facilitate textual or conceptual queries in large image repositories and thus to classify the image data into semantic classes. However, different users' perception of image contents and the lack of standards among different annotation tools make it necessary to develop methods for the unification and integration of different annotation schemes. In this chapter we present a graph approach as a representation technique for the complex semantic annotation space which is generated by the transformation of the subjective perceptions into a unified knowledge base. Our technique bridges the discrepancy between users' vocabulary and the several levels of abstraction at which content descriptions are assigned. Based on examples, we show how to integrate our method into probabilistic approaches to (semi-) automatic image annotation.

1 Motivation

The *Semantic Gap* [1], known as the difference between the low-level features and the high-level human perception of image contents [2], represents the most disruptive factor in semantic retrieval and the management of large image data repositories. To perform content-based image retrieval (CBIR) at feature level, numerous heuristic and machine-learning methods [2–4] with relevance feedback have been developed in recent years. However, there are still many unresolved issues in content-based systems. A big drawback is the fact, that similarity between images is typically determined by applying a distance metric on a feature space, where only low-level features like color, texture, or shape [5] are considered. These methods do not take into account that an advanced and fully functional retrieval system would require support for queries at the semantic level. Furthermore, CBIR retrieval methods are mostly restricted to particular application fields (e.g. medicine, geographic information systems) causing the assignment to heterogeneous image collections to

J. Vompras and S. Conrad: *Management and Processing of Personalized Annotations in Image Retrieval Systems*, Studies in Computational Intelligence (SCI) **93**, 137–155 (2008)
www.springerlink.com

fail in terms of accuracy. To facilitate image queries at conceptual and semantic level, several approaches, like [6–9], have been proposed to minimize the mentioned difficulties by combining automated feature extraction approaches with concept-based or annotation-based techniques. Their main objective is to (semi-)automatically attach images with appropriate descriptions and thus support image retrieval by a keyword-based search. The annotations can be composed of free text keywords or instances from complex ontologies which allow the specification of objects and actions depicted in an image. Beyond the reduction of the high manual effort required in creating annotations there is the need for methods which effectively manage the increasing annotation data and possess the ability to automatically discover differences in interpretations of image contents or inconsistencies in keyword assignments among indexers [2]. The latter problem has been addressed by using existing ontologies to make use of concept search instead of keyword search and thus allow the user to specialize or generalize a query with the help of a concept hierarchy [10–12].

Furthermore, traditional retrieval systems should be expanded by personalization components in order to make heterogeneous data collections accessible to an increasingly heterogeneous population of users [13]. In our work, image annotations are understood as an accumulation of strongly personalized information given by users which have different standards of knowledge and act in different contexts. This subjectivity in image annotation may appear in several points, for example at the querying step in form of users' preferences and skills or the background knowledge during annotation of images [14]. This form of information overload complicates the search facilities and makes the retrieval of relevant information an exhausting task. Hence, the most important requirement for annotation-based systems is the flexibility to accommodate differing semantic views of the same image and the dynamics to handle the advances in the areas of image processing as well as the evolution of application domains [15]. Furthermore, it is desirable that an image retrieval system will be able to adapt itself continuously to the changing requests of the user [12] by adjusting the changing mappings between image data and its annotations (e.g. by relevance feedback). The new idea in our approach is to integrate hierarchical multi-level information that is contained in annotations into an image annotation and retrieval framework. Our method incorporates the relations between annotation terms, their contexts, and the frequency they are assigned, resulting in the unification of different users' views.

This chapter is structured as follows. In Sect. 2 we briefly review the properties of the *Image Annotation Process* and the encountered problems. After an introduction of basic definitions in Sect. 3, a detailed description of the *Multi-level Annotation Component* is given with its functionalities for analyzing and encoding different abstraction levels in annotations. Afterwards, we demonstrate the application of the resulting annotation space for the automatic image annotation. A summary of related work is given in Sect. 4. Finally, Sect. 5 concludes our approach and gives further research directions.

2 Image Annotation Process

Users' interpretations can be summarized by means of *terms* or *keywords* describing the recognized semantic concepts. The association of these keywords with images for capturing their semantic contents and enriching them by additional information is known as *Image Annotation*. At the same time, the annotation should assign the image data to one or more of the predefined categories resulting in a semantic classification of the underlying data collection. Ambiguous interpretations can be avoided by using a lexicon-based knowledge (e.g. an ontology) which serves as a source of semantic types and their relations. In order to combine the high-level tasks of *scene recognition* and *user interpretation* with traditional CBIR systems, the manual annotation is performed by users. Figure 1 illustrates a course of image annotation according to human perception ability and the corresponding image data model used for modeling content information. Accordingly, the image annotation process includes the following steps:

1. Application of visual analysis of the image contents in order to identify relevant objects or regions and their relations.
2. Determination of a set of candidate keywords for the annotation of the image by using an application-specific lexicon. These textual keywords are supplemented by attribute based meta data, such as creator, dates, genre, file type, size, etc.
3. Assignment of a set of keywords to the image at different abstraction levels, for example by describing the recognized objects, their relations, and the overall classification of the scene. To perform clustering at semantic level, information about the low level features, like color, texture, and (primitive) shape within the image have to be associated with the recognized semantic concepts.

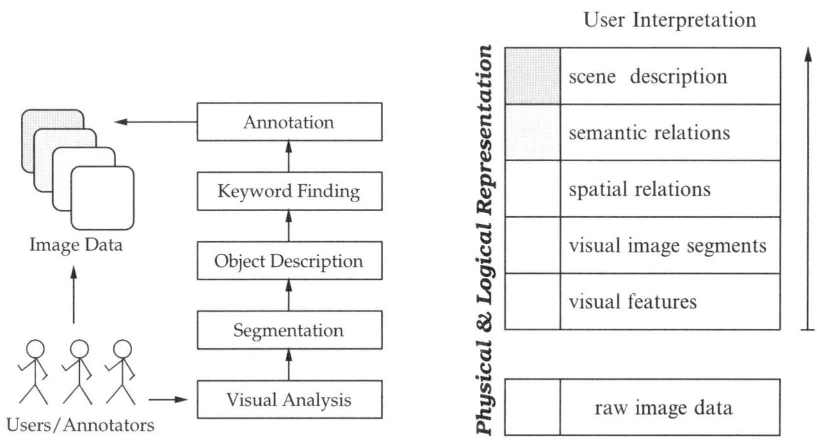

Fig. 1. The workflow of semantic annotation and the image data model

Although this described process appears to be a straightforward task which seems to succeed error-free, it is afflicted with uncertainties. Beginning with the selection of an appropriate set of keywords and the abstraction level, it turns out to be a complex task. Particularly, to make manual annotations reusable and integrate them into semi-automatic annotation and retrieval systems, more than incoherent keyword descriptions are needed. The mostly encountered problems during the annotation process are:

Multiple levels of abstraction. Annotations are assigned by different users in different contexts and in different points of view. In addition, the type of information and levels of abstraction may often depend on the application domain. Some annotations may work well with one application, but by exchanging the context they may turn out to be useless or unsuitable for reuse.

Incompleteness. In retrieval systems using semi-automated approaches, the correlations between image features and their annotations are discovered. The incompleteness of annotation data and the biased assignment of annotations leads to a small recall value in search results.

Non-uniform word distribution and word sparseness. The term frequency of used words for the annotation is not uniformly distributed. Only a small number of words appears very often in annotations and most words are used only by a few images. Consequently, word co-occurrence frequencies within a set of annotated images cannot be determined. The problem of word sparseness can be overcome by incorporating additional knowledge such as annotation ontologies that explicitly identify the relationships between words and their meanings.

Hard-to-describe objects Complex and hard-to-describe objects in images or objects occluding other objects can complicate semantic annotation. The extraction of semantic concepts is difficult because images may contain multiple semantic concepts and different objects corresponding to different concepts. In addition, images differ from each other in the number of objects and their area, resulting in different-sized annotations for the same semantic category of images.

Users' perception. Users' perception proves to be highly subjective and leads to inconsistent annotations among indexers. In addition, users' views may change over time, that means that different interpretations could be assigned to the same images or the same annotations could be given to different image contents.

3 Multi-level Annotation Model

This section describes the basic definitions which are involved in the retrieval component and the modeling of annotations included in our approach. In general, we want to provide a unique framework to define those definitions formally which can serve as a solid foundation for the theory of personalized annotation-based image retrieval. Afterwards, we give a detailed description of the multi-level annotation model and its usage.

*Semantic concepts.*We define a set $\Phi = C_1, C_2, \ldots, C_n$ of semantic concepts arranged in a concept hierarchy. The subset relation $\subseteq_\Phi$ between two concepts $(C_i, C_i) \in \Phi \times \Phi$ is a partial order between concepts $(C_i \subseteq_\Phi C_j)$, which denotes that C_i ia a sub-concept of C_j. The set of the concepts is not known apriori and is dynamically extended according to the appearance of a new instance of semantics.

Representative features. Let $\mathcal{D} = \{d_1, d_2 \ldots d_n\}$ be a set of application domains and $\mathcal{F}_{d_k}$ a set of representative visual features for a domain d_k.

Image data set. A database $\mathbf{D}$ includes a set of images $\mathcal{I} = \{I_1 \ldots, I_N\}$ which are characterized by their feature vectors $\mathbf{f}_{I_1} \ldots \mathbf{f}_{I_N}$.

Segmentation set. Let $\mathcal{R}$ be the set of manually or automatically segmented ROIs (regions of interest). We define a function $\pi_\mathcal{R} : \mathcal{D} \rightarrow 2^\mathcal{R}$ so that $\pi_\mathcal{R}(d_k)$ is the set of representative regions of interest of a domain d_k.

Image annotations. Let be $\mathcal{K} = \{k_1, k_2, \ldots, k_n\}$ a set of keywords. The subset $\mathcal{K}_d \subseteq \mathcal{K}$ is a sequence of keywords or semantic labels $\{k_j, k_{j+1}, \ldots, k_m\}$ which are used in an application domain $d \in \mathcal{D}$. An annotation Γ_a of an image I is a set of keywords from $\mathcal{K}_d$ which are arranged in a multi-level structure. The number of images attached with the annotation Γ_i is denoted by $\mathcal{N}(\Gamma_i)$.

Semantic category. Images containing particular visual concepts $\Phi_\mathcal{S} \subseteq \Phi$, where $\Phi_\mathcal{S} = \{\ldots, C_{n-1}, C_n, C_{n+1}, \ldots\}$ are assigned to a semantic category $\mathcal{S}$. The number of sematic categories is not fixed, and is expanded during annotation and retrieval.

Annotation mapping. Let $\Gamma_1, \Gamma_2 \ldots, \Gamma_n$ be annotations used to describe a set of images. Than the mapping into the *Annotation Space* is created by arranging the annotations in a multi-graph structure consisting of a set of nodes $V = \{v_1, \ldots, v_n\}$ corresponding to the image annotations and a set of edges $e_i \in E$ $(E \subset V \times V)$ connecting the annotations.

Domain-dependent annotation ontology. A domain-dependent ontology o_d provides a collection of concepts and their interrelationships providing an abstract view of the application domain d.

Table 1 summarizes the possible instances of the introduced conceptualization. Let us consider the image in Fig. 2. The image $I_1 \in \mathbf{D}$ is represented by an n-dimensional feature vector $\mathbf{f}_{I_1}$. The selection of the n features from the set $\mathcal{F}_{d_k}$ depends on the membership of the image to an application domain d_k. A domain (e.g. medical applications, geographic applications, etc.) reflects the objectives and demands on CBIR methods for the detection of particular patterns in images, in this case, outdoor images, which are described by several semantic concepts like 'sky', 'building' or 'tree', each represented by an individual set of representative features.

Table 1. Examples

Representative features	Color, histogram, texture features
Semantic category	Beach images, historic photographs, sightseeing
Annotation	Textual description, e.g. city, building, *London*
Image segments	Segmentation of an image into information-bearing contents e.g. extracting objects from background
Application domain	Medical, geographic, face detection, cell detection
Annotation ontology	Conceptualization of objects and their relations, for example entities like '*library* `is-a` *bulding*'.

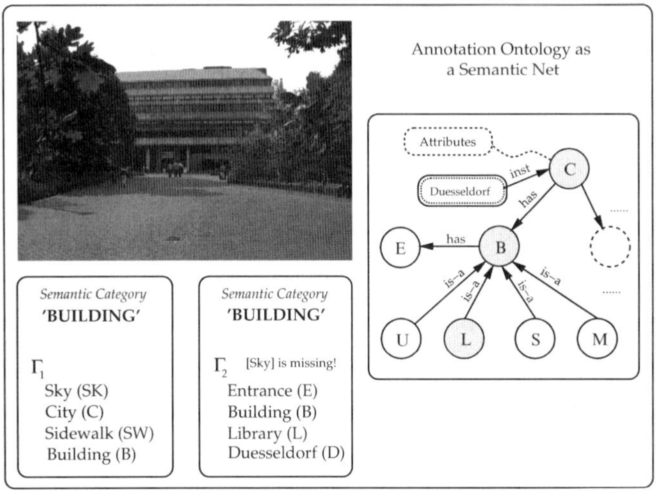

Fig. 2. Two annotations Γ_1 and Γ_2 and the corresponding annotation ontology

3.1 Subjectivity in Image Annotations

Variations in user's contextual knowledge, resulting in different quality and preciseness of content descriptions, leads to problems when retrieval is performed on annotations.

This fact is demonstrated in Fig. 2 by means of two annotations Γ_1 and Γ_2 which have been assigned to an image illustrating a building which is surrounded by greenery by different system users with varying background knowledge. The first annotation Γ_1 is a *flat* annotation only composed of keywords which are not semantically related. The second annotation is a *semantically (partially) structured* annotation which links the keywords according to their semantic meaning. For the calculation of the similarity between the two annotations, the annotation ontology, the structure of the annotations, and the types of the containing keywords have to be considered. By considering an excerpt of the ontology (Fig. 2, right) describing the concept 'building' (B) with its subconcepts 'university' (U), 'library' (L), 'school' (S), and 'museum' (M) and its superconcept 'city' (C), it is obvious that the keywords are related to each other and require specific rules to compute the extent to which they share similar semantic contexts.

Furthermore, this subjectivity in annotations also provides advantages, because the annotations contain contextual information derived from the annotators view on the images. Although this subjectivity might cause some mismatches between the users' intentions and retrieval system behavior, such contextual information embedded in annotations is sometimes useful for interpreting images. For the most part, subjective context (keywords *laughing* and *children*) is accessible only by the annotation words assigned to images. Thus, subjectivity may enhance semantic retrieval when there exist methods to understand and interpret the characteristics of the resulting annotations.

3.2 Components of the Annotation Analysis Framework

Since in the majority of cases the application domain in which the annotations will be used in the future is unknown at the annotation time, methods for understanding and interpreting different annotations are required. The development of a model for the *Multi-Level Annotation* (MLA) is an essential step to the unification and integration of different annotation schemes. The thus obtained annotation provides a semantically consistent description of the data which results in a higher precision and recall in image retrieval. For this purpose, a statistical approach combined with lexical analysis is used to find correspondences between the used keywords and visual concepts.

The four main functionalities the multi-level annotation component, demonstrated in Fig. 3 are the following:

Unification of annotations and inference. The unification of annotations which have been inconsistently created by different users, and the determination of redundant information and the detection of different views.

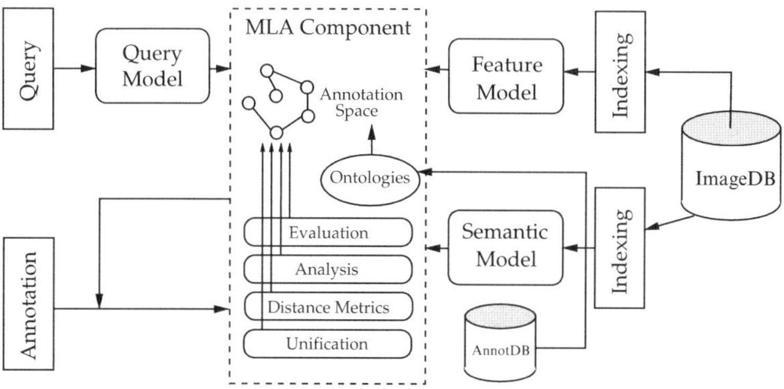

Fig. 3. Multi-level annotation component

Annotation analysis. By the analysis of assigned annotations for a specific concept, we can determine suitable abstraction levels and the optimal length of the annotation sequence.

Context-based distance functions. Distance functions for annotations should take into consideration the different views and relations between annotations and the context they appear in.

Statistical evaluation. The evaluation of annotation co-occurrences allows the creation of annotation sets which are used to switch between different users' views without negatively influencing the retrieval results. Moreover, using associations between image low-level data (features) and the assigned keywords an appropriate propagation of new annotations can be performed.

Example 1. Unification of annotations and inference. As a first step, the conceptual distance between a set of image descriptions (e.g. annotation A) and a structured global annotation ontology is computed in order to determine the amount of information they share. Secondly, the unification is done by finding a covering of the annotation terms with the given ontology. Assume that we have

- Annotation A: {*skyscrapers, New York*},
- Ontology with concepts 'city' and 'building',
- Relations, e.g. (*skyscraper* **is-a** 'building') and ('building' **is-part** 'city'),

then the rule (*skyscraper* **is-part** 'city') can be inferred. The unification provides an annotation $\bar{A}$: {→building:*skyscrapers*, →city.inst:*New York*} with pointers (→) to the respective concepts in the ontology.

3.3 Graph Representation for Multi-level Annotations

The main objective of this work is to extract and unify the information from multi-level annotations. In order to fulfil the mentioned requirements, annotations are not only considered as a collection of semantically independent keywords. For this purpose, we introduce a general multi-level annotation structure, which is presented in Fig. 4.

Accordingly, the MLA consists of *annotation units* and their *structure* which reflects the composition and relations between the annotations units. The annotations consist of textual descriptions (descriptors) which are either linked to a part of the image data (*segment*) or unattached (implicit descriptors). The structural information consists of an *object layer* and a *description layer.* At the object layer *annotation relations* describe the 'visual' relations between annotations, e.g. the position of an object, whereas at the description layer annotations are linked to each other or to other objects, for example to feature an optional description for the same content or to describe other relational properties. The annotations' properties of reusability and generality are warranted by their flexible structure: *annotation types* define the kind of

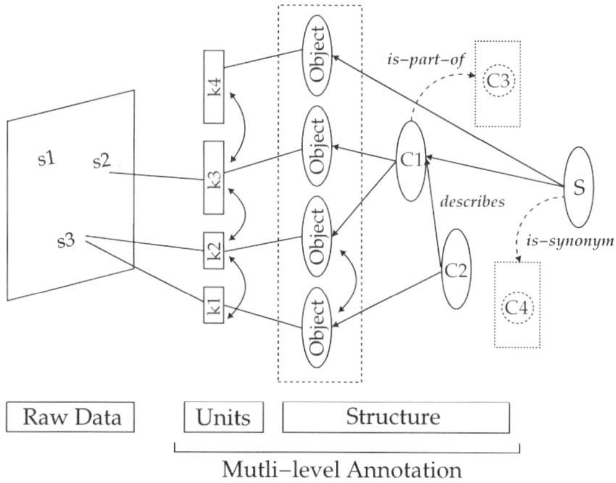

Fig. 4. Example of the multi-level structure of annotations

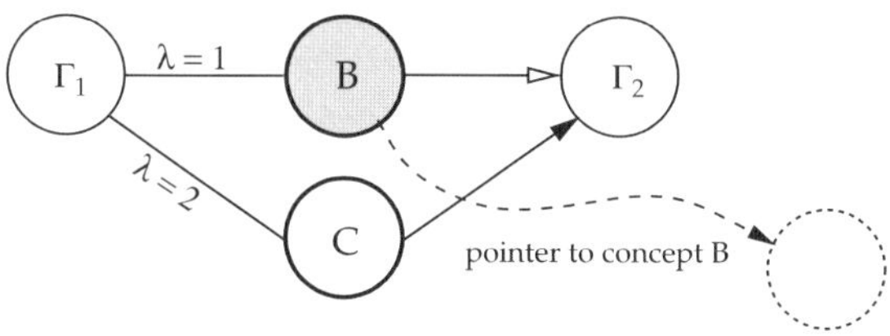

Fig. 5. Representation of specialization/genaralization as a multi-graph

content held (e.g. *object*, *action* or *event*) by annotations. A type possesses a name and the types of possibly connected annotations. Multiple annotation types in one image annotation are used to describe the variety of user views. Further information about an annotation relation is specified by a *relation type* which describes the type of the objects associated by a relation and defines the types and the number of participating annotations. For example, the type of *action* could represent that an object invokes operations on other objects.

In order to facilitate semantic retrieval at multiple abstraction levels, annotations are not strictly assigned to semantic categories, but are arranged in an internal weighted representation to encode the hierarchical annotation information and to express the relations and similarity between the underlying images. Thus, using an existing annotation ontology a space of annotations (*Annotation Space*) is built to derive connections between image annotations using specific rules. Figure 5 visualizes a small example of the semantic network constructed for the image annotations presented in Fig. 2.

Formally, the network consists of nodes $V = \{v_1, \ldots, v_n\}$ generally labeled with one or more semantic features which correspond to the image annotations and a set of edges $\{e_1, \ldots, e_m\} \in E \subset V \times V$ connecting the annotations. For each concept X two annotations have in common, their nodes are connected by an edge $e[\Gamma_1, \Gamma_2]^{[X]}$ which passes through the concept node X. There is a distinction between two types of edges:

- *subsumption edge* e_{sub}: denotes the stronger specificity of the respective concept in the annotation. The direction of the arrow is the more specific annotation.
- *expansion edge* e_{ext}: expands the annotation by a new concept which represents additional information derived from the annotation ontology.

Consequently, the stronger specificity of the concept B in the annotation Γ_2 is visualized by a subsumption edge (white arrowhead) $e_{sub}[B]$, because the concept 'building' (B) is more general than 'library' (L) according to the annotation ontology. By using the *expansion edge* $e_{ext}[C]$ (black arrowhead) the semantic annotation is expanded by a new concept. For example, the fact that the entity 'Duesseldorf' is connected with the concept 'city' (C) by the `is-inst` relation is used to derive this additional information.

Syntactic relations, like synonymy, where the meaning one of two or more terms is considered to be the same, are connected by a super-concept, they were assigned to (see Fig. 6). Example: *notebook* ← 'computer' → *laptop*. Abbreviated terms and their full forms are also treated as synonyms.

Special features describing an image content more detailed can be expressed by attributes, which are attached to the nodes in the annotation graph (Fig. 7). An object annotated with keyword k_1 is characterized by additional descriptive attributes such as k_1.**color**:*orange* or expressing an action performed by the agent, for example, k_1.**action**:*eating* or k_1.**action**:*laughing*. The arrow indicates the direction of the relationship between nodes, in this case from the general to more specific node content.

In addition, the network (see Fig. 5) is parameterized with the value $\lambda \in [0, \nu]$, denoting the *level* of the semantic relation between the annotations which is inferred from the ontology hierarchy. For example, Γ_2 is extended by

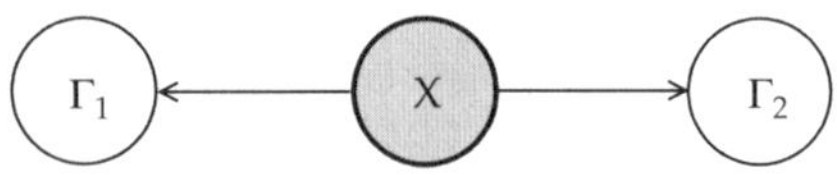

Fig. 6. Syntactic (synonymy) relation between annotation terms

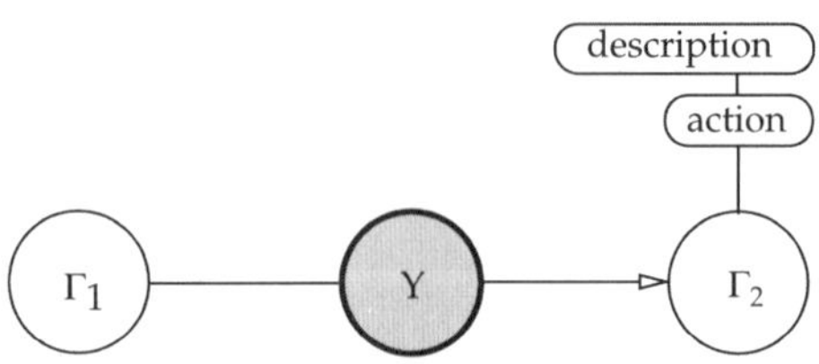

Fig. 7. Representing descriptive features

the concept 'city' (C) which is situated at a higher semantic level according to the hierarchy. The corresponding level is represented by the maximum distance between the individual keywords which are associated with the concept 'city'. Formally, the value of λ for an edge $e[\Gamma_a, \Gamma_b]^{[X]}$ is computed by the following formula:

$$\lambda(e[\Gamma_a, \Gamma_b]^{[X]}) = \max\left(\sum\nolimits_{[c_i,c_j]^{[X]}} \mathrm{dist}_o(c_i, c_j)\right), \tag{1}$$

where $[c_i, c_j]^{[X]}$ denotes the path between concepts $c_i \in \Gamma_a$ and $c_j \in \Gamma_b$ via the node X and $\mathrm{dist}_o(c_i, c_j)$ represents the distance between two concepts c_i and c_j according to the annotation ontology o. Thus, the overall similarity of two annotations depends on the number of their connections and the levels λ between the used concepts.

However, for retrieval purposes only the distance between two annotations does not predicate their expressivity and quality. For this purpose, two further measures are introduced in Table 2 to indicate the goodness of an annotation.

The *specificity* σ is based on the fact, that annotations with a high number of subsumption edges describe more specifically the image contents and the more specifically described is an image the fewer images with this content will exist in the data collection. It is quantified to a positive real number $\sigma(\Gamma_a) \in \mathbb{R}^+$ where Γ_a is an annotation. An annotation with a large number of edges entering the node (subsumption edges) describes more specifically the image contents with additional context knowledge delivered by the user, and the more specifically described is an image, the fewer images with this content will exist in the database. For example, if there are only two images with a particular annotation, then we can assume, that the annotation is very specific. Therefore, $\sigma(\Gamma_a)$ is computed by dividing the number of subsumption edges by the number of images attached with this annotation.

The second measure reflects the *visual expressiveness* ϵ of an annotation, specifying to what extent the used annotations have visual characteristics. The smaller is this value, the more discriminative power at feature level is provided by the annotation. This information is important for image annotation, especially for (semi-)automatic image annotation, since not all concepts are related to visual contents. This characteristics firstly depends on the number of used keywords which have been assigned at low level to image segments. If this information is unavailable, it can be intuitively concluded that concepts described by annotations which are close to each other in the annotation space and whose images have similar visual characteristics have more discriminative

Table 2. Measures for annotation quality

Measure	
$\sigma(\Gamma_a) = \lvert e_{sub}\rvert \times \frac{1}{N(\Gamma_a)}$	*preciseness*
$\epsilon(\Gamma_a) = \frac{1}{n} \sum_{\Gamma_i \in N(\Gamma_a)} \mathrm{dist}(I_a, I_i) \cdot \Lambda(\Gamma_a, \Gamma_i)$	*visual expressiveness*

properties than similar annotations specifying images with high discrepancy at feature level. The value of ϵ is therefore computed by the formula presented in Table 2, where $\Lambda(\Gamma_a, \Gamma_i) = \exp\left(-\frac{\|\Gamma_a - \Gamma_i\|^2}{2\sigma^2}\right)$. $\|\Gamma_a - \Gamma_i\|$ denotes the distance between the annotations, dist$(\cdot)$ the images' distance at feature level, n the number of similar annotations within the neighborhood N and σ the circumference of N. Thus, annotations in the neighborhood of Γ_a describing similar image contents are weighted according to their distance. In this case, if an annotation is close to Γ_a the "penalty" of visual dissimilarity is high; in contrast, if the corresponding annotations are far away from the reference annotation Γ_a, the penalty will decreased to zero, according to the *Gaussian* neighborhood function.

The advantages of the new representation are the following. First, implications about the similarity of annotations can be determined by considering the incoming and outgoing edges in the multi-graph structure. In addition, relations like specialization can be determined by considering the degree of the hierarchical distance. Thus, the resulting multi-graph structure is used to support the semantic retrieval at different levels of abstraction.

3.4 Relations Between Terms of Annotation

Relations between annotations are needed to describe the content at multiple levels and to create structured and consistent annotations. During the annotation process, the user either defines relations between keywords according to the relation catalog (an extract is shown in Fig. 8) or if available the relational information is extracted from the annotation ontology, which is used to define semantic and lexical relations when they cannot be inferred automatically from the image's content. The resulting structural composition of the annotations is used to extract the information needed to compare different annotations at semantic level.

A relation is composed of a *relation id*, its *type* and the *levels* describing the semantic level of two participating annotation terms. For example, **left-of** describes a structural relation which denotes the spatial arrangement of

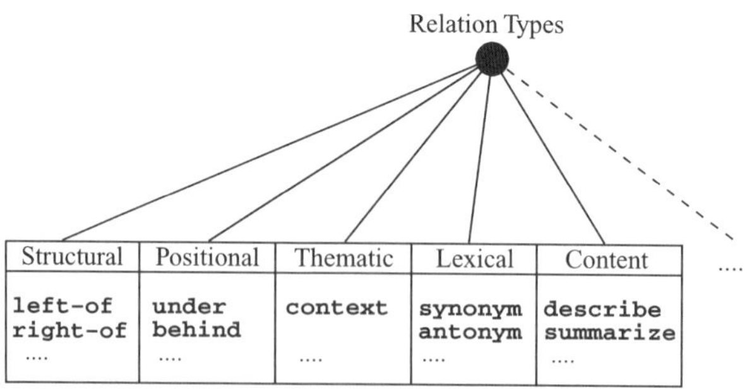

Fig. 8. Annotation relations catalog

two objects. At a higher semantic level, there are positional relations, like **under** or **behind** which perception is more influenced by the user. Thematic relations, which represent a subgroup of semantic relations, connect verbal concepts with nominal concepts which preferably occur as their complements. For example, the verbal concept *write* should have pointers to the concept *person*. Another relations, e.g. lexical, are used to to mitigate synonymy and polysemy problems in the retrieval process. By providing a such finite catalog, the possible relations between concepts are constrained which reduces the amount of annotation errors and moreover simplify the evaluation of the relations. By the way, the inference making process can be used to discover hidden relationships.

3.5 Extending the Probabilistic Annotation by Multi-level Annotations

According to the probabilistic annotation approach, where keywords' relevance or importance for an image is determined by the hypothesis that similar images may share the same keywords, the set of keywords for annotating I_q is determined by the following three steps:

1. Calculation of the k most similar images $I_1, \ldots, I_k$ based on their low-level features.
2. Statistical identification of the frequent annotations associated with the k images.
3. Extension of the results by taking into account the multi-level properties of annotations.

The set of images which are similar to a target image I_q is computed by applying the k-Nearest Neighbors algorithm (kNN). The detected images satisfy the criteria $sim(I_q, I_x) < \varepsilon$, where $sim(\cdot)$ $(0 \leq sim(\cdot) \leq 1)$ is the distance metrics computing the dissimilarity between two low-level feature vectors. The most suited annotations for the image I_q can be simply determined based on the annotations of its similar images. Assuming that we have an underlying probability distribution $P(\cdot|I_x)$ for each image $I_x \in \mathbf{D}$, which can be thought as a vector that contains the low-level features of the image, as well as all keywords $\{k_1, k_2, \ldots, k_n\}$ that appear in the annotation of I_q. Due to the *Probabilistic Model* the probability $P(k_j|I_q)$ that a keyword k_j is suited for the annotation of the image I_q is defined as [6]:

$$P(k_j|I_q) = \frac{w_j}{\sum\limits_{j'=1\ldots n} w_{j'}}, \tag{2}$$

where w_j is the weight of keyword k_j, which is computed as following:

$$w_j = \sum_{\forall i} sim(I_q, I_i) \times \beta_{ij}, \tag{3}$$

where $sim(I_q, I_i)$ represents the similarity value between the images I_q and I_i and β_{ij} $(0 \leq \beta_{ij} \leq 1)$ defines the importance of the keyword k_j to the image I_i. This importance can be estimated by a modification of the *tf.idf* weighting, namely by the frequency of the word k_j in annotations of similar images multiplied by the inverse frequency of this keyword in other annotations. Nevertheless, this approach does not consider the fact that keywords are related to each other and can be assigned at several levels. In addition, some keywords describing emotions or actions (e.g. 'driving') are difficult to be associated with visual features.

To alleviate the deficiencies mentioned previously, another third step is needed to enhance the results of the automatic annotation. Now, the created semantic annotation space and the annotation ontology is used to evaluate the connections between keywords and related concepts (Fig. 9). By examining the relations between several annotations containing a keyword the importance of this keyword for different concepts can be inferred. Thus, the probability of a given keyword k_j in a given context l (or abstraction level) – abbreviated by $P([k_j, l]|I_q)$ – to be accurate for the annotation of the image I_q is defined by:

$$P([k_j, l]|I_q) = \frac{w_j^{new}}{\sum_{j'=1 \ldots n} w_{j'}^{new}}, \tag{4}$$

where w_j^{new} is computed as following:

$$w_j^{new} = \sum_{i=1 \ldots q} H[C_i, k_j, l] \times (\sum_{j'=1 \ldots n} w_{j'} \times H[C_i, k_{j'}, l]), \tag{5}$$

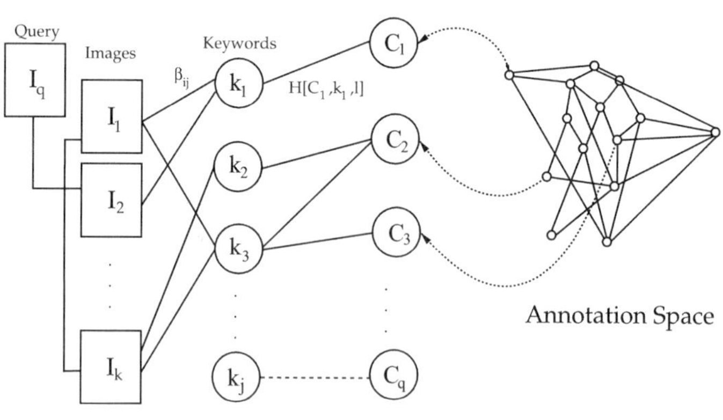

Fig. 9. Linking keywords to the annotation space

and represents the weighted sum of the concept weights. The w_j's are the weights computed in (3), and $H[C_i, k_j, l]$ denotes the relevance of the keyword k_j for the description of the concept C_i depending on a given context l.

This context can be defined by adding a supplementary concept which will be preferable considered in the computation of the distance. The relevance $H[C_i, k_j, l]$ is a number between [0..1] which mirrors the contextual distance between the keyword k_j and the concept C_i depending on the annotation space and the given domain dependent ontology. The used method will be described in Sect. 3.6. The idea behind this multi-level extension is, that the accuracy of a keyword is determined depending on its context. For example, if a keyword k had a low probability determined by the classical approach, it can be increased by the fact, that it is frequently used in annotations of similar images or by frequently having a relation to other frequent annotations within the same concept.

3.6 Example of Context-Based Similarity

The advantages of the graph representation are the following. First, implications about the similarity of annotations can be determined by considering the incoming and outgoing edges in the multi-graph. In addition, relations like specialization can be determined by considering the degree of the hierarchical distance. The way the resulting graph structure is used to support the semantic retrieval by distinguishing between different abstraction levels will be demonstrated in an example. Semantics is commonly defined as the meaning of data, and the task of evaluating the extent of semantic matching between different annotations should be based on their meanings. Since in the most cases the *meaning* of a piece of data cannot be expressed by only one concept, the graph approach is used to determine the *context-based* similarity between annotations by examining the contexts in which the keywords appear. By introducing the relevance $H[C_i, k_j, l]$, denoting the importance of a keyword k_j for the description of the concept C_i depending on a given context l, we can estimate the location of frequent occurrences of the keyword in the annotation space and thus discover its relations to other annotations.

Example 2. Figure 10 explains the the usage of the graph for the distance computation. Assuming that we have two users U_1 and U_2 which are searching

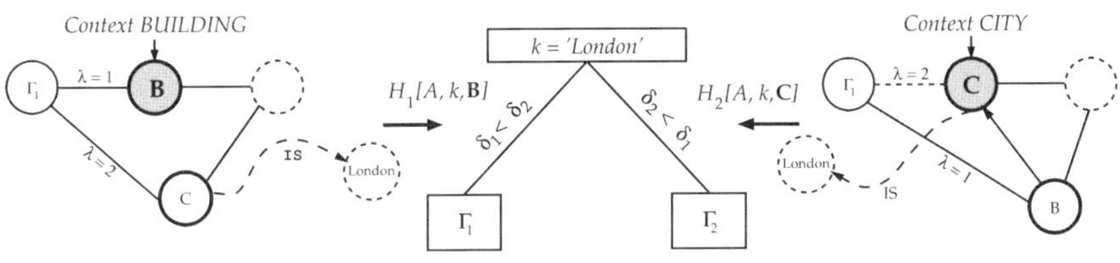

Fig. 10. Context-based distance between semantic concepts

for photographs of the city of London using the keyword k = '*London*'. The several abstraction levels appearing in image collections are simulated by two images, the first one displays the *Tower Bridge* and is annotated with Γ_1, and the other one presents the *Coat of Arms* (annotation Γ_2) of the City of London. The two context-based distances $\delta_1 = H_1[A, k, \mathbf{B}]$ and $\delta_2 = H_2[A, k, \mathbf{C}]$ between the used keyword k = '*London*' and the concept A = CITY:London are computed in different contexts $\mathbf{B}$ = *building* and $\mathbf{C}$ = *city*. In order to fulfill the *contextual* requirements, the distance δ_1 between Γ_1 and the new keyword k should be smaller than δ_2 when the user's search is restricted to the context of *building* and vice versa.

For the determination of δ_1 and δ_2, the underlying annotation graph has to be considered. Here, $h(k_j, C_i)$ denotes the relevance of keyword k_j for the concept C_i and $\mathrm{p}(k_j, l)$ the path between concepts k_j and l in the annotation ontology. The factor η is an evaluation measure applied to the annotation space which describes the relevance and the sematic distance between the concept assigned to the keyword k_j and the context l. In the above example, the annotation Γ_1 is connected to the concepts 'city' and 'building' in the annotation graphs, on the left side the context is *building* ($\mathbf{B}$) which shares with Γ_1 (with its keyword 'bridge') the same hierarchical layer in the annotation ontology. This results in a low η and thus strengthens the relevance of the keyword in the context of *buildings*. On the opposite side, in the context *city*, where Γ_1 is only indirectly connected to $\mathbf{C}$ by an expansion edge, the relevance for this keyword is lower.

The *context-based* methodology incorporates different sematic contexts in which keywords can appear to image retrieval systems. By applying suitable graph transformation rules, annotations can be efficiently organized with the objective to provide a global semantic overview of the underlying image collections.

4 Related Work

An architecture for semi-automated image annotation is proposed in [7], which integrates keyword-based search, content-based image retrieval, and user feedback. This approach is presented from the perspective of enriching the image data by keywords, which are extracted by considering results of the feedback cycle performed by the user. The initial search keywords are automatically added to the images that receive positive feedback and facilitate keyword-based image retrieval in the next iteration. Another tool for semi-automatically annotating image regions is presented in [16]. It is based on manual selection of positive and negative examples performed by users and then uses texture similarity to propagate annotations. In several work, the choice of appropriate annotation terms is supported by existing ontologies [10]. We also found several work on relevance feedback [12, 17] using learning methods and model inference to find correspondences between the high-level concepts users perceive and the low-level features extracted from the images. Because of the automatic query reformulation by the system, the user does not need to explicitly specify weights for different computational features. Several approaches in the area of semantic information retrieval incorporating the mapping of local features into *w*ords of an image have been proposed [18, 19]. The approaches are based on the creation of a partial taxonomy for home photos, modeling of high-level information like events, and the definition of visual keywords to describe semantic concepts. The main difference to our approach lies in the fact that we place more emphasis on the nature of annotations. Our work differs from the mentioned approaches through its focus on users' subjectivity which creates special problems, such as the detection of equal content descriptions at different abstraction levels. The presented approach returns useful information about the underlying data collection, that cannot be found in the annotation ontology. Through this preprocessing of semantic information, the mappings of the low level features into semantic concepts can be improved, leading to an increase of precision in semi-automatic annotation methods and image retrieval.

5 Summary and Future Work

In this chapter we have demonstrated existing problems in the field of image retrieval supported by semantic annotation. In the main part we have introduced the multi-level annotation component which analyzes and evaluates the assigned multi-level annotations at both feature level and semantic level. The resulting semantic information is transformed into a multi-graph representation, which encodes the complex structure of both hierarchical semantic relations and similarities between differently annotated images. The information derived from this representation can be easily utilized to supplement existing annotation models and to allow a context-based similarity evaluation between keywords and different annotations. For future work, we plan retrieval performance experiments for semantic queries and the investigation of the accuracy of the semi-automatic annotation using our approach. Another promising aim is to automatically detect annotation inconsistencies within image collections or use our approach for the creation of correctly annotated image data corpora which are the basis for the evaluation of annotation-based retrieval systems.

References

1. Rong Zhao and William I. Grosky. Bridging the Semantic Gap in Image Retrieval. In *Distributed Multimedia Databases: Techniques & Applications*, pages 14–36, Hershey, PA, USA, 2002. Idea Group Publishing.

2. T. Huang, Y. Rui, M. Ortega, and S. Mehrotra. Relevance Feedback: A Power Tool for Interactive Content-Based Image Retrieval. *IEEE Transactions on Circuits and Systems for Video Technology*, pages 25–36, 1998.
3. Y. Rui, T. Huang, and S. Mehrotra. Relevance Feedback Techniques in Interactive Content-Based Image Retrieval. In *Storage and Retrieval for Image and Video Databases (SPIE)*, pages 25–36, 1998.
4. Y. Rui, T. Huang, and S. Mehrotra. Content-Based Image Retrieval with Relevance Feedback in MARS. In *Proceedings of the 1997 International Conference on Image Processing (ICIP '97)*, pages 815–818, 1997.
5. Wayne Niblack, Ron Barber, William Equitz, Myron Flickner, Eduardo H. Glasman, et al. QBIC Project: Querying Images by Content, using Color, Texture, and Shape. In *Proceedings of Storage and Retrieval for Image and Video Databases (SPIE)*, volume 1908, April 1993.
6. Pu-Jen Cheng and Lee-Feng Chien. Effective Image Annotation for Search using Multi-level Semantics. In *Proceedings of International Conference of Asian Digital Libraries*, pages 230–242. Springer, 2003.
7. L. Wenyin, S. Dumais, Y. Sun, H. Zhang, M. Czerwinski, and B. Field. Semi-Automatic Image Annotation. In *Proceedings International Conference on Human–Computer Interaction (INTERACT'01)*, pages 326–333, 2001.
8. P. Duygulu, Kobus Barnard, J. F. G. de Freitas, and David A. Forsyth. Object Recognition as Machine Translation: Learning a Lexicon for a Fixed Image Vocabulary. In *ECCV '02: Proceedings of the 7th European Conference on Computer Vision-Part IV*, pages 97–112, London, UK, 2002. Springer, Berlin Heidelberg New York.
9. Jose Torres, Alan Parkes, and Luis Corte-Real. Region-Based Relevance Feedback in Concept-Based Image Retrieval. In *Proceedings of the 5th International Workshop on Image Analysis for Multimedia Interactive Services*, Lisboa, Portugal, 2004.
10. L. Hollink, G. Schreiber, J. Wielemaker, and B. Wielinga. Semantic Annotation of Image Collections. In *Proceedings of the K-CAP 2003 Workshop on Knowledge Markup and Semantic Annotation*, 2003.
11. A. Th. Schreiber, Barbara Dubbeldam, Jan Wielemaker, and Bob Wielinga. Ontology-Based Photo Annotation. *IEEE Intelligent Systems*, 16(3):66–74, 2001.
12. Rosalind W. Picard, Thomas P. Minka, and Martin Szummer. Modeling User Subjectivity in Image Libraries. In *IEEE International Conference On Image Processing*, volume 2, pages 777–780, Lausanne, Switzerland, 1996.
13. Micheline Beaulieu, Pia Borlund, Peter Brusilovsky, et al. Matthew Chalmers. Personalisation and Recommender Systems in Digital Libraries. *Joint NSF-EU DELOS Working Group Report.* Technical Report, May 2003.
14. Masashi Inoue. On the Need for Annotation-based Image Retrieval. In *Workshop on Information Retrieval in Context (IRiX)*, pages 44–46, Sheffield, UK, 2004.
15. James Griffioen, Rajiv Mehrotra, and Rajendra Yavatkar. An Object-Oriented Model for Image Information Representation. In *CIKM '93: Proceedings of the Second International Conference on Information and Knowledge Management*, pages 393–402, New York, NY, USA, 1993. ACM Press.
16. Rosalind W. Picard and Thomas P. Minka. Vision Texture for Annotation. In *Multimedia Systems*, volume 3, pages 3–14, 1995.
17. Takio Kurita and Toshikazu Kato. Learning of Personal Visual Impression for Image Database Systems. In *Second International Conference on Document Analysis and Recognition*, pages 547–552, 1993.

18. Joo-Hwee Lim. Building Visual Vocabulary for Image Indexation and Query Formulation. In *Pattern Analysis and Applications (Special Issue on Image Indexation)*, volume 4, pages 125–139, 2001.
19. Joo-Hwee Lim, Qi Tian, and Philippe Mulhem. Home Photo Content Modeling for Personalized Event-Based Retrieval. *IEEE MultiMedia*, 10(4):28–37, 2003.

An Incremental Framework for Adapting the Hypermedia Structure of Multimedia Documents

Sébastien Laborie and Jérôme Euzenat

INRIA Rhône-Alpes and LIG
Montbonnot, France
Sebastien.Laborie@inrialpes.fr, Jerome.Euzenat@inrialpes.fr

Summary. The multiplication of presentation contexts (such as mobile phones, PDAs) for multimedia documents requires the adaptation of document specifications. In an earlier work, a semantic approach for multimedia document adaptation was proposed. This framework deals with the semantics of the document composition by transforming the relations between multimedia objects. In this chapter, we apply the defined framework to the hypermedia dimension of documents, i.e., hypermedia links between multimedia objects. By considering hypermedia links as particular objects of the document, we adapt the hypermedia dimension with the temporal dimension. However, due to the non-deterministic character of the hypermedia structure, the document is organized in several loosely dependent sub-specifications. To preserve the adaptation framework, we propose a first straightforward strategy that consists of adapting all sub-specifications generated by the hypermedia structure. Nevertheless, this strategy has several drawbacks, e.g., the profile is not able to change between user interactions. Hence, we propose an incremental approach which adapts document sub-specifications step by step according to these interactions. To validate this framework, we adapt real standard multimedia documents such as SMIL documents.

1 Introduction

A multimedia document may be played on different devices with different capabilities: phones, PDAs, desktop computers, setup boxes, etc. These devices introduce different constraints on the presentation itself. For instance, keyboard limitations (e.g., portable multimedia player) may prevent selecting hypermedia links concurrently. Other constraints may also be introduced by user preferences, content protection or terminal capabilities [20]. The constraints imposed by a client constitute its profile.

To satisfy profiles, multimedia documents must be adapted, i.e., transformed into documents compatible with target contexts before being played.

S. Laborie and J. Euzenat: *An Incremental Framework for Adapting the Hypermedia Structure of Multimedia Documents*, Studies in Computational Intelligence (SCI) **93**, 157–176 (2008)
www.springerlink.com

Several kinds of adaptation are possible, such as local adaptation (adaptation of media objects individually) and global adaptation (adaptation of the document structure). This chapter focuses on the latter.

In [6], a framework for adapting multimedia documents based on the qualitative semantics of the documents and constraints was proposed. This approach transforms the relations between multimedia objects and ensures two main properties (1) that adaptation constraints are satisfied and (2) that the adapted document is as close as possible to the initial document. As an example, for the temporal dimension of documents, this work has been applied to descriptions based on the Allen algebra [2].

In this chapter, we apply this adaptation framework to the hypermedia dimension of documents, i.e., hypermedia links between multimedia objects. By considering hypermedia links as particular objects of the document which belong to multimedia objects, we adapt the hypermedia dimension with the temporal dimension (Sect. 4).

However, due to the non-deterministic character of the hypermedia structure, the document is organized in several loosely dependent sub-specifications. To preserve the adaptation framework, we propose a first straightforward strategy that consists of adapting all sub-specifications generated by the hypermedia structure (Sect. 5). Nevertheless, we show that this strategy has several drawbacks, e.g., the profile is not able to change between user interactions. Hence, we propose an incremental approach which adapts document sub-specifications step by step according to these interactions.

For a smooth implementation of this adaptation framework, we provide an algorithm which computes adapted multimedia document specifications (Sect. 6). Moreover, we validate our adaptation approach by adapting real standard multimedia documents such as SMIL documents [21] (Sect. 7).

Let introduce first a multimedia document example (Sect. 2) that will be used in this chapter and a current proposed adaptation framework (Sect. 3).

2 Multimedia Document Specification

Multimedia documents are composed of various multimedia objects such as texts, images, sounds and videos. These multimedia objects are organized into several dimensions [15]:

- *Temporal*: multimedia objects are synchronized.
- *Spatial*: some multimedia objects have a spatial layout.
- *Logical*: some multimedia objects can be assemble in groups.
- *Hypermedia*: the user is able to interact with some multimedia objects.

Figure 1 shows the temporal and the hypermedia dimensions of a movie trailer presentation composed of different multimedia objects like a movie Poster, a Trailer, an Abstract, the Characters information and the upcoming Dates.

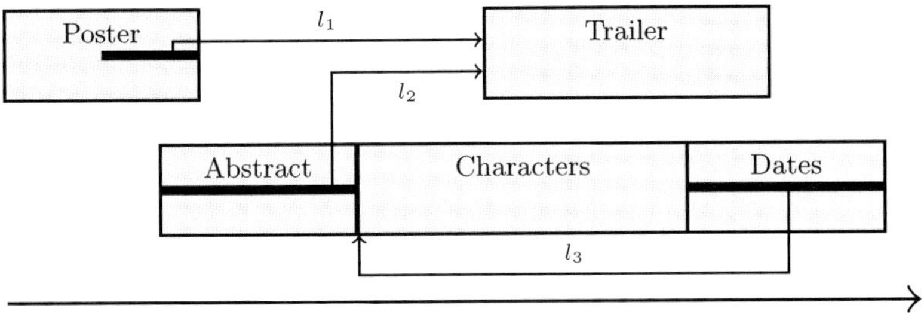

Fig. 1. A multimedia document example

The hypermedia dimension of multimedia documents is characterized by hypermedia links between multimedia objects. For example in Fig. 1, the Poster and the Abstract are linked to the beginning of the Trailer thanks to the hypermedia links l_1 and l_2, respectively. Note that, the hypermedia dimension is not independent from the other dimensions like the temporal one. In Fig. 1, suppose the user never selects a hypermedia link, each multimedia object of the presentation is executed. However, if the user selects l_1, the presentation "jump forward" in time to the beginning of the Trailer and the beginning of the Characters information is not presented.

There are several languages or formats for specifying multimedia documents such as SMIL [21], SVG [19], Madeus [9], etc. Making the adaptation format-dependent requires an adaptation strategy for every single format. Instead, we propose an abstraction layer which allows to hide the format specific syntax and details, and capture the essence of a given document with respect to its main dimensions. We call this document abstraction a multimedia document specification.

Definition 1 (Multimedia document specification). *A multimedia document specification $s = \langle O, C \rangle$ is made of a set of multimedia objects O and a set of constraints C between these objects. In the remainder, the constraints will be considered as binary.*

The advantages of such an abstraction and adaptation scheme are twofold. First, it allows to reuse the same strategy for the different languages or formats. Second, the abstract representation provides more flexibility for the adaptation since the relations between multimedia objects can be described qualitatively.

3 Semantic Adaptation of Multimedia Documents

In [6], a semantic approach for multimedia document adaptation is defined. This approach interprets each document as the set of its potential executions (i.e., related to the initial document) and a profile as the set of its possible executions. In this context, "adapting" amounts to find the set of potential executions that are possible. When none is possible, the goal of adaptation is to find possible executions close to potential executions that satisfy the profile.

We consider both the multimedia document specifications and the profiles as a set of relations holding between multimedia objects. The potential and possible executions are ideally represented by relation graphs.

Definition 2 (Relation graph). *Let O be a set of multimedia objects and C a set of constraints between the elements of O. A multimedia document specification $s = \langle O, C \rangle$ relative to a set of executions can be represented as a complete directed labeled graph $g_s = \langle O, E, \lambda \rangle$ called a relation graph. The elements of O are nodes, E is the set of edges and $\lambda : E \rightarrow 2^{\mathcal{R}}$ is a total labeling function from the edges to a set of relations of $\mathcal{R}$ such that for each $x\ r\ y \in C$, $r \in \lambda(\langle x, y \rangle)$.*

Figure 2 presents two relation graphs. Each node corresponds to a multimedia object and each edge is labeled by a set of relations. In this example, the Allen algebra of temporal interval relations $\mathcal{R}$ [2] is used to describe the temporal dimension. These relations are presented in Table 1.

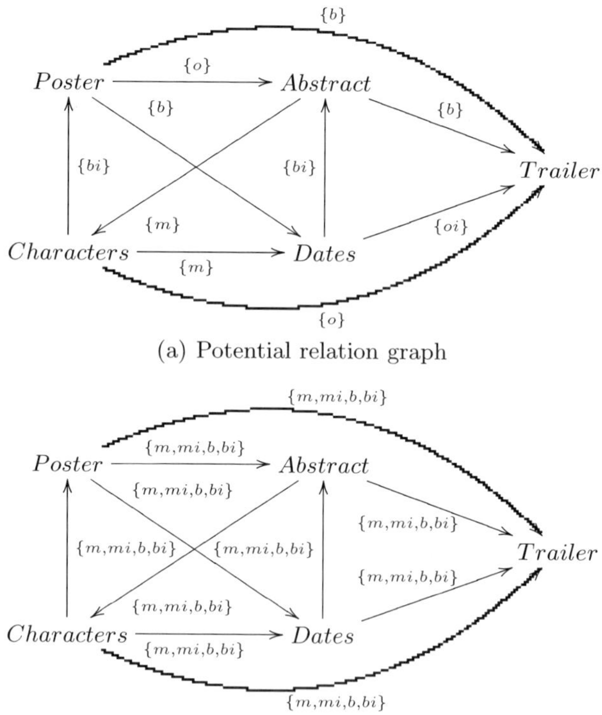

(a) Potential relation graph

(b) Possible relation graph

Fig. 2. Potential and possible executions

Table 1. The thirteen Allen relations

Relation (r): x r y	x/y	Inverse: y r^{-1} x
before (b)		(bi) after
meets (m)		(mi) met-by
during (d)		(di) contains
overlaps (o)		(oi) overlapped-by
starts (s)		(si) started-by
finishes (f)		(fi) finished-by
equals (e)		(e)

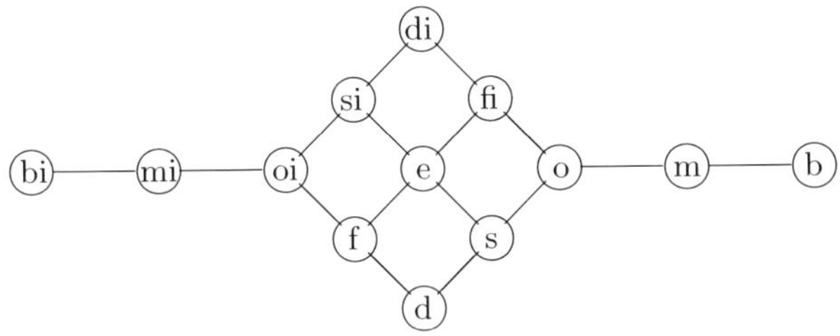

Fig. 3. Allen conceptual neighborhood graph

The potential relation graph (Fig. 2a) includes, in particular, the execution of Fig. 1. It corresponds to the initial document. The possible relation graph (Fig. 2b) corresponds to the following profile: Impossible for multimedia objects to be played concurrently. It is important to note that it may occur that some potential relations are not possible (e.g., between *Poster* and *Abstract*, and between *Characters* and *Trailer*).

In this context, adapting consists of finding a set of relation graphs corresponding to the possible relation graph (i.e., satisfying the profile) at a minimal distance from the potential relation graph (i.e., the initial document specification). This set of relation graphs is thus called adapted relation graph solutions.

We consider that the proximity between two relation graphs depends on the proximity between relations borne by the same edge in both graphs. This proximity relies on the conceptual neighborhood between these relations and is measured by the shortest path distance in the corresponding conceptual neighborhood graph (Fig. 3, [8]).

Thus, a distance d between relation graphs is obtained by summing up all the conceptual distances δ between relationships used in both graphs (Definition 3).

Definition 3 (Conceptual distance between two relation graphs).

$$d(\lambda, \lambda') = \Sigma_{o_1,o_2 \in O} \mathrm{Min}_{r \in \lambda(\langle o_1,o_2\rangle),\ r' \in \lambda'(\langle o_1,o_2\rangle)} \delta(r, r')$$

Figure 4a presents an adapted relation graph solution computed from Fig. 2b. Its distance from the initial relation graph of Fig. 2a is $d = 4$ because

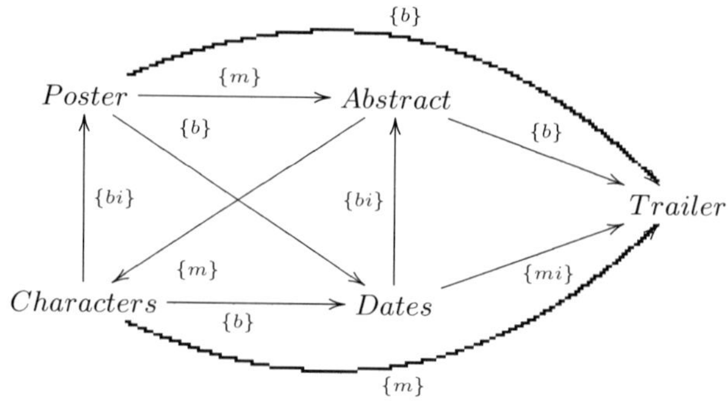

(a) An adapted relation graph solution

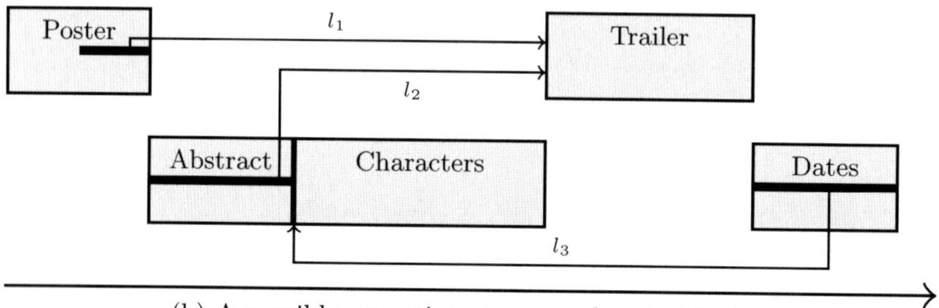

(b) A possible execution corresponding to Fig. 4a

Fig. 4. An adapted relation graph solution and a corresponding execution

the relations between *Poster* and *Abstract*, and between *Characters* and *Trailer* change from o to m ($\delta(o, m) = 1$), the relation between *Characters* and *Dates* changes from m to b ($\delta(m, b) = 1$) and the relation between *Dates* and *Trailer* changes from oi to mi ($\delta(oi, mi) = 1$).

A possible execution corresponding to Fig. 4a is presented in Fig. 4b. As we may notice, the profile is satisfied and this adapted document is close to the initial one.

This approach has been fully defined for the temporal [6], spatial [11] and spatio-temporal [10] cases. Our goal is to extend this framework to the hypermedia dimension.

4 Adaptation of the Hypermedia Dimension

As shown in Fig. 1, each hypermedia link of a multimedia document is related to a multimedia object. Moreover, hypermedia links possess their own execution time inside their related multimedia object. For example in Fig. 1, l_1 is a hypermedia link belonging to *Poster* and is active during a part of its execution, while l_2 belongs to *Abstract* and is active during all its execution.

Hence, hypermedia links can be considered as particular objects of the document specification. Thus, we enhance in the following definition the multimedia document specification of Definition 1 with hypermedia links.

Definition 4 (Multimedia document specification with hypermedia). *Let* O_m *be a set of multimedia objects and* O_l *a set of hypermedia links.* $s = \langle O, C \rangle$ *is a multimedia document specification enhanced with hypermedia links with* $O = O_m \cup O_l$ *the set of objects of the document and* C *the set of constraints between the elements of* O*. To express that a link* $l \in O_l$ *belongs to a multimedia object* $m \in O_m$*, the constraint between* l *and* m *should express such property, e.g., for the temporal case* l *is active only when* m *is executed.*

Figure 5 presents a subpart of the document presented in Fig. 1.

The upcoming Dates object is voluntary omitted for visibility reason.

Figure 5b is the potential relation graph corresponding to the execution presented in Fig. 5a. Note that the relation graph is enhanced with hypermedia links, i.e., l_1 and l_2. Moreover, the relation between l_1 and *Poster* and the one between l_2 and *Abstract* state that a hypermedia link is active only when its corresponding multimedia object is executed.

Thanks to he multimedia specification extension defined in Definition 4, the adaptation framework described in Sect. 3 can be used to adapt the hypermedia dimension of documents. Let adapt Fig. 5 with the following profile: impossible to select hypermedia links concurrently, e.g., because the device has only one button.

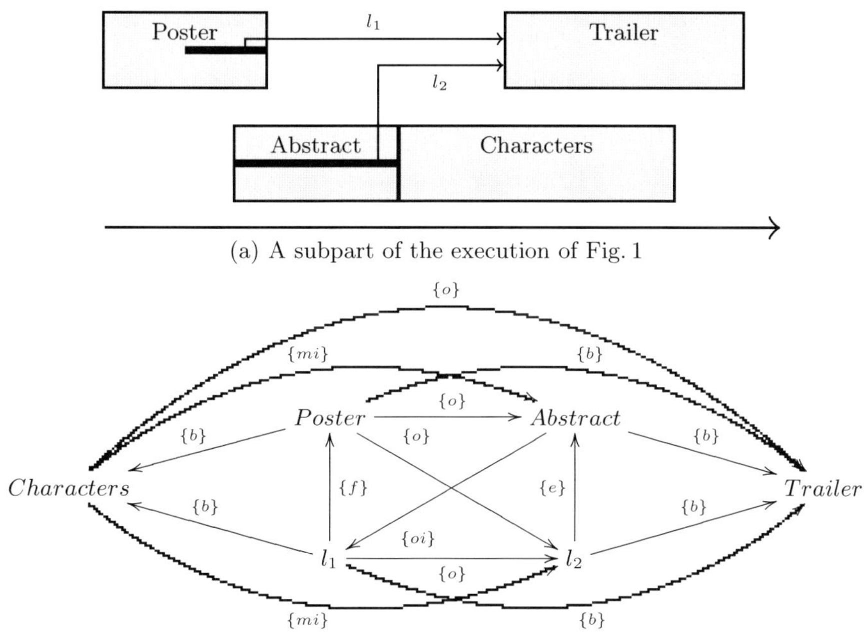

(a) A subpart of the execution of Fig. 1

(b) The potential relation graph of Fig. 5a

Fig. 5. A subpart of the document of Fig. 1

Figure 6a presents the possible relation graph corresponding to this profile. The set of relations $\{m, mi, b, bi\}$ between l_1 and l_2 is due to the profile. When no adaptation constraint is expressed in the profile between two objects,

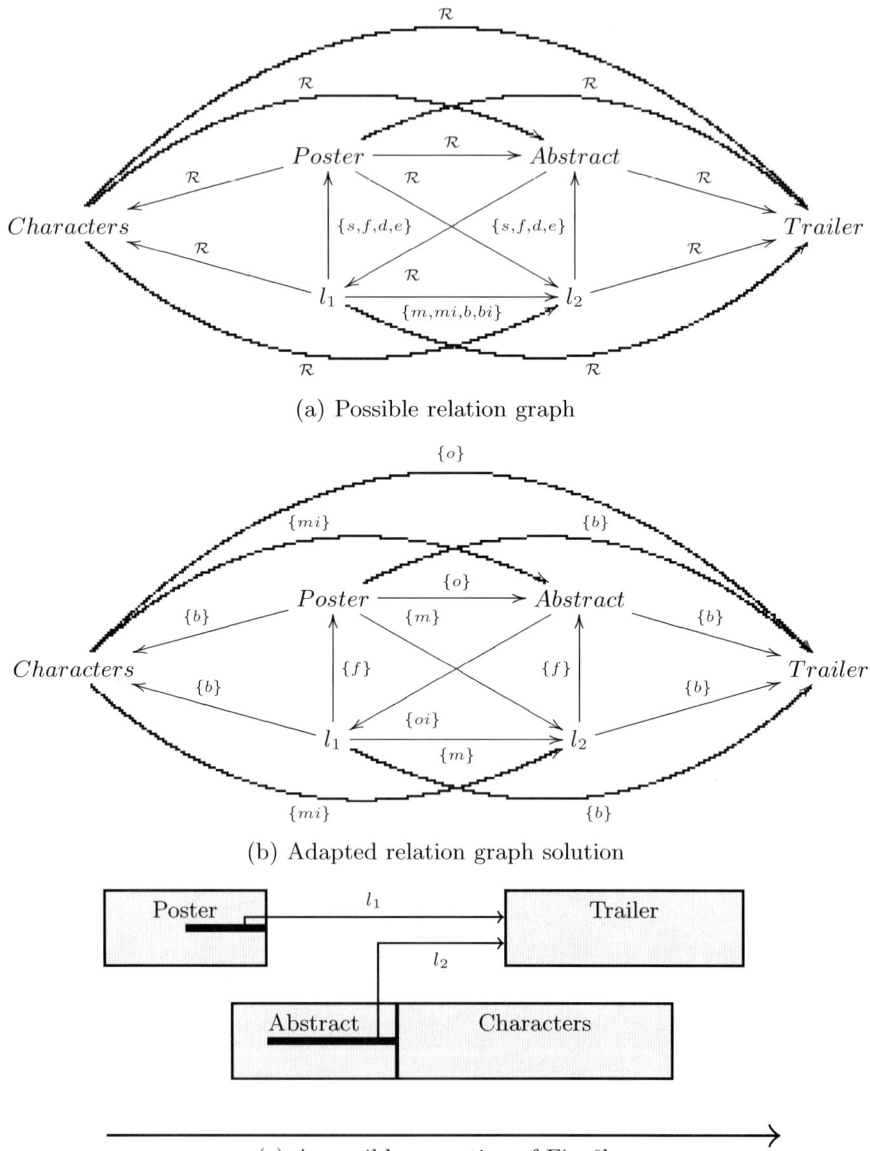

(a) Possible relation graph

(b) Adapted relation graph solution

(c) A possible execution of Fig. 6b

Fig. 6. Adaptation of Fig. 5 with the following profile: impossible to select hypermedia links concurrently we consider that all relations $\mathcal{R}$ are possible. For example, no adaptation constraint is expressed in the profile between *Poster* and *Abstract*, thus all relations $\mathcal{R}$ are possible between these two objects.

Moreover, in the relation graph of Fig. 6a the constraint between a hypermedia link and its related multimedia object is satisfied. For example, the hypermedia link l_2 belongs to *Abstract*, and the only allowed relations in time between l_2 and *Abstract* are starts (s), finishes (f), during (d) and equals (e), i.e., inclusive temporal relations.

Figure 6b presents an adapted relation graph solution computed from Fig. 6a (cf. Sect. 3). Its distance from the initial relation graph of Fig. 5b is $d = 3$ because the relation between l_1 and l_2 changes from o to m ($\delta(o, m) = 1$), the relation between l_2 and *Abstract* changes from e to f ($\delta(e, f) = 1$) and the relation between *Poster* and l_2 changes from o to m ($\delta(o, m) = 1$).

A possible execution corresponding to Fig. 6b is presented in Fig. 6c. As we may notice, the profile is satisfied and this adapted document is close to the initial one.

In this section, we adapt with the proposed adaptation framework of Sect. 3 the hypermedia dimension of multimedia documents by considering hypermedia links as particular objects which belong to multimedia objects. We mix this dimension with the temporal one providing a temporal-hypermedia adaptation. However, due to the non-deterministic character of the hypermedia structure involved by hypermedia links, several loosely dependent sub-specifications have to be considered. In the next section, we propose an incremental approach for adapting these sub-specifications according to the user interactions.

5 Adaptation Based on the Hypermedia Structure

A multimedia document may be composed of several hypermedia links. Each hypermedia link points to a particular time in the presentation. For example, in Fig. 1 the hypermedia links l_1 and l_2 point to the beginning of the *Trailer*, while l_3 points to the beginning of *Characters*.

Hence, several multimedia document sub-specifications can be identified. They correspond to the hypermedia structure of the document. For example, Fig. 7 presents the hypermedia structure of Fig. 1. Each node represents a multimedia document sub-specification. In this figure, they are three sub-specifications, namely s_1, s_2 and s_3. For a better understanding, for each sub-specification we show a possible execution. Each edge is labeled by a set of hypermedia links which leads from one sub-specification to another.

We propose, in Sect. 5.1, a method which build from an initial document specification several sub-specifications corresponding to its hypermedia structure. Thereafter, in Sect. 5.2, we propose a first straightforward strategy that consists of adapting at once all sub-specifications generated by the hypermedia structure. However, several drawbacks are identified. Hence, in Sect. 5.3,

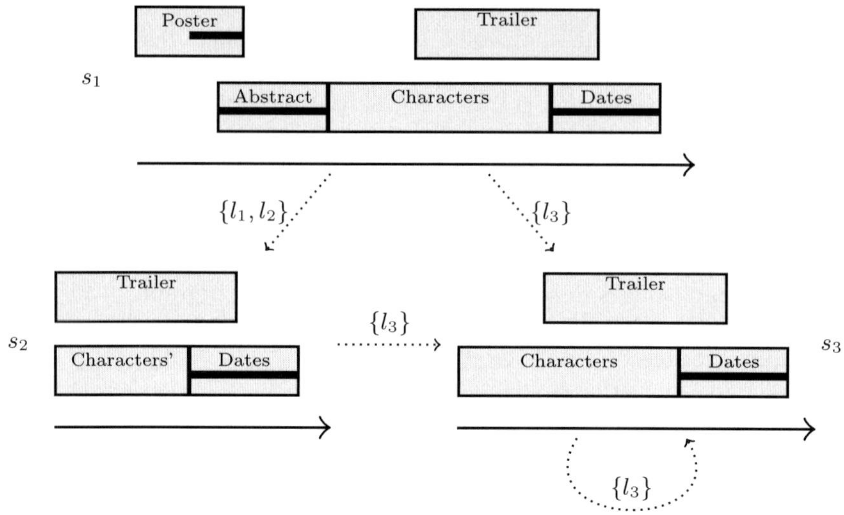

Fig. 7. The hypermedia structure of Fig. 1

we propose an incremental adaptation which adapts sub-specifications step by step according to the user interactions.

5.1 Generating the Multimedia Document Sub-Specifications According to the Hypermedia Structure

From an initial document specification, several sub-specifications can be computed according to its hypermedia links. For example, Fig. 1 is an initial document specification. If the hypermedia link l_1 is selected, the presentation "jump" in time to the beginning of the *Trailer*. It is thus useless to preserve in the future sub-specification all objects that are before this moment.

Figure 8 presents how s_2 is constructed from the initial document specification of Fig. 1. As we may notice, when l_1 is selected, the objects *Poster*, *Abstract*, l_1 and l_2 are useless in the future sub-specification. Moreover, the multimedia object *Characters* is splitting into two parts, i.e., one useless and one presented. In this case, this object is transformed into a partition of *Characters*, called *Characters'*, and starts the present time (cf. s_2 in Fig. 7).

This process is done for each hypermedia link of the initial document specification. In the next section, we propose to adapt these sub-specifications.

5.2 Adapting All Sub-Specifications at Once

Each document sub-specification of the hypermedia structure does not contain the same number of objects and the same constraints between them.

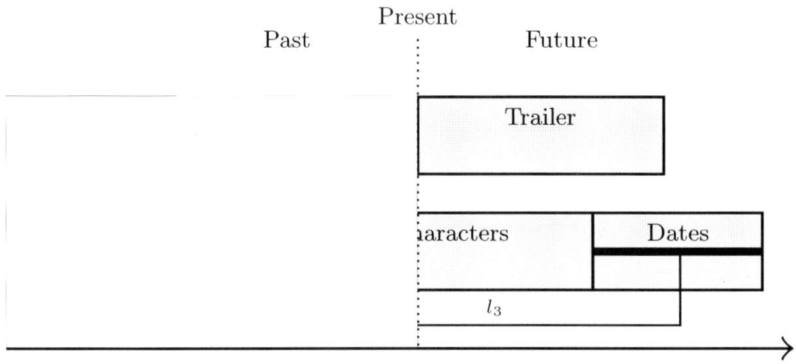

Fig. 8. Building the sub-specification s_2

Moreover, the adaptation framework presented in Sect. 4 never suppresses objects. Thus, it is difficult to group in one global document specification all sub-specifications generated by the hypermedia structure without any modification of our adaptation approach.

Hence, to preserve our framework, defined in Sect. 4, we propose the following straightforward strategy:

Suppose p a given profile,
for each sub-specification s_i of a hypermedia structure, $Adapt(s_i, p)$.

For example, suppose the following profile: impossible to play hypermedia links concurrently with other objects (we do not consider the multimedia object related to the hypermedia link). Figure 9 presents the adapted hypermedia structure of Fig. 1. In the adapted sub-specification s'_1, the hypermedia link duration of l_1, l_2 and l_3 is reduced, while in s'_2 and s'_3 only the hypermedia link duration of l_3 is reduced.

This strategy effectively produces an adapted document where all its sub-specifications satisfy the profile and are close to the initial ones. However, this strategy has several drawbacks:

- Some models are adapted even if they are not executed. For example in Fig. 7, if the user never selects a hypermedia link, it would be more efficient to adapt only s_1.
- If the profile p changes between each user interaction, all sub-specifications s_i have to be adapted again.
- Parts of several sub-specifications which are identical are adapted several times. For example in Fig. 7, s_1 and s_3 have identical parts and are adapted twice (see s'_1 and s'_3 in Fig. 9).

Hence, we propose in the following section to adapt the sub-specifications of the hypermedia structure step by step according to the user interactions.

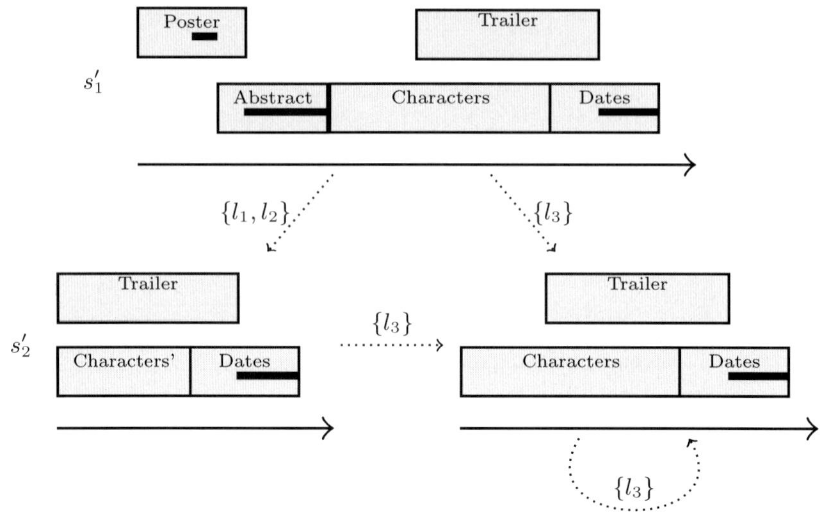

Fig. 9. An adapted hypermedia structure of Fig. 1

5.3 Adapting Sub-Specifications Step by Step

Euzenat et al. [6] proposed to compute adapted documents close to the initial one which satisfy the profile. Based on the hypermedia structure, we propose in this section to compute adapted sub-specifications close to the previous adapted one if the profile do not changed, and close to the initial one if the profile changed. Our aim is to adapt only subparts of the document that is going to be presented. Let call this adaptation incremental adaptation.

Initially, we compute and adapt a first initial sub-specification. We choose the one executed when the user never selects a hypermedia link. This leads to a first adapted sub-specification. For example, s_1 (Fig. 7) is the first initial sub-specification and p a given profile (the same as in Sect. 5.2). We adapt s_1 according to the profile p, i.e., $Adapt(s_1, p) = s'_1$ (Fig. 10 top).

If a hypermedia link is selected by the user and the profile p do not changed, we compute the next corresponding sub-specification according to the previous adapted sub-specification. For example in Fig. 10, suppose during the execution of s'_1 the user selects l_3, we compute s_3 according to s'_1. This computation propagates the adapted information of s'_1 into s_3. Thereafter, we adapt s_3 according to the profile p, i.e., $Adapt(s_3, p) = s'_3$ (Fig. 10 middle).

If a hypermedia link is selected by the user and the profile p changed into p', e.g., impossible to play multimedia objects concurrently, we initialize the corresponding sub-specification according to the initial document specification and adapt it. For example in Fig. 10, suppose during the execution of s'_3 (computed previously) the user selects l_3, we initialize s_3 according to the initial document specification. Thereafter, we adapt s_3 according to the profile

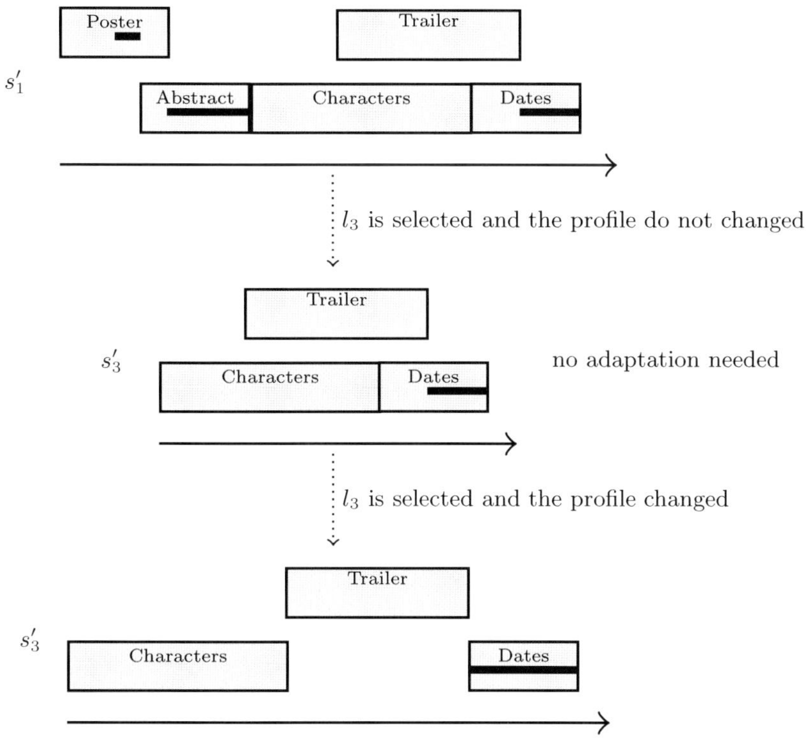

Fig. 10. An incremental approach for adapting the hypermedia structure of Fig. 1

p', i.e., $Adapt(s_3, p') = s'_3$ (Fig. 10 bottom). Note this time that l_3 is not affected by the adaptation, thus to be close to the initial document.

This strategy is repeated each time a hypermedia link is selected.

With this incremental adaptation, we only adapt sub-specifications which are executed, the profile is able to change between each user interaction, and we propagate the adapted information between each sub-specification.

6 Implementation

For a smooth implementation of our adaptation framework, we present in this section the *Adapt* algorithm (Algorithm 1) which computes adapted relation graph solutions. This algorithm is Nebel's backtracking algorithm [14], which enumerates consistent relation graphs, enhanced with a Branch and Bound optimization [12] for adaptation.

Algorithm 1: Adapt

Input: An initial matrix $I_{i,j}$ corresponding to potential executions and a matrix $P_{i,j}$ corresponding to possible executions.
Data: $\mathcal{S}$ is a set of adapted relation graphs; Min is a current minimum computed distance (initialized with a high value).

```
pathConsistency(P);
if P does not contain an empty relation then
    Choose an unprocessed label P_{i,j} and split P_{i,j} into r_l = r_1, ..., r_k;
    if no label can be split then
        tmp ← d(I,P);
        if tmp < Min then
            Min ← tmp; S ← {P};
        if tmp = Min then
            S ← S ∪ {P};
    else
        for all labels r_l (1 ≤ l ≤ k) do
            P_{i,j} ← r_l;
            if d(I,P) ≤ Min then
                Adapt(I,P);
```

The relation graph corresponding to potential executions (i.e., the initial document specification) and the one corresponding to possible executions (i.e., satisfying the profile) are encoded into two matrix $I_{i,j}$ and $P_{i,j}$, respectively.

Before the *Adapt*$(I_{i,j}, P_{i,j})$ call, we first sort each label of the matrix $P_{i,j}$ according to the distance δ (cf. Sect. 3) from each label of the matrix $I_{i,j}$.

Thereafter, Algorithm 1 computes consistent possible relation graphs with help from the *pathConsistency* function [2]. Each time consistent possible relation graphs are computed, we select the minimal ones in $\mathcal{S}$ thanks to the distance d defined in Definition 3. When Algorithm 1 stops, $\mathcal{S}$ is the set corresponding to adapted relation graph solutions and Min the minimal distance from the initial document specification.

7 Application to Concrete Multimedia Documents

We want to validate our adaptation framework on real standard multimedia documents such as SMIL documents [21]. We present, in Sect. 7.1, a strategy which generalizes a multimedia document, adapts it with the adaptation framework defined previously, and translates it back to an adapted document. This strategy can be applied to any multimedia document language. In Sect. 7.2, we present an interactive prototype which allows users to create SMIL documents, specify adaptation constraints and adapt these documents.

7.1 Concrete Multimedia Documents Adaptation

The natural way of using our adaptation framework, presented in the previous sections, for adapting actual documents, consists in taking the equivalent qualitative representation of the document (α), adapting it and translating back to an adapted document (β). Figure 11 presents this straightforward strategy.

The strategy presented above, if it is effective, does not apply to a particular multimedia specification language. It must be precised for each particular language. We apply it to SMIL documents.

SMIL is an XML-based language for specifying multimedia presentations [21]. Figure 12 presents a SMIL specification of Fig. 1. As we may notice, the temporal behavior of a SMIL document is not expressed with Allen relations, as used in the previous sections, but time containers and attributes for objects synchronization.

Hence, the goal of the α function is the extraction of the relational structure involving objects of the document and its encoding as a relation graph on which to apply the previous defined adaptation operations. This extraction is obtained by the following procedure:

1. Extract all multimedia objects and hypermedia links, and make them a node in the potential relation graph
2. Add an edge between each pair of nodes (they are labeled by a set of relations $\mathcal{R}$)
3. Extract the relations implied by the multimedia objects and hypermedia links synchronization, e.g., the time containers, the objects attributes, etc.

Thanks to the qualitative representation built from α, it is thus possible to adapt it with the adaptation framework described in the previous sections, e.g., for adapting hypermedia links. Thereafter, we need to inject the adapted information inside the document.

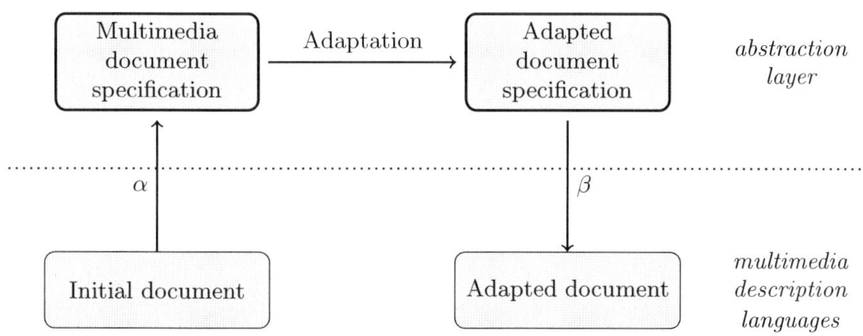

Fig. 11. Concrete multimedia documents adaptation

```
<smil xmlns="http://www.w3.org/2001/SMIL20/Language">
  <head>
    <layout>
      <root-layout background-color="#000000" height="400" title="Movie" width="700"/>
      <region height="400" id="Poster" left="0" top="0" width="700" z-index="1"/>
      <region height="400" id="Abstract" left="0" top="0" width="700" z-index="2"/>
      <region height="200" id="Trailer" left="175" top="0" width="350" z-index="3"/>
      <region height="200" id="Characters" left="0" top="200" width="700" z-index="3"/>
      <region height="400" id="Dates" left="0" top="0" width="700" z-index="4"/>
    </layout>
  </head>
  <body>
    <par>
      <a href="#trailer" begin="3s" dur="3s">
        <img id="poster" dur="6s" region="Poster" src="poster.png"/>
      </a>
      <video id="trailer" begin="15s" end="43s" region="Trailer" src="trailer.avi"/>
      <seq begin="4s">
        <a href="#trailer">
          <img id="abstract" dur="6s" region="Abstract" src="abstract.png"/>
        </a>
        <video id="characters" region="Characters" src="character.avi"/>
        <a href="#characters">
          <img id="dates" dur="6s" region="Dates" src="dates.png"/>
        </a>
      </seq>
    </par>
  </body>
</smil>
```

Fig. 12. A SMIL specification of Fig. 1

For that purpose, the β function can be defined in a straightforward manner:

1. Propagate all adapted constraints from the qualitative adapted representation to the document, i.e., update the multimedia objects and hypermedia links synchronization (e.g., the time containers, the objects attributes).

7.2 Prototype

The presented adaptation framework has been implemented in an interactive adaptation prototype. Its architecture is based on Fig. 11 and used Algorithm 1 to compute adapted relation graph solutions. Figure 13 presents a screenshot of the prototype.

The author can edit his SMIL document via the editor. Then, the relation graph corresponding to his document can be computed by the system. Moreover, it is possible to indicate to the system adaptation constraints (i.e., specify the profile).

On one hand, if the multimedia document does not need to be transformed, the adapted SMIL document is identical to the initial one. On the other hand, if the document needs to be adapted, the system presents all adapted relation graph solutions, i.e., those which are at the minimal distance from the initial relation graph and satisfy the profile.

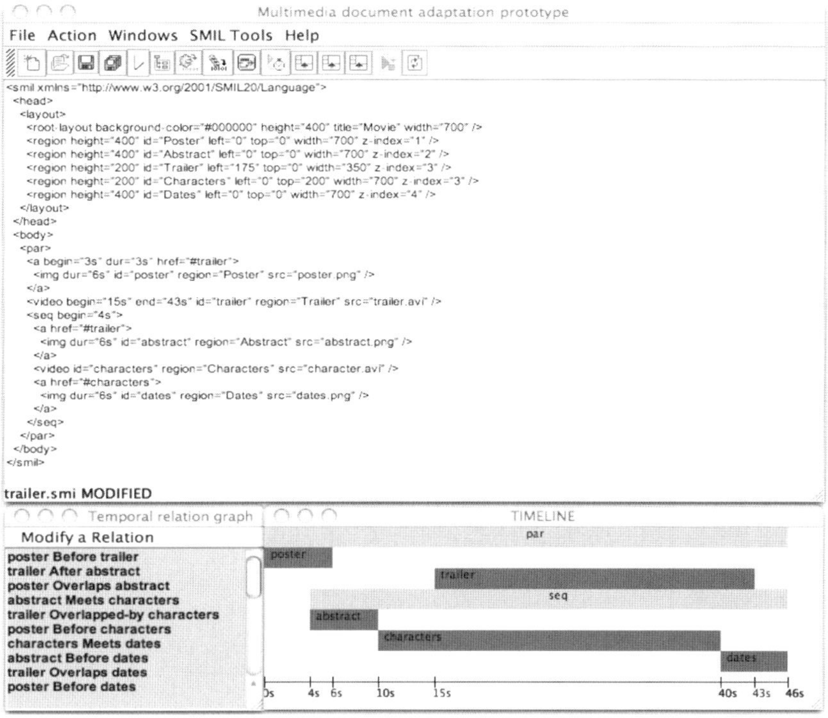

Fig. 13. A screenshot of our prototype

Once an adaptation solution is chosen, the system provides an adapted SMIL document and the user can execute it.

8 Related Work

The SMIL language [21] defines a `switch` element that allows authors to specify a set of presentation alternatives from which only the first acceptable one is chosen. Nevertheless, it is necessary to know in advance the different target profiles. Furthermore, authors should order the alternatives from the most desirable to the least desirable.

De Bra et al. [4] and De Bra and Stash [5] consider the adaptation of the hypermedia dimension of multimedia documents. They describe methods and techniques for adapting the hypermedia structure. Note that the latter considers SMIL hypermedia adaptation. However, these frameworks only perform link annotation and/or link hiding.

Francisco-Revilla and Shipman [7] define a spatial hypermedia adaptation framework. Like our adaptation approach they use an abstract representation of documents. Nevertheless, they do not include in their adaptation approach other dimensions like the temporal one.

Adam et al. [1] present a Petri net model for adapting hypermedia presentations. Like our hypermedia adaptation framework presented in this chapter, they mix the temporal and the hypermedia dimensions. However, they only drop multimedia objects to satisfy profiles.

The works described in [3, 13] adapt SMIL documents. Asadi and Dufourd [3] adapts the temporal and spatial structure using semantic dependencies between multimedia objects. Lemlouma and Layaïda [13] seeks to compute efficiently an adapted document satisfying a limited form of adaptation constraints, namely the translation of SMIL documents between the different language profiles (i.e., a form of syntactic restrictions of the different SMIL dialects [21]). Our adaptation approach differs in that it operates at a semantic level and is therefore capable of adapting documents in the same language or between different languages or formats. Moreover, our adaptation process produces documents which are as close as possible from the initial ones.

Other approaches for multimedia document adaptation are based on specialized specifications for generating documents such as [17, 18]. However, this requires the content to be initially described with these specifications instead of adapting existing documents.

9 Conclusion

In this chapter, we applied a semantic adaptation framework to the hypermedia dimension of documents by considering hypermedia links as particular objects of the presentation. Moreover, to adapt the several document sub-specifications generated by the hypermedia structure and preserve our approach, we proposed an incremental adaptation which takes into account the user interactions. This adaptation framework does not apply to a particular multimedia description language but is generic. In order to validate it we implement a prototype that relies on adapting SMIL documents.

In the future, we want to refine the adaptation by selecting suitable adapted solutions. Indeed, several adapted relation graph solutions can be computed. Currently, it is difficult to select a particular solution from this set because all solutions are close to the initial document. Using the rhetorical structure of multimedia documents, as [16] did for generating multimedia content, could be a promising direction for refining our adaptation framework.

Moreover, if the profile is highly-constrained, the adaptation may produce no solution. In such a case, we have to delete or transform (in terms of transcoding) objects of the document, thus requiring new proximity measures. Hence, it would be possible to modify the hypermedia structure of documents.

To conclude, we also plan to extend our adaptation framework to all multimedia document dimensions, thus providing a spatio-temporal-hypermedia adaptation.

As discussed above, there remains more work to be carried out for covering all the aspects of multimedia documents and for deepening the specification of documents and adaptation constraints so that the adaptation produces quality results.

References

1. Nabil R. Adam, Vijayalakshmi Atluri, Igg Adiwijaya, Sujata Banerjee, and Richard Holowczak. A dynamic manifestation approach for providing universal access to digital library objects. *IEEE Transactions on Knowledge and Data Engineering*, 13(4):705–716, 2001.
2. James Allen. Maintaining knowledge about temporal intervals. *Communications of the ACM*, 26(11):832–843, 1983.
3. Mariam Kimiaei Asadi and Jean-Claude Dufourd. Knowledge-based and semantic adaptation of multimedia content. In Paola Hobson, Ebroul Izquierdo, Yiannis Kompatsiaris, and Noel E. O'Connor, editors, *Knowledge-Based Media Analysis for Self-Adaptive and Agile Multimedia Technology*, pages 285–293, 2004.
4. Paul De Bra, Peter Brusilovsky, and Geert-Jan Houben. Adaptive hypermedia: from systems to framework. *ACM Computing Surveys*, 31(4), 1999.
5. Paul De Bra and Natalia Stash. Multimedia adaptation using AHA! In *Proceedings of the World Conference on Educational Multimedia, Hypermedia and Telecommunications*, pages 563–570, 2004.
6. Jérôme Euzenat, Nabil Layaïda, and Victor Dias. A semantic framework for multimedia document adaptation. In *Proceedings of the 18th International Joint Conferences on Artificial Intelligence*, pages 31–36. Morgan Kauffman, 2003.
7. Luis Francisco-Revilla and Frank Shipman. MASH: A framework for adaptive spatial hypermedia. In *Proceedings of the Third Workshop on Spatial Hypertext, ACM Hypertext Conference*, 2003.
8. Christian Freksa. Temporal reasoning based on semi-intervals. *Artificial Intelligence*, 54(1–2):199–227, 1992.
9. Muriel Jourdan, Nabil Layaïda, Cécile Roisin, Loay Sabry-Ismaïl, and Laurent Tardif. Madeus, an authoring environment for interactive multimedia documents. In *Proceedings of the 6th ACM Multimedia conference*, pages 267–272, 1998.
10. Sébastien Laborie. Spatio-temporal proximities for multimedia document adaptation. In *Proceedings of the 12th International Conference on Artificial Intelligence: Methodology, Systems, Applications*, pages 128–137, 2006.
11. Sébastien Laborie, Jérôme Euzenat, and Nabil Layaïda. A spatial algebra for multimedia document adaptation. In *Poster Proceedings of the First International Conference on Semantic and Digital Media Technologies*, pages 7–8, 2006.
12. A. H. Land and A. G. Doig. An automatic method for solving discrete programming problems. *Econometrica*, 28(3):497–520, 1960.
13. Tayeb Lemlouma and Nabil Layaïda. The negotiation of multimedia content services in heterogeneous environments. In *Proceedings of the 8th International Conference on Multimedia Modeling*, pages 187–206, 2001.
14. Bernhard Nebel. Solving hard qualitative temporal reasoning problems: Evaluating the efficiency of using the ORD-horn class. In *Proceedings of the 12th European Conference on Artificial Intelligence*, pages 38–42, 1996.

15. Cécile Roisin. Authoring structured multimedia documents. In *Proceedings of the Conference on Current Trends in Theory and Practice of Informatics*, pages 222–239, 1998.
16. Lloyd Rutledge, Brian Bailey, Jacco van Ossenbruggen, Lynda Hardman, and Joost Geurts. Generating presentation constraints from rhetorical structure. In *Proceedings of the 11th ACM on Hypertext and hypermedia*, pages 19–28. ACM Press, 2000.
17. Ansgar Scherp and Susanne Boll. mobileMM4U – framework support for dynamic personalized multimedia content on mobile systems. In *Proceedings of Techniques and Applications for Mobile Commerce (TaMoCO)*, pages 204–215, 2004.
18. Jacoo van Ossenbruggen, Frank J. Cornelissen, Joost Geurts, Lloyd Rutledge, and Lynda Hardman. Cuypers: a semi-automatic hypermedia generation system. Technical Report INS-R0025, CWI, Amsterdam (NL), 2000.
19. W3C. *Scalable Vector Graphics (SVG)*, 1999. http://www.w3.org/Graphics/SVG/.
20. W3C. *Composite Capability/Preference Profiles (CC/PP): Structure and Vocabularies*, 2001. http://www.w3.org/TR/CCPP-struct-vocab/.
21. W3C. *Synchronized Multimedia Integration Language (SMIL 2.0) Specification*, 2001. http://www.w3.org/TR/smil20/.

Improving the Adaptiveness of an e-Learning System

Ioannis Hatzilygeroudis[1], Constantinos Koutsojannis[2], and Nikolaos Papachristou[1*]

[1] Department of Computer Engineering and Informatics, School of Engineering, University of Patras, Greece
[2] Departmen of Nursing, School of Health Sciences, Higher Technological Educational Institute of Patras, Greece

Summary. Education of health-related professions such as medicine, nursing, dentistry, pharmacy, etc. represents a major category of adult training. Cognitive flexibility theory which emphasizes a case study based approach involving context-dependent and realistic situations applies directly to health care education. COFALE is an adaptive e-learning environment supporting cognitive flexibility. In this chapter, we present EX-COFALE, an extension to COFALE. COFALE, although offers facilities for adaptive content presentation, adaptive use of pedagogical devices and adaptive communication, it lacks facilities for adaptive student assessment. EX-COFALE remedies this deficiency of COFALE by allowing for automated test creation and assessment based on the students' knowledge information. To this end, COFALE has been modified to allow for representation of associations between test questions and learning concepts. Also, assessment is made at two levels, the concept and the goal level, taking into account the difficulty level of the questions. To achieve the above, a rule-based intelligent component is used. To evaluate EX-COFALE, we implemented a course concerning fundamental aspects of radiation protection, which was used by a number of students in a nursing department and evaluated for usability of the e-learning system and educational content administration. The results are quite promising.

1 Introduction

Most health care institutions and educational departments must typically schedule radiation protection training classes for selected groups of their stuff or their students. Computers can provide alternative radiation safety training in the form of a computer-based training program that can be accessed or downloaded via the Internet. Some University Radiation Safety Programs are already using the Internet as a means to provide computer based radiation safety training to radiation workers employed at their facilities [10, 11].

*The order is alphabetical

I. Hatzilygeroudis et al.: *Improving the Adaptiveness of an e-Learning System*, Studies in Computational Intelligence (SCI) **93**, 177–198 (2008)
www.springerlink.com

1.1 Web-Based Intelligent Education Systems

Recently, there has been a large research activity on web-based intelligent educational systems (WBIESs) [19]. WBIESs use Artificial Intelligence (AI) Techniques in order to adapt mainly to student needs for self-study. As WBIESs we consider either web-based intelligent tutoring systems (ITSs) [13] or adaptive hypermedia education systems (AHESs) incorporating intelligent techniques [5]. Adaptiveness refers to the fact that a system can automatically adapt to user needs. On the other hand, e-learning environments provide facilities mainly for helping course generation and management and refer to both the tutors and the students. Adding facilities (intelligent or not) for tutors in WBIESs make them a kind of intelligent e-learning systems (IELSs) [8,9], so that they offer some type(s) of adaptivity.

According to [4], a system, in order to facilitate adaptive support, should be designed to meet the following operational criteria for adaptability:

- *Adaptive presentation of learning content.* At any given time during the learning process, the learner is provided with learning contents that are appropriate to his or her present competence.
- *Adaptive use of pedagogical devices.* At any given time during the learning process, the learner is encouraged to do learning activities that are appropriate to his or her present competence.
- *Adaptive communication support.* For any learning discussion, the learner is suggested with peers who are appropriate to help him or her overcome his or her own difficulties.
- *Adaptive assessment.* At any given time during the learning process, the learner is provided with assessment problems and methods that are appropriate to his or her present competence.
- *Adaptive problem-solving support.* For any problem solving session, the learner is supported with appropriate feedback (e.g. appropriate hints) to solve the problem effectively.

1.2 Medical Education

Education of health-related professions such as medicine, nursing, dentistry, pharmacy, etc. represents a major category of adult training. Because of the high costs and consequences of poor performance, a great deal of attention has been devoted to health care education [26,36]. One of the characteristics of medical knowledge and its technology-depended part is that it is immense and constantly changing. Health professionals must acquire and remember a tremendous number of details, making memory processes critical. Theories of learning that focus on memory are therefore especially relevant. Cognitive flexibility theory, which emphasizes a case-based study approach involving context-dependent and realistic situations, applies directly to health care education. In addition, the use of learning strategies to maximize study behaviour is also very important. Certain cognitive processes and skills are critical in medical practice, e.g. decision-making, reasoning, and problem-solving. Problem solving, in particular, has been the basic pedagogy for many medical curricula [3,12,26]. The medical environment is very stressful, making anxiety and arousal important issues. Many topics of health care (e.g. surgery, radiology, dentistry) require high levels of sensory-motor ability. Health care workers are frequently required to make important value judgements, so research on attitudes is also relevant. Health care education extends over the lifetime of the individual. Health professionals must be self-directed in their learning activities and be able to relate new information to their own needs and experiences. For this reason, theories of adult learning that emphasize self-directed and experiential learning are highly pertinent. Furthermore, theories of instruction that are based upon self-study or use of media [3] are also significant to health care education.

1.3 Cognitive Flexibility Theory

Cognitive flexibility theory focuses on the nature of learning in complex and ill-structured domains [33,34]. The theory is largely concerned with transfer of knowledge and skills beyond their initial learning situation. For this reason, emphasis is placed upon the presentation of information from multiple perspectives and use of many case studies that present diverse examples. The theory also asserts that effective learning is context-dependent, so instruction needs to be very specific. In addition, the theory stresses the importance of constructed knowledge; learners must be given an opportunity to develop their own representations of information in order to properly learn. Cognitive flexibility theory builds upon other constructivist theories and is related to the work of Salomon in terms of media and learning interaction. Cognitive flexibility theory is especially formulated to support the use of interactive technology (e.g. videodisc, hypertext). Its primary applications have been literary comprehension, history, biology and medicine. Jonassen, Ambruso & Olesen [21] describe an application of cognitive flexibility theory to the design of a hypertext program on transfusion medicine. The program provides a number of different clinical cases which students must diagnose and treat using various sources of information available (including advice from experts). The learning environment presents multiple perspectives on the content, is complex and ill-defined, and emphasizes the construction of knowledge by the learner. The principles of this theory are:

1. Learning activities must provide multiple representations of content.
2. Instructional materials should avoid oversimplifying the content domain and support context-dependent knowledge.
3. Instruction should be case-based and emphasize knowledge construction, not transmission of information.
4. Knowledge sources should be highly interconnected rather than compartmentalized.

Learning strategies refer to methods that students use to learn. This ranges from techniques for improved memory to better studying or test-taking strategies. For example, the method of loci is a classic memory improvement technique; it involves making associations between facts to be remembered and particular locations. In order to remember something, you simply visualize places and the associated facts. Some learning strategies involve changes to the design of instruction.

1.4 The Motivation

In order to create a course on "radiation safety", we have chosen COFALE [6] as the development IELS platform, mainly because it supports "cognitive flexibility" as its learning theory, which, given the above analysis, is suitable for such type of courses. COFALE (cognitive flexibility in adaptive learning environments) is an open-source adaptive e-learning environment supporting cognitive flexibility [7]. It is concerned with transfer of knowledge and skills beyond their initial learning situation. Emphasis is placed upon the presentation of information from multiple perspectives and use of many case studies that present diverse examples; learners must be given an opportunity to develop their own representations of information in order to properly learn. Self-study skills are promoted and cultivated through cognitive flexibility [34]. Health professionals must be self-directed in their learning activities and be able to relate new information to their own needs and experiences. Medical training that focuses on self-directed and experiential learning, as indicated above, is highly pertinent [21].

As far as adaptiveness is concerned, COFALE, apart from supporting common adaptive techniques, such as student modeling and adaptive content presentation, also provides means for adaptive use of pedagogical devices and adaptive communication support. COFALE manages to successfully accommodate the first three requirements [4, 5] of those presented in Sect. 1.1. So, COFALE lacks functionalities related to adaptive student assessment/evaluation. In order to help tutors to be able to create courses with adaptive assessment capabilities, we modified and extended it to provide such functionalities.

The chapter is organized as follows. In Sect. 2, a short overview of COFALE with an emphasis to its adaptive capabilities is presented. Section 3 deals with the proposed extensions to COFALE, mainly concerning adaptive assessment, whereas Sect. 4 deals with implementation aspects. Section 5 presents related work while Sect. 6 presents the evaluation results and finally Section 7 concludes the chapter.

2 The COFALE Environment

COFALE is based on ATutor, an open-source, web-based learning content management system (LCMS) designed and maintained by ATRC [1]. Compared to contemporary adaptive learning systems, COFALE seems to fulfill all the needed criteria for cognitive flexibility. COFALE gives the tutor the ability to implement student models. Student model refers to a classification depending on student's knowledge state and learning preferences. This information is vital for the system to adapt to the student's needs. There are many possible student characteristics that can be taken into account in the student model and choosing the most appropriate ones can be a problem. If the student model is incomplete, the system's adaptivity will be unsuccessful, whereas if it is complex, the system's operation will be encumbered [20].

For example, in COFALE one can choose and implement two types of student model, "novice" and "expert". For the purpose of the discussion, we shall assume that a "novice" first semester learner (George), who has not learnt anything about the basic physics and biological effects of radio activity, and a fourth semester expert learner (Anna), who is acquainted with the above, use Ex-COFALE to learn Radio-Safety; a tutor and a number of other learners (peers) also participate in the same learning experience. The learning content in COFALE can be decomposed into quite primitive content (or learning) units, so that the system can present each student different content units. For example, simpler examples for a "novice" learner and advanced for an "expert" one. This implements adaptive presentation of learning content. In our example, Ex-COFALE present to George analytically all the basic theory about the properties, forms and biological effects that radio activity has, whereas Anna is presented only with pinpoints for the same chapters.

At the end of each content page, the student is encouraged and guided to do a number of learning activities, depending on his/her current "mental model" about the concept of study. This means that COFALE allows for a second level of student modeling, that of "mental models" of the students, which are related to the type of activities that a student prefers and is able to manage. Given the type of first-level student model (novice, expert), certain types of mental models may be excluded. So, COFALE may suggest activities based on simpler mental models to a "novice" learner, but based on more complex models to an "expert" one. This implements adaptive use of pedagogical devices. For example, George is given with simpler activities like tests and participation in collaborative tasks through the first chapters when Anna is given homework like making a research about a topic, or being the team leader about the collaborative tasks of the same chapters.

Moreover, while learning with COFALE, students can use a tool to search for peers who could help them to overcome difficulties about acquiring the concept of recursion. For example, COFALE may suggest some "expert" students to a "novice", so that he/she can ask them questions about problems, or may suggest an "expert" student to another "expert" student, so that they can exchange ideas about advanced concepts or activities. That implements adaptive communication support. George is able to communicate only with expert peers like Anna in order to get help about his learning problems, when on the other hand Anna is able to communicate both with novice and expert peers, in order to decide how they will split up the class and make teams for the collaborative tasks that concern them.

3 Extending COFALE

To make COFALE meet our needs, we made some modifications and extensions to it, presented in the following subsections. We call the extended system EX-COFALE (EXtended COFALE).

3.1 Domain Knowledge

COFALE uses a tree-like structure to represent domain knowledge. COFALE can alter the domain knowledge tree presented to the user depending on the user model. In Fig. 1, a partial such domain knowledge tree, which we constructed, implementing a course on "radio safety" from the health care domain [10, 11], is presented. The root of the tree is the above field. The field is divided into subjects (like "radiation") and the subjects into topics (like "X-rays"). Each topic deals with a number of concepts which are the leaves of the tree. Topics may have common concepts. Topics typically constitute learning goals and related concepts their learning concepts.

Actually, what COFALE can do is to hide certain subtrees, which are not appropriate for a certain user. EX-COFALE goes a step further. It can re-arrange the branches of the tree, based on the user's model, thus achieving something like concept sequencing. For example, in Fig. 1a, part of the domain knowledge tree of the "radiation safety" course is presented, whereas in Fig. 1b, the same tree adapted to a specific user is presented. Adaptation here concerns not only hiding part of the tree, but also changing the order of the branches in a subtree, thus suggesting a new topic sequence, adapted to the user.

3.2 Test Creation

First, we modified the test manager. We added more functionality as far as test construction is concerned:

(a) The tutor can associate a test to a specific learning goal and
(b) The system can now automatically create a test.

The tutor should only create and store questions in the system's database. Also, he/she may define the number and the difficulty levels of the questions

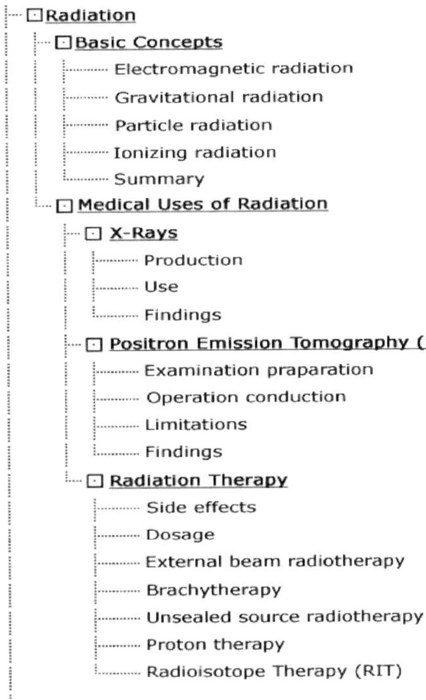

(a)

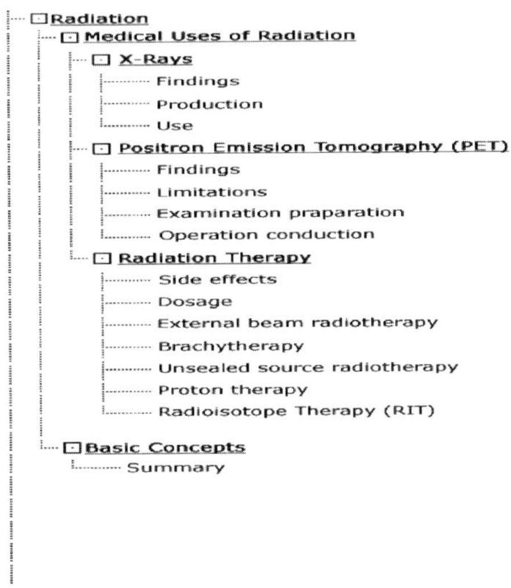

(b)

Fig. 1. (**a**) Part of domain knowledge tree. (**b**) Adapted domain knowledge tree (concept sequencing) to be included in a test for each concept. Automatic creation is done via a rule-based intelligent system.

As far as creation of test questions is concerned, we added the capability of defining associations between a learning concept and corresponding questions. This way a test has each question associated with a specific learning concept. More than one question may refer to the same concept. Questions may have different difficulty levels assigned to them. The tutor is able to insert, delete or modify all the parts/attributes of a question (i.e. the body of a question, its answers, possible help hints, the associated concept, the difficulty level, etc.). We must note here that two types of questions, multiple-choice and true-false, can be automatically marked. There is a third one, open-end questions, which are manually marked. The questions are created once, while in COFALE the tutor had to create the same question for different tests more than one time.

The process of a test creation by the tutor for a specific learning concept or goal, assigned to different student models is quite straightforward. For example, in the course of "radio safety" there are different questions, with different difficulty levels associated with the concepts of "production", "use", and "findings" of the learning goal (topic) "X-rays".

Creation of a test for the topic "X-rays" is possible through random selection of the questions concerning the above concepts. The only thing the tutor has to do is to select the concepts for which the system will collect questions from the database and generate a test. He/she also has to assign the test to a specific student model. A revision test can be made by the tutor, after he/she selects all the concepts he/she wants the system to generate a test for. For example, a revision test for the subject of "nuclear medicine", in the course of "radio safety", is generated after selecting the concepts of "PET", "radiation therapy", "radiosurgery", "proton therapy" and "Brachytherapy"(Fig. 2).

Students and tutor have view of the total test result and of each test question result, too. Students can open a completed test and see for each question the number of tries and the number of successes. If the same question is answered elsewhere all tests with this question are updated. For example, in Fig. 3, in Test 6 a student has answered the easy question correctly, while in a summary test he answered this question wrong, whereas he answered the difficult question in both tests correctly. Tutors apart from opening a test for each student, can also open a general view of the test like in Fig. 3, and see the total number and percentage of tries and successes of each test question for the whole class.

This is a flexible way of monitoring the student performance, both for the student and tutor. Tutor is provided with a tool which helps him not only to monitor the student efficiency analytically in the course but also to reconsider each concepts test questions and difficulty, taking into account the class performance at each of them.

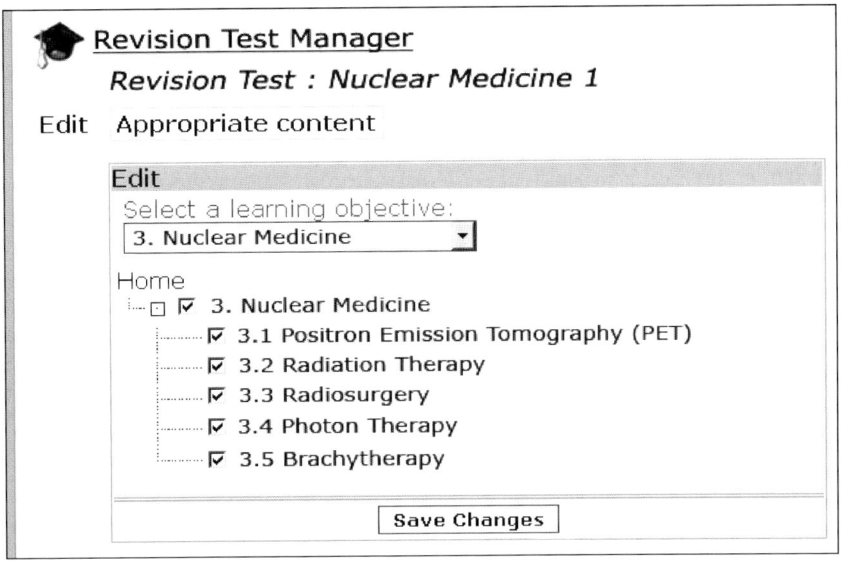

Fig. 2. Revision test manager of EX-COFALE

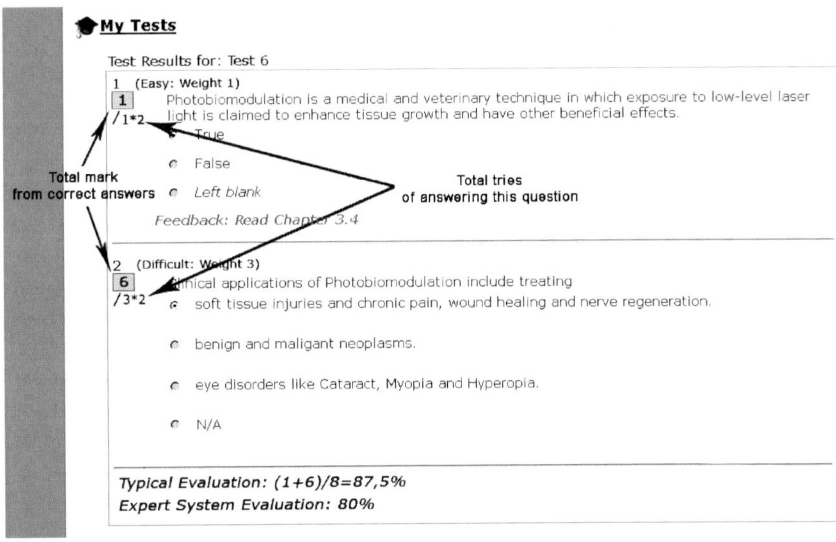

Fig. 3. Test results in EX-COFALE

3.3 Student Assessment

One of the most important functions of an intelligent e-learning system is student evaluation (or assessment). Student evaluation refers to the evaluation of the knowledge level of a student after having dealt with a learning goal. In other words, how well a student has learnt the concepts related to a learning goal. Student evaluation is important for both the students and the tutor. COFALE allows for evaluation of students based on the tests provided for each learning goal. It actually allows for a learning goal level evaluation based on tests with predefined indistinguishable questions. By "indistinguishable" we mean that there is no explicit representation of which concept each question refers to (or examines). So, the system cannot be aware of those associations to use them. In EX-COFALE, a student is evaluated at two levels: (a) the concept-level and (b) the goal-level.

The concept-level evaluation deals with the level of understanding of the individual concepts of a learning goal test, whereas the goal-level evaluation deals with the level of understanding of a learning goal as a whole. Furthermore, EX-COFALE allows for on-line test creation, even if a student has not completed the study of all of the concepts related to a learning goal. This is achieved via the above mentioned rule-based intelligent system. For example, on the topic of "X-rays", if the student hasn't completed the study of concept "Findings", the intelligent system will select questions only from the two other concepts of the topic, "Production", "Use". The intelligent system is used for this purpose because a future enhancement could be the automatic creation of tests with question difficulty level adaptation. Tests which adapt not only to the material studied but also to how well the student has performed to this material. Depending on how many concepts of a learning goal are covered and how well the student performed for each concept, the next tests generated about these concepts will bring suitable difficulty questions, appropriate to his/her current knowledge level as well [18].

The knowledge level of a student, as far as a concept is concerned, is classified in one of the following three categories: (a) low (0–49), (b) medium (50–70) and (c) good (71–100), whereas, as far as a learning goal is concerned, in one of the following five categories: (a) low (0–30), (b) average (31–49), (c) good (51–70), (d) very good (71–85) and (e) excellent (86–100) (within the parentheses are the corresponding ranges of the marks to be achieved). The knowledge level of a student for a concept, called concept level (CL), is calculated via the following formulas:

$$CL = \sum_{i=1}^{3} Qm_i * qw_i,$$

$$Qm_i = \frac{\sum_{j=1}^{n} Qm_j^i}{n},$$

where i represents the difficulty level ($1 \rightarrow$ easy, $2 \rightarrow$ medium, $3 \rightarrow$ difficult), Qm_j^i represents the answer to question j (which is 100 if it's correct and 0 otherwise), n is the number of the questions of the same difficulty associated with the concept, Qm_i is the average mark of the answers to questions of the same difficulty level related to the concept and qw_i is the question's weight.

Table 1. Question difficulty level contribution to student assessment

Easy (E)	Medium (M)	Difficult (D)	Question (s) weight (s) (/100)
1	1	1	20 (E), 50 (M), 30 (D)
1	1	0	40 (E), 60 (M), 0 (D)
1	0	1	40 (E), 0 (M), 60 (D)
0	1	1	0 (E), 60 (M), 40 (D)

The weight of a question is related to the difficulty level of a question and the composition of the set of questions used for testing the concept. Table 1 presents the corresponding weights. It is assumed that at least two questions for each examined concept exist in a test. If $CL \geq 0.5$ then the student has an acceptable level of knowledge about the corresponding concept.

As an example, suppose we have four questions that are used to evaluate the knowledge of a student about a concept (e.g. "X-rays production") and that two of them are "easy" and from the other two one is "medium" and one is "difficult". If the student gives right answers to one of the two easy questions and the medium one, but wrong to the other easy question and the difficult one, then the $Qm_1 = (100 + 0)/2 = 50$, $Qm_2 = 100$ and $Qm_3 = 0$. So, the corresponding $\mathrm{CL} = 50 * 0.2 + 100 * 0.5 + 0 * 0.3 = 60$.

The knowledge level of a student for a learning goal, called *goal level* (GL), is calculated from the following formula:

$$GL = \frac{\sum_{i=1}^{n} CL_i}{n},$$

where CL_i is the value of the knowledge level of the student for concept i (i.e. the achieved mark before it is classified in one of the three levels) and n is the number of concepts that constitute the learning goal. Again, if $\mathrm{GL} \geq 0.5$ then the student has an acceptable level of knowledge about the corresponding learning goal, given that each $CL_i \neq 0$, that is there is no concept not studied at all.

Let's consider the learning goal "X-rays", which consists of the following concepts: "X-rays production" (C1), "X-rays use" (C2), "X-ray findings" (C3) and "X-rays diagnostical value" (C4). Also, let's suppose that a student has achieved the following scores from the study of the individual concepts: CL1 = 60, CL2 = 40, CL3 = 80 and CL4 = 50. Then his knowledge level for "X-rays" will be GL = (60 + 40 + 80 + 50)/4 = 57.5. Usually, apart from local tests, there is also a summary test at the goal level, including questions for all related concepts. Let us suppose that there are four questions, one easy, two medium and one difficult, about the concept "X-ray production", and the student fails at the difficult one. Then, the CL for this concept is re-evaluated: $Qm1 = 100$ $Qm_2 = (100 + 100)/2 = 100$, $Qm3 = 0$, $\mathrm{CL1} = 0.2 * 100 + 0.5 * 100 + 0.3 * 0 = 20 + 50 + 0 = 70$. Finally, GL is

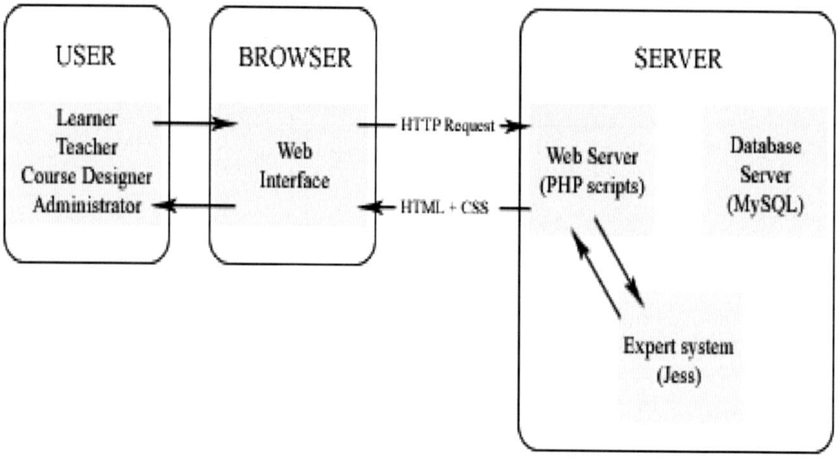

Fig. 4. EX-COFALE architecture

re-calculated: GL $= (70+40+80+50)/4 = 60$ (given that the rest of concept levels remain the same). So, evaluation is done in two levels.

4 Architecture and Implementation Aspects

Figure 4 shows the implementation architecture of EX-COFALE (as a modification of that of COFALE). The functionalities of the architecture are as follows:

- The user uses a Web browser to log into the system and make a request.
- Taking into account the user's request, the browser sends a HTTP request to the Web server in which a set of PHP scripts were installed.
- Depending on the kind of the HTTP request, the Web server creates new data or update existing data or retrieve existing data by connecting the MySQL database server in which all data of the user, learning content, tests, forums, and so on are stored and indexed. Then, the Web server formulates a HTML file including a CSS format and sends it back to the browser.
- On the basis of the HTML file and the CSS format received from the Web server, the browser creates a Web page and presents it to the user.
- For the tests' evaluation a rule-based intelligent system is used, which is implemented in Jess, a java based tool for development of rule-based intelligent systems [14]. PHP scripts make all the appropriate communication between the intelligent system and the browser. PHP scripts also bridge the intelligent system with the MySQL database server to store the results of the tests evaluation.

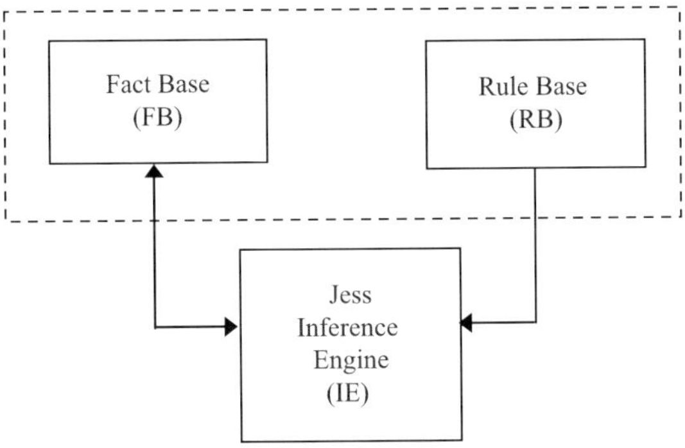

Fig. 5. The intelligent system structure

The structure of the intelligent system is illustrated in Fig. 5. It consists of the fact base (FB), the rule base (RB) and the Jess inference engine (JESS IE). FB contains facts, which are created from the problem data, whereas RB contains the rules used by the IE. The intelligent system processes the facts via the rules of RB, according to the Jess IE instructions, to (a) select the questions from the database according to the students' knowledge levels and tutor's settings and (b) deduce the knowledge level values of the students for the concepts involved in the delivered test, based on the test results.

The process of estimating the knowledge level, after a user has been through a test is as follows:

1. Test results are recorded to the database
2. Those results generate appropriate facts for the Jess fact base
3. Jess is triggered, and after taking into account the fact base, deduces the knowledge level
4. The system reads the result and record it to the proper place.

The following templates are used to represent corresponding facts:

```
(deftemplate answer
           (slot concept)
           (slot easy) (slot medium) (slot difficult)
           (slot ans1) (slot ans2) (slot ans3))
(deftemplate concept-level
           (slot concept)
           (slot level))
```

So, after a test has been performed, one of the generated facts (related to the example used in Sect. 3) is:

```
(answer (concept ''X-rays production'') (easy 1) (medium 2)
(difficult 1) (ans1 1) (ans2 2) (ans3 0))
```

The following function is used within the next rule to compute the concept knowledge level on the current concept, when the first case of Table 1 holds.

```
(deffunction cl-case1 (?d1 ?d2 ?d3 ?x1 ?x2 ?x3)
      (bind ?q1 (/(* 100 ?x1) ?d1))
      (bind ?q2 (/(* 100 ?x2) ?d2))
      (bind ?q3 (/(* 100 ?x3) ?d3))
      (bind ?cl (+(* 0.2 ?q1) (* 0.5 ?q2) (* 0.3 ?q3)))
      (return ?cl))
```

The following rule investigates whether the current fact is of the first case of Table 1. If it is, the rule calls the above function to calculate the knowledge level of the student on the current concept and assert the corresponding fact.

```
(defrule rule1
        ?x<- (answer (concept ?c) (easy ?d1) (medium ?d2)
                     (difficult ?d3) (ans1 ?x1) (ans2 ?x2)
                     (ans3 ?x3))
        (and (test (> ?d1 0)) (test (> ?d2 0)) (test (> ?d3 0)))
        =>
        (assert (concept-level (concept ?c)
                 (level (cl-case1 ?d1 ?d2 ?d3 ?x1 ?x2 ?x3))))
        (retract ?x))
```

After the execution of the above rule, the following fact is added to the fact base:

```
(concept-level (concept ``X-rays production'') (level 70))
```

A core prototype of the system has been implemented, which does not offer at the moment all designed functionalities and is currently accessible at http://150.140.142.79/.

5 System Evaluation

We have made an indirect evaluation of some aspects of our system. That is, we implemented a course on "radiation protection" for students in a Nursing Department using EX-COFALE and evaluated it. The first version of the course was released in December 2006 and used by the class of the "medical informatics" course, in a Nursing Department, which consisted of seventy students. The students had not been taught about "radiation protection" during the course lectures. Evaluation aimed to uncover students' preferences and points of view regarding the technical, pedagogical and teaching efficiency of the course implemented with EX-COFALE. We also aimed at getting their feeling on the correctness or fairness of the followed assessment methodology.

The students were instructed to use the system as follows: login at least five times and make at least (a) three sets of exercises from subject 1, (b) two sets of exercises from subject 2 and (c) two exercises from the summary evaluation test created using their knowledge and prior assessment information. Then, they were asked to fill in a questionnaire, including questions for evaluating usability and learning [2, 35]. The questionnaire was created on the basis of well known systematic evaluation procedures for interactive multimedia and Intelligent E-Learning Systems with Authoring and Assessment Mechanisms for Education and Training [23, 28, 37].

The questionnaire included eighteen questions (see Table 2) concerning strong and weak points or problems faced. They can be distinguished in four groups:

(a) Questions 1–11, aimed to evaluate the interface of the web site
(b) Questions 12–15, aimed to evaluate the created course
(c) Questions 16–17, aimed to evaluate the student assessment method and
(d) Question 18, aimed at getting ideas for further improvement of both, the system and the course.

Table 2. Questionnaire results

Q	Questions	Answers (%)				
		1	2	3	4	5
1	I think I would like to use this website frequently	0.0	8.1	29.7	41.1	20.3
2	I thought the website was easy to use	0.0	5.4	24.3	44.6	25.7
3	I found the website unnecessarily complex	29.7	35.9	28.3	6.8	0.0
4	I think I would need Tech Support to be able to use this site	29.7	36.5	14.9	10.8	8.1
5	I thought there was too much inconsistency in this website	35.1	29.1	35.1	0.0	0.0
6	I would imagine that most people would learn to use this website very quickly	0.0	5.4	21.6	70.3	2.7
7	I found the website very cumbersome to use	25.7	23.0	17.6	24.3	9.5
8	I felt very confident using the website	4.1	8.0	29.2	38.4	20.3
9	I need to learn a lot about this website before I could effectively use it	8.1	39.2	27.5	23.8	1.4
10	The course objectives are clear to me	0.0	10.8	1.3	73.0	14.9
11	The course activities stimulated my learning	0.0	0.0	25.7	58.1	16.2
12	EX-COFALE was essential in the course	17.6	6.8	32.4	27.0	16.2
13	The test(s) in this course were accurate and fair	0.0	1.4	27.0	58.1	13.5
14	The difficulty level of this course is appropriate	6.8	9.5	23.0	60.8	0.0
15	Ex-COFALE was a good way for me to learn nursing radiation protection	0.0	10.0	21.6	41.4	27.0
16	The performance assessment method gave me a good guidance for self study, in order to revise and comprehend the learning items in this course	0.0	1.4	23.0	30.4	35.2
17	Evaluation results agree with my personal feeling about my performance	0.0	6.8	27.6	40.4	25.2

Questions 1–17 are based on the Likert scale (1: agree not at all, 5: agree very much), but a number of them included a "please explain" request too [10]. Question 18 was a multiple-choice one. The results are presented in Table 2.

Their answers showed that the students in general consider that the system was easy to use (Q2). Most of them reported that they felt very confident using the system (Q8) and that they didn't need to learn a lot before they could effectively use it (Q9).

On the other hand, the students agreed that the system helped them in learning about radiation protection, because the course objectives are clear (Q10), activities stimulated learning (Q11), the test(s) in this course were accurate and fair (Q13). According to the students, EX-COFALE was a good environment to learn nursing radiation protection (Q15).

Finally, students found the assessment method used in EX-COFALE helpful for self study purposes and close to their personal impression about how well they did in the course(Q16), (Q17).

The multiple-choice question revealed that the system needs some improvements. For example, the use of more videos and images throughout each of the modules could provide helpful visual aid to the training experience and reinforce the topics.

6 Related Work

There have been a number of e-learning systems (or environments) that can be used to produce adaptive courses. However, a few of them provide facilities for adaptive student assessment.

ALE (adaptive learning environment) is an e-learning environment (implemented in the context of the WINDS project [17, 25], that integrates an intelligent tutoring system, a computer instruction management system and a set of cooperative tools [24,29]. ALE can produce individualized courseware for students, based on their current state of knowledge, their preferences and learning styles. So, it supports adaptive content presentation. However, it does not support any facility for student assessment management. EX-COFALE provides tutor with the ability to evaluate the learner's performance and monitor his/hers comprehension of the concepts and chapters independently.

aLFanet (Active Learning For Adaptive interNET) is an e-learning platform created in the context of an IST project [30,31]. It integrates adaptive techniques from intelligent tutoring systems (ITSs), adaptive hypermedia systems (AHSs) and computer supported collaborative learning (CSCL). It consists of three subsystems. The authoring subsystem provides facilities for the creation of instructional-based courses and optionally the possibility to define adaptive questionnaires. Course administration subsystem includes the users' management, learners/tutors assignment to the courses, permissions management and users' data privacy. It also provides facilities for the definition of new presentation layouts. Finally, LMS eLearning instruction subsystem includes an adaptive course presentation depending on the learner profile and preferences, dynamic user modeling (learner profiles refinement) and tools for learning process (collaborative frameworks). Again, aLFanet, although provides facilities for test/questionnaire creation, it does not seem to provide any facilities for student assessment. It does not provide any sensible reports to authors to help them to evaluate how learners are dealing with course activities.

In [32] a system that integrates course authoring and student assessment capabilities is presented, focusing on the assessment issue. The presented assessment method is based on SPC (student-problem-course) table. Assessment is based on the correct answers and the time taken to deliver it. Although it offers automatic assessment, through the SPC table, it does not seem to offer automatic test creation. An on-line exam will be held after instructor select problems in the exam scope. Also, it uses weights associated with course units, which are taken into account in student evaluation. However, it does not take into account the neither the difficulty of the questions nor the student level. All students must attend the same course and the questions' difficulty is calculated by the average performance of all the students to each of these questions. Also, students get a secondary evaluation by the total study time of a course unit compared to the minimum requirement time which was preset by the instructor for that particular course unit. Except that, the object navigation frequency must be over the threshold which was defined by the instructor. We implemented a more flexible and tolerant technique for student evaluation, where more experienced learners might have good test results without having to have a thorough attendance to some learning concepts. Finally, the system in [32] does not provide any collaborative tools for students inter-assessment, in contrast to EX-COFALE, which provides a number of such tools, like e-mail, forums, chat rooms and learners' hyperspace area for team solving situations.

The system presented in [8, 9] is really a good ILES. It provides content presentation and student assessment adaptivity alongside extensive authoring facilities. It uses three levels of assessment (paragraph, subchapter, and chapter). Tests can be adaptively (i.e. dynamically) created. It uses only multiple choice questions, where wrong answers are taken into account for the assessment. Questions are not distinguished by difficulty level; all are considered as equally contributing to the final mark. Finally, there is a mark threshold associated with each learning item that should be overtaken by the student in order to consider that he/she has successfully passed it. The system is rather examination-oriented. Although it does not luck a tool for communication of the tutor and the learners, EX-COFALE offers a variation of such tools. EX-COFALE offers the tutor a more efficient tool, to understand how different types of learners comprehend and what difficulties they face in studying each learning concept.

As far as the implemented course is concerned, according to our knowledge there is no other system teaching radiation protection for nurses developed on aspects of tele-education in Greece. Worldwide there are in use only a few systems for radiation workers. For example at Princeton University, the Office of Environmental Health and Safety developed an instructive computer-based radiation safety training program, accessible to anyone via the Internet (http://www.princeton.edu/~ehs/), in order to train many of their laboratory radioactive source users [10]. Their web site, which focuses mainly on the fundamentals of radiation safety and physics, is used only by radiation workers at the University as an adjunct to the traditional classroom radiation safety training they receive [10]. Written in a web page format, the program was authored using the readily available Netscape Composer.

Training modules were developed for several common subjects in medical radiation safety relating to hospital facilities that are licensed to use radioactive devices and sources. Since the training modules were written primarily for medical radiation technologists, the most basic information on radiation physics was not included. For most technologists, this basic information should be well understood from previous academic studies and experience. Most of the written material in each module was referenced from several regulatory guides [11,12].

7 Conclusions

Hospital management is continually challenged by the increasing costs of managed care and the formidable task of providing the best quality care to patients. One important aspect of this continuing effort requires well-trained workers in medical radiation safety. A practical easy-to-use computer-based radiation safety-training course could prove to be an instrumental and economic alternative in the training of medical radiation workers while making a significant impact on radiation safety awareness.

To create such a course, we chose COFALE, an existing open-source, web-based intelligent e-learning system, as the development platform, because it is based on the "cognitive flexibility" theory, which is suitable for the learning tasks required in the above kind of domains. To make COFALE more adaptive, we extended it to EX-COFALE. COFALE, although offers facilities for adaptive content presentation, adaptive use of pedagogical devices and adaptive communication, it lacks facilities for adaptive student assessment, one of the requirements for such systems. In EX-COFALE, we introduce automated test creation and assessment based on the students' knowledge information. To this end, COFALE has been modified to allow for representation of associations between test questions and learning concepts. Also, questions are distinguished in three levels of difficulty. Assessment is done at two levels, the concept and the goal level. In the assessment process, the difficulty level of questions is taken into account, which is not the case in existing systems. To technically achieve the above, a rule-based approach is used. Very few e-learning environments provide facilities for adaptive assessment. In this vein, extending existing open source tools seems to be an interesting idea.

We also developed a web-based e-learning course using EX-COFALE for training nursing stuff on "radiation protection" aspects. We used it in a nursing department and evaluated it via a questionnaire. That was an indirect evaluation of EX-COFALE, as far as usability, pedagogy and the assessment method are concerned. The results are promising, although a more systematic evaluation is required. For example, at least two different groups of students could be used, one attending the course in the traditional way and the other through EX-COFALE. Then their knowledge should be tested to assess the effectiveness of the system. Furthermore, its adaptive assessment method should be tested for its validity.

Although EX-COFALE at its present design offers capabilities for adaptive assessment, it does it in a degree that can be improved. For example, the categories of the concept and goal knowledge of a student have no sharp borders in reality. So, a representation based on fuzzy sets would improve their effectiveness. Also, answers to questions are marked as correct or wrong, i.e. by two concrete values. This may not assess correctly the knowledge level of students. A more fine grained marking would improve it. To this end, other factors related to student interaction could be taken into account (e.g. number of tries, whether system help were used, etc.).

Furthermore, assessment is not parameterized as far as the intelligent rule-based system is concerned. An authoring unit related to the rules of the intelligent system could be another direction for improvement. This would give the tutor the possibility of changing the way that student assessment is done, e.g. by changing the contributions of the difficulty levels.

Finally, adapting the difficulty level of the questions of a test to student's current knowledge level is another aspect for strengthening assessment adaptation. To this end, other intelligent technologies, like, e.g. genetic algorithms could be used [18].

Acknowledgement

This work was partially supported by the European Social Fund (ESF), Operational Program for Educational and Vocational Training II (EPEAEK II), 2.2.2.a (Nurse Department of the Technological Educational Institute of Patras, Greece).

References

1. Adaptive Technology Resource Center, A tutor learning content management system (http://www.atutor.ca/), 2004.
2. Avouris, N., Tselios, N., Fidas C. and Papahristos E. (2003). Website evaluation: a usability-based perspective, In: Y. Manalopoulos et al. (ed.). LNCS No. 2563, pp. 217–232. Springer Verlag, Berlin Heidelberg New York.
3. Barrows, H.S. and Tamblyn, R.M. (1980). Problem-Based Learning: An Approach to Medical Learning. Springer, New York.

4. Brusilovsky, P. (1999). Adaptive and intelligent technologies for web based education, In: C. Rollinger and C. Peylo (eds.). Special Issue on Intelligent Systems and Teleteaching, Künstliche Intelligenz 4, pp. 19–25.
5. Brusilovsky, P. (1998). Methods and techniques of adaptive hypermedia. In: Brusilovsky, P., Kobsa A., and Vassileva, J. (eds.). Adaptive Hypertext and Hypermedia. Kluwer, Dordecht.
6. Chieu, V.M., Anh, D.T.V. and Hung, P.K. (2006). An operational approach for analyzing ICT-based constructivist and adaptive learning systems. Fourth IEEE International Conference on Computer Sciences: Research, Innovation and Vision for the Future (RIVF'06), February 12–16, Hochiminh City, Vietnam, 1–10.
7. Chieu, V.M. and Milgrom, E. (2005). COFALE: an adaptive learning environment supporting cognitive flexibility, The Twelfth International Conference on Artificial Intelligence in Education, pp. 491–498.
8. Christea, P.D. and Tuduce, R., Test authoring for intelligent E-learning environments, First International Workshop on Authoring of Adaptive and Adaptable Educational Hypermedia, 2004, http://wwwis.win.tue.nl/~acristea/WBE/416-805_WBE-PCristea_RTuduce_6pg.pdf.
9. Christea, P.D., Tuduce, R., Savescu, I.A., Grogorin, C.A., Tomozei, D.-C., Gradinescu, V.R. and Rangu, C.M. (2004). Prototype implementation of an intelligent e-learning system, Proceedings of the IASTED International Conference on Web-Based Education (WBE-04), Feb. 16–18, Innsbruck, Austria. Acta Press, pp. 441–446.
10. Hamilton, D.D., Peck, M.M., Yu H., and Kearfott, K.J. (2000). Computer-based radiation safety training for hospital radiation workers. Health Physics 78(Suppl. 1):S4–S8.
11. Dupree, S.M. (1999). Development of web-based radiation safety training to meet the needs of a small academic institution. Abstracts of papers presented at the Forty-Fourth Annual Meeting of the Health Physics Society, Philadelphia, Pennsylvania. Health Physics 76:5146.
12. Elstein, A., Shukman, L. and Sprafka, S. (1978). Medical Problem Solving. Harvard University Press, Cambridge, MA.
13. Freedman, R. (2000). What is an Intelligent Tutoring System? Intelligence 11(3):15–16.
14. Friedman-Hill, E. (2003). Jess in Action: Rule-Based Systems in Java. Manning Publishing.
15. Hatzilygeroudis I. (Guest Editor) (2004). Special Issue on AI Techniques in Web-Based Educational Systems. International Journal on AI Tools (IJAIT) 13(2).
16. O'Neil, H.F. and Spielberger, C. (1979). Cognitive and Affective Learning Strategies. Academic Press, New York.
17. Hóttenhain, R., Klemke, R., Kravcik, M., Pesin, L., and Specht. M. (2002). Adaptive learning environment in WINDS, In: Proceedings of ED-MEDIA 2002, Denver Colorado. AACE Press, Charlottesville, VA, pp. 1846–1851.
18. Hatzilygeroudis, I., Koutsojannis, C., and Papavlasopoulos, C. (2006). Knowledge-Based Adaptive Assessment in a Web-Based Intelligent Educational System.
19. Hatzilygeroudis, I., Prentzas, J. and Garofalakis, J. (2005). Personalized Learning in Web-Based Intelligent Educational Systems: Technologies and Techniques.
20. Intelligent Educational Systems: Technologies and Techniques.

21. Jonassen, D., Ambruso, D. and Olesen, J. (1992). Designing hypertext on transfusion medicine using cognitive flexibility theory. Journal of Educational Multimedia and Hypermedia 1(3):309–322.
22. Kazi, S.A. (2004). A conceptual framework for web-based intelligent learning environments using SCORM-2004", Proceedings of the IEEE ICALT-2004, Aug. 30–Sept. 1, 2004, Joensuu, Finland, IEEE Computer Society, pp. 12–15.
23. Knussen, C., Tanner, G.R. and Kibby, M.R. (1991). An approach to the Evaluation of Hypermedia. Computers Education 17(1).
24. Kravcik, M. and Specht, M. (2004). Authoring adaptive courses-ALE approach, Proceedings of the IASTED International Conference on Web-Based Education (WBE-04), Feb. 16–18, 2004, Innsbruck, Austria. Acta Press, pp. 396–401.
25. Kravcik, M., Specht, R., Pesin, M. and Klemke, R. (2001). Authoring adaptive educational hypermedia in WINDS, Online Proc. ABIS 2001 Workshop (Adaptivitat und Benutzermodellierung in interaktiven Softwaresystemen), Dortmund, Germany (Available at http://www.kbs.uni-hannover.de/~henze/ABIS_Workshop2001/ABIS_2001.html).
26. McGuire, C.H. et al. (1983). Handbook of Health Professions Education. Jossey-Bass, San Francisco.
27. Norman, G.R. and Schmidt, N. (1992). The psychological basis of problem-based learning. A review of the evidence. Academic Medicine 67:557–565.
28. Reeves, T.C. (1994). Systematic evaluation procedures for interactive multimedia for education and training, Reisman, S. (ed.). Multimedia Computing: Preparing for the 21st Century.Idea Group, Harrisburg, PA.
29. Klemke, R., Kravcik, M., Pesin, L. and Specht, M. Authoring adaptive educational hypermedia in WINDS, Online Proc. ABIS 2001 Workshop (Adaptivitat und Benutzermodellierung in interaktiven Softwaresystemen), Dortmund, Germany, 2001 (Available at http://www.kbs.uni-hannover.de/~henze/ABIS_Workshop2001/ABIS_2001.html).
30. Santos, O.C., Boticario, J.G. and Barrera, C. (2005). ALFANET: an adaptive and standard-based learning environment built upon DotLRN and other open source developments, Calvo, R.A., Ellis, R.A. and Peters, D. Internationalisation and ElearningSystems: LRN Case Studies. In: Delgado Kloos, C. and Boticario, J.G. (eds.), Proceedings of Foro Hispano, LRN, Madrid, 2005.
31. Santos, O.C., Gaudioso, E., Barrera, C. and Boticario, J.G. (2003). "LFANET: an adaptive e-learning platform, Second International Conference on Multimedia and ICTs in Education (m-ICTE2003).
32. Shih, T.K., Lin, N.H., and Chang, H.-P., An intelligent e-learning system with authoring and assessment mechanism, Proceedings of the 17th International Conference on Advanced Information Networking and Applications (AINA'03).
33. Spiro, R.J. and Jehng, J. (1990). Cognitive flexibility and hypertext: theory and technology for the non-linear and multidimensional traversal of complex subject matter. D. Nix and R. Spiro (eds.), Cognition, Education, and Multimedia. Erlbaum, Hillsdale, NJ.
34. Spiro, R.J., Feltovich, P.J., Jacobson, M.J., and Coulson, R.L. (1992). Cognitive flexibility, constructivism and hypertext: random access instruction for advanced knowledge acquisition in ill-structured domains. In: Duffy, T., and Jonassen, D. (eds.), Constructivism and the Technology of Instruction. Erlbaum, Hillsdale, NJ.

35. Tullis, T.S. and Stetson, J.N. (2004). A comparison of questionnaires for assessing website usability", UPA 2004, Human Interface Design Department, Fidelity Center for Applied Technology, Fidelity Investments.
36. Tosteson, D.C. (1979). Learning in medicine. New England Journal of Medicine 301(13):690–694.
37. Psaromiligkos, Y. and Retails, S. (2003). Re-evaluating the effectiveness of a web-based learning system: a comparative case study. JI of Educational Multimedia and Hypermedia Barker, P. and King, T. (1993). Evaluating interactive multimedia courseware – a methodology. Computers and Education 2:4.

The Context Oriented Architecture: An Augmentation of Context Awareness and Reactivity into Web Services

Aisha Mohamed-Salama Elsafty, Sherif Gamal Aly, and Ahmed Sameh

Department of Computer Science, The American University in Cairo, 113 Kasr El Aini St., Cairo, 11511, Egypt aisha@aucegypt.edu, sgamal@aucegypt.edu, sameh@aucegypt.edu

Summary. Standardization promotes web services as a very promising candidate for successfully integrating disparately heterogeneous systems. As such, web services prove themselves to be very suitable candidates for supporting the predominantly heterogeneous pervasive environments. The ability of web services however to sense their surrounding context and effectively react to it is still a matter of research. In this work, we introduce a new architecture, an architecture solely built on open standards, that supports the development of context aware and context reactive applications that use web services as building components. We describe in detail the various components of this architecture, along with their supporting interactions. Furthermore, we describe the expansion of the OWL-S ontology language, namely expanding both the profile and the process model ontologies, to allow for encoding context behavior of both web services and clients. We eventually illustrate validation scenarios for this architecture, and demonstrate an application example built using this architecture that adapts to ambient security requirements.

1 Introduction

Today's computer systems meet very pressing challenges. Users are increasingly not expected to adapt to existing systems, but it is the systems themselves that are challenged with the need to adapt to the ever changing and ever heterogeneous requirements of users. Today's systems become increasingly nomadic, in environments that are very ubiquitous and pervasive. In order to continue adapting to user needs, applications must now be knowledgeable about ambient information, not only that which surrounds them, but also information that surrounds the users themselves. In other words, applications must be equipped with the ability to know the context surrounding both them, and their users. In effect, not only should applications become context-aware, but they should also be adaptive in the way they react to changing

A.M.-S. Elsafty et al.: *The Context Oriented Architecture: An Augmentation of Context Awareness and Reactivity into Web Services*, Studies in Computational Intelligence (SCI) **93**, 199–219 (2008)
www.springerlink.com

context information, and in a way that best satisfies user needs. As such, users expect applications to be both accessible and effectively functional while the resources and context of the application and the user change continuously.

The World Wide Web, through its continuous growth has become an invaluable resource to both users and applications. The web became a very successful candidate for the provision of customized and personalized services, and in specific, services that are available everywhere, all the time, and through standardized usage. Web services proved themselves to be the prominent solution for this kind of revolution in the provision of services in a very heterogeneous environment such as the web. They became the solution for interoperability between applications and service providers. Web service description languages on the other hand, were introduced to support automatic integration and composition. The web service ontology language, OWL-S is a very successful example of such language. However, such language does not support context awareness in itself.

Even existing infrastructures for supporting context awareness suffer from problems of openness and scalability. Applications still depend on very thick communication layers to perform multiple tasks including managing and monitoring resources, user activities, and changes in devices. Scalability on the other hand is still challenged by the existing discovery models. The discovery layers still pose themselves as bottlenecks during context monitoring and component discovery against the ability of applications to properly scale in size.

Context aware support was so far provided via middleware layers or as complete platforms starting from the operating system layer upwards [1–3]. Although such efforts contributed towards specifying some behavior that is expected of context aware applications, they are still far away from standardization and deployment. There is no consensus on a recommended architecture, such as those that exist for real time systems and workflow applications, which can act as a guideline for the building of context aware and adaptive applications.

This research investigated and developed what we call the Context Oriented Architecture: an architecture for building adaptable and context-aware systems [16], based on the existing, already standardized, web infrastructure, and using web services as building blocks. The specification of the OWL-S ontology language was extended to describe the context needs and adaptive behavior of web services within the underlying architecture. Furthermore, operational scenarios were used to validate aspects of the architecture.

In Fig. 1, we demonstrate the relation of the context oriented architecture with respect to the service oriented architecture, and the object oriented architecture. While the object oriented architecture stressed defining the interface of an object to enhance its reusability, service oriented architectures aimed at integrating multiple services for interoperation, and thus stressed the formalization of service interfaces or contracts. The context oriented architecture acknowledges the relation between the system and its user and deployment

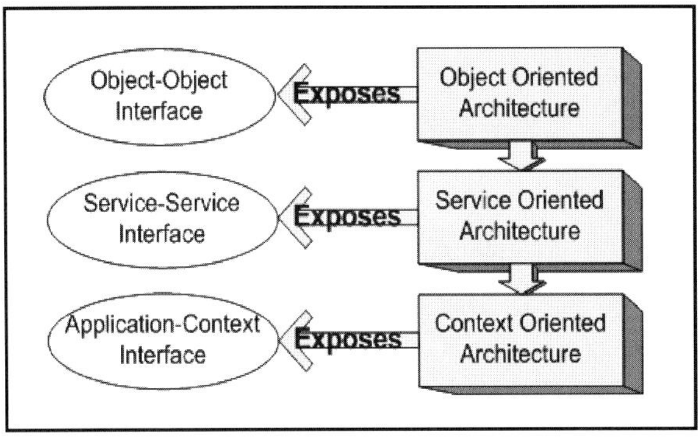

Fig. 1. Evolution of architecture concerns

environment, and thus emphasizes the definition of the interface between the application and its context.

2 Related Work

In order to build a web-based architecture for systems that are context aware, we primarily rely on two domains, namely context awareness and web services. Context awareness, as defined below in more details has been utilized in many domains, most recently, in the domain of pervasive systems; systems that diffuse into the lives of their users and adapt to ever changing user requirements. In the subsections hereafter, we define the notion of context as it appears in literature. Of particular interest, we also describe the OWL-S web service ontology language, which we eventually extended to support the building of this architecture. Furthermore, we also describe existing context aware infrastructures, as well as previous attempts to make web services context aware.

2.1 Context Defined

Several definitions for context exist in literature. We use Dey and Abowd's definition for context as: "Any information that can be used to characterize the situation of entities (i.e. whether a person, place or object) that are considered relevant to the interaction between a user and an application, including the user and the application themselves" [4]. Context information is collected by sensors, which can be classified into physical, virtual and logical sensors. Physical sensors collect data that can be sensed physically, for example temperature, light intensity and humidity. Virtual sensors are software based, like motion detection applications. Logical sensors combine two or more sensors to provide higher level conclusions [5].

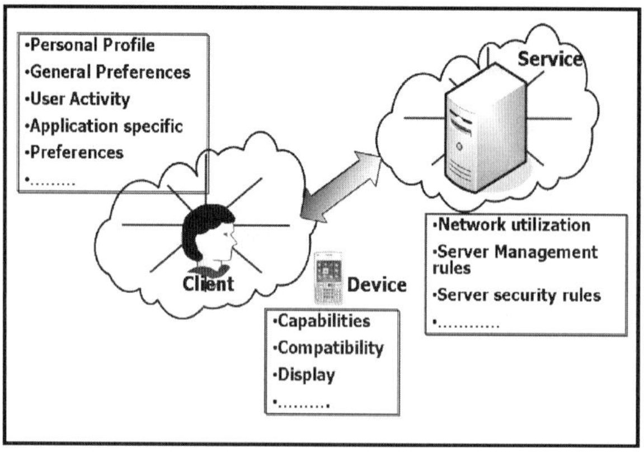

Fig. 2. Context types and examples

Our classification of context depends on the entity the context describes. In Fig. 2 we categorize context into user context, service context and device context. User context includes personal data like name, age, address, profession and schedule. User context can also include preferences; like the preferred language, alerting mechanism and payment method. Service context can include: security level required by the application, minimum bandwidth, storage capacity and compatible platforms. Service context thus defines non functional characteristics primarily related to deployment and runtime environments. Tightly related to both user and service is the device used in the transaction. Device context includes the device type, display capabilities, memory and processor capacity, connectivity, compatibility and data format among other attributes.

The variance in context types and attributes necessitates a structure for context data that is comprehensive and extensible. For this aim, several efforts have targeted standardizing a hierarchy of context types. For example, CONON is a context ontology that represents context entities typical to pervasive environments [6]. CONON divides context into upper and lower ontologies. The upper ontology includes the basic entities: person, activity, computational entity and location and it can be extended by domain specific lower ontologies. CONON is encoded in the web ontology language (OWL) enhancing its portability. Another OWL ontology that encodes vocabulary of pervasive environments is SOUPA (Standard Ontology for Ubiquitous and Pervasive Applications). SOUPA is compatible with existing ontologies using OWL's equivalentClass and equivalentProperty attributes allowing for reusing existing ontologies [7]. The core ontology consists of nine ontologies: person, activity, space, time, policy, action and geographical measurements. The SOUPA extension ontologies defined complementary ontologies to satisfy the needs of an intelligent meeting room.

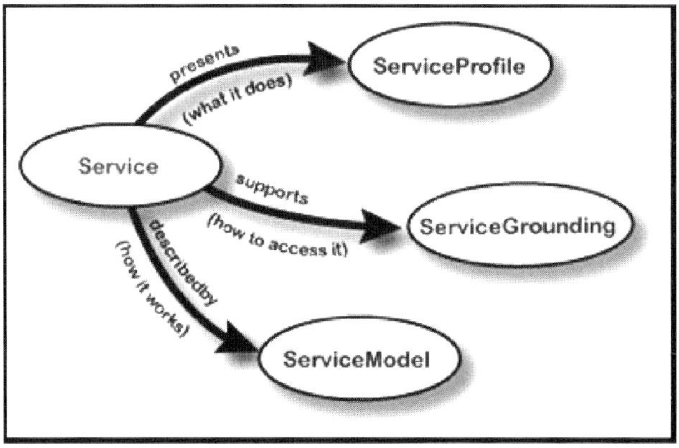

Fig. 3. The service ontology in OWL-S (Figure from [14]: Martin et al. 2003)

2.2 OWL-S: The Semantic Web Service Ontology Language

The semantic web effort aims at building the specifications and tools necessary for the data representation on the web to better suit the automatic fetch and interpretation of data. The semantic web ontology language (OWL-S) is the latest development in semantic web markup languages, after DAML+OIL and OWL. OWL-S provides a computer interpretable description of a web service. Thus, OWL-S eases the discovery of services according to their specifications, allows the automatic invocation of services, provides for the interoperation of services and service composition and provides monitoring of the service progress [13]. In OWL-S, a service is viewed as a process which has inputs, outputs, parameters and any number of effects on the state of the world. Figure 3 shows the service ontology in OWL-S, which consists of three sub ontology: the profile ontology used to advertise the service and the providing business entity, the grounding ontology used to provide physical access information like WSDL URI and the process model which describes the interoperation of the service. The OWL-S process model differentiates between atomic and composite processes. An atomic process has one transaction with the user; triggered by an input; an atomic process executes and provides an output. The user has no access to the details of execution inside an atomic process. OWL-S provides for a set of constructs used to link other services in one composite service, called control constructs. These are: sequence, spilt, split+join, If-Then-Else, Repeat-while and Repeat-until. OWL-S specifies ontologies for time and resources, to allow for representing the interaction between service and the environment and its resources.

OWL-S uses OWL for describing the entities and objects the service uses. For example, while describing a bookstore service in OWL-S one expects to use the OWL description of entities like book and author. OWL has three types, Lite, Description Logic (DL) and Full. OWL-S builds on OWL DL, which is expressive and powerful while still computationally complete. Another OWL-S implementation is in the OWL-S2UDDI Converter [15], which creates a mapping between the profile sub-ontology of OWL-S and the UDDI registry. The generated UDDI description can then be published in any UDDI registry, allowing for UDDI to make use of the semantic description of the service.

2.3 Context Awareness Infrastructures

A notable effort towards the standardization of the context representation is the context toolkit which represents the most integrated work for support of context-awareness [5]. It was first introduced by IBM Watson research center in 2002. In the context toolkit, the main provider of context information is a Context Widget, which provides applications with a standardized interface for context types. Applications would subscribe to a widget and be notified on context changes; alternatively an application can query the widget. The toolkit provided intermediate context processing, like interpreters that provided logical context through mixing different context data and applying different inference techniques, and aggregators that dealt with the distributed nature of sensors by providing a central access to logically related context types through an aggregator interface. Also, services objects provided reusable building blocks for the response of an application to a change in context, i.e. sending a message or turning on lighting in a room. Finally, discoverers allowed applications to find out what interpretation and aggregation services are available in the environment [4].

The Context Broker Architecture (COBRA) uses OWL to represent, infer and validate context information. The context broker is central to the architecture; other components are assumed to have knowledge of the broker location and to delegate context processing to it when resources are limited. The broker consists of a knowledge base, reasoning engine, acquisition module and management module. The context knowledge base stores context information allowing for validating its consistency. Our architecture uses context information about an entity available on the entity itself, allowing sharing of such information when desired by the entity itself. Figure 4 shows the main components of the architecture: the reasoning engine infers advanced context information and validates the consistency of context information. The acquisition module shields the sensors acquiring the low level context data, similar to our unified interface for context sensors. The management module provider rule based policy for sharing context among applications. The privacy concern is reduced in our architecture as we rely on a distributed architecture where the centralized repository of context information is absent.

2.4 Context Aware Web Services

Integrating context information into the protocols, interactions and structure of web services has taken two main paths. First, embedding context data within the existing web services protocols and devising an architecture

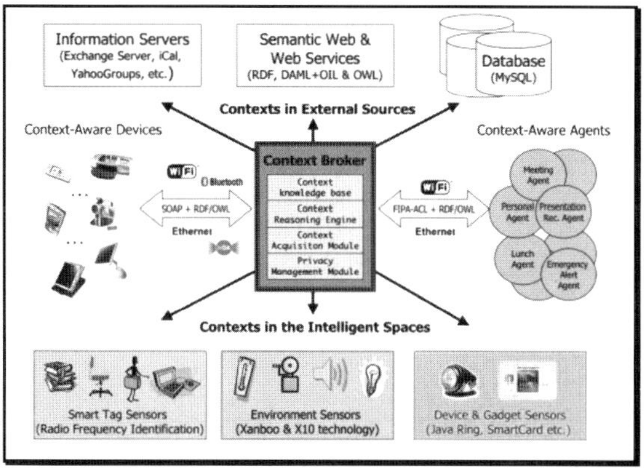

Fig. 4. The CoBrA context architecture (Figure from [9]: Chen, 2004)

component that processes this information. Or alternatively, to define web services that are simple and fine grain that act as building blocks for the context aware application. Later at run time, and according to the active context, these fine grain services can be composed according to the runtime context.

SOAP Based Context Aware Web Services

Keidl and Kemper integrate context information in the interaction among web services and between the client and the service [10]. The SOAP header is extended to include a context header block which the web service can optionally process. Context information includes user identity, location, preferences and the type of device in use. The processing of this context data can be done inside the web service or externally in a proxy as shown in Figure 5. In the latter case, only the request and response of the web service can be manipulated according to the context rather than its internal control flow as in the former case.

Keidl and Kemper thus maintain a static notion of context as set of attributes representing user and environment state, the response to context state is restricted to updates in the context state again. While this restriction opens up the security issues of which parts of the user context are accessible and can be modified by the service, it is unrealistic to assume that response to context change will be limited to the same context attribute. Our architecture acknowledges the need for dynamic response to context, allowing the designer to specify routines that execute in response to context states. These routines vary from simple value updates to invoking other web services.

Planner Based Context Aware Web Service Composition

Vukovic and Robinson used the SHOP2 planner to device a composition of web services to achieve user target [11]. The service is represented as an operator

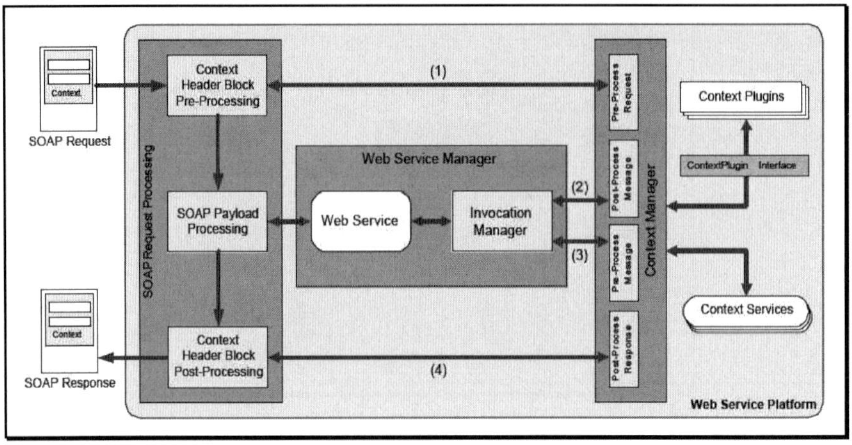

Fig. 5. SOAP based context aware architecture (Figure from [10]: Keidl and Kemper, 2004)

in the planner, which can achieve certain results in the world state; this is the output of the service. The input to he planner is the desired output of the system, whether it is streaming a certain video or viewing client email. With this target, the planner is fed the context information, whether the client is using his mobile, or his pc etc. The planner is able to device a composition of web services that can achieve the desired target, according to the context state.

The output of the planner is a composition of web services, encoded in the BPEL4WS language. A simplification of the web service details is necessary to realize it as an operator in the planner; it was argued that representing web services as operators entails an amount of simplification that contradicts the detailed nature of business domain most web services target.

3 The Context Oriented Architecture

This section provides a description of the proposed context oriented architecture. The various subsystems, along with their interactions are described herein. The context oriented architecture as we describe it is an infrastructure that transparently monitors context surrounding both clients and services within a given environment. Not only does the architecture support monitoring of context, but it also allows for the development of custom responses designed by service developers to be triggered when conditions in the monitored context are met. A primary leverage of this architecture is that it relies solely on web service technologies and standards, that, which immediately supports openness and scalability, similar to the Internet standards it uses.

This architecture has two basic requirements for both the sensing of context and its related response. With regards to the first requirement, each application using this architecture as a reference, and interested in becoming context sensitive must initially declare what is called the application's context profile. Such profile includes the types of services that the application monitors, and how such monitoring is achieved, for example, through the utilization of various types of sensors for this purpose. As relates to the second requirement, each application must also declare what action to perform in reactivity to changes in the monitored context.

As this architecture relies on existing standards, we built upon the existing SOUPA context ontology and added a hierarchical, yet extensible definition of context types. The context profile of the service defines the location in which the context data is collected from, and the frequency of collection. The response to context changes can have different forms. One response could involve the invocation of another web service. Another could be in the form of alteration of the internal flow of the service, and a third form may be in the form of the invocation of a local routine. The ultimate decision of the kind of reactivity to adopt in response to contextual changes is within the hands of the application architects and designers. The architecture in itself is open to allow for further types of responses other than web service invocation, alteration of internal flow, and invocation of local routines. However, such three response types were initially presented in the architecture as an example of typical commonly used responses. For each response type, certain information and resources need to be defined in the context profile of the service. We define the components of the architecture as shown in Fig. 6.

Seven primary components form the skeleton of the service oriented architecture, namely, the negotiator, context sensors, the monitor, context response, the response manager, the component registry, and the context repository. A usage scenario is further illustrated below. Each of the seven components is described in detail below.

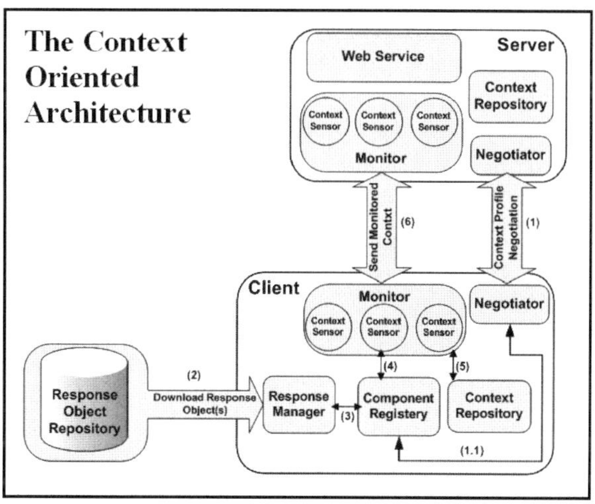

Fig. 6. The context oriented architecture

3.1 The Negotiator

Reaching an agreement between both the clients and services as relates to the types of contexts to be monitored, the sensors to be used for monitoring, and the response components to be used for reacting to contextual changes is an integral functionality of this architecture. The negotiator on both the client and the server is responsible for negotiating the context profile to be used by the client and the service in the transaction. While the service is designed to be sensitive to certain context types, the client may have privacy or efficiency concerns against allowing the service to monitor some of these context types. During the negotiation phase, the context types and profile for each type are negotiated between the client and service. The objective is to find context sensitivity objects on the service side that do not violate the requirements of the client profile. Negotiation ends when the client accepts a proposed context profile or when either the client or the service ends the negotiation pre maturely, causing the transaction to start in the context unaware mode.

3.2 The Context Sensors

Context sensors are responsible for collecting context information from the environment. Sensors are the unified interface to the various physical and logical sensing entities. As indicated earlier, sensors need not only be physical, such as temperature, motion, sound, and light sensors, but can also be logical deriving information from underlying physical sensors. Each sensor type has a unique identifier. Sensors can exist on both the client and server sides.

3.3 The Monitor

The monitor on both the client and server manages the sensors running to collect context data. The monitor is responsible for managing the processing of sensed data together with the initiation of response triggering requests. It is the monitor that controls the frequency of sensing activities that need to be performed. Each sensor may have different needs for data reading frequencies. The monitor also is responsible for testing the collected context data against the conditions required for triggering responses to contextual changes.

3.4 The Context Response

Context responses are considered to be the materialization of an application's need to adapt to its surrounding context. It is those responses that finally make applications adaptable to contextual changes. The service designer specifies the service behavior in response to context changes, encapsulated as response components. Responses can be as simple as changes in the internal flow of a service, or as involving as triggering a search and invocation of another web service. Responses can execute on either the client side or server side, and not necessarily in the same location the context data was initially collected. For example, detecting a security threat on the user device can trigger the use of virus scanning on the server side the processing request received from this client. Furthermore, the detection of a client's linguistic preferences can invoke a translation web service to translate any required text to the client's own preferences.

It is expected that context response to the same context type will be reused across applications. It is thus desirable to isolate the response routine from the main service. This way the context model of a service could be changed independently from the service logic.

3.5 The Response Manager

As indicated earlier it is the monitor that checks whether the conditions are met or not to trigger appropriate responses to contextual changes. It is the response manager however that is responsible for locating and initiating the response objects for each context type monitored in the interaction. Together with the negotiator, the response manager determines which response objects are to be downloaded from the response object repository. The response manager also handles communication with remote response objects.

3.6 The Component Registry

In an attempt to keep track of all sensor and response objects, as well as their relevant details such as name, functionality, and location, a registry is needed. The component registry includes a list of local components that function as sensor or response objects, their locations, unique identifiers and invocation semantics. The component registry is used by the monitor and response mangers to elect sensor and responses for specific context types. The component registry is also used by the negotiators to determine which components are available locally and which components need to be obtained from the response objects repository, possibly affecting the success or failure of negotiation.

3.7 The Context Repository

As usual, there always exists a need to store contextual information that is persistent by nature. The context repository includes context information that persists across the transaction with a web service and beyond the life time of a single message. Data stored in the repository are not persistent among transactions with multiple web services, thus relieving privacy concerns. The repository includes the persistent profile as well; this is user profile information that persists across applications, they are stored locally and the monitor is responsible for applying user privacy rules. For example, gender, language preferences and contact profile. The persistent profile includes the meta-preferences specifying which context types the user is willing to share together with the sensors and monitors permitted to execute locally on the clients side.

3.8 Usage Scenario

As shown in Fig. 6, when the client invokes the context aware web service, a negotiation phase is initiated. According to client preferences, a consensus on monitoring certain context types is reached. If the client lacks some of the sensors or response objects indicated in the context profile agreed on in the negotiation phase, these components are downloaded from the response object repository and registered in the local component registry. Typically if a sensor or response object is needed on the server side it will be already deployed at the service deployment time. As the transaction starts, the monitor and the response manager are responsible for obtaining the sensor and response object handles from the component registry and activating them. The monitor handles collecting the context data and testing it for meeting the condition for triggering a response. The response manger handles activating the response object when notified by the monitor.

Figure 7 shows a modified version of the web service interaction scenario after adding interactions related to context exchange and processing. The phase of negotiating the service syntax and semantics is extended to allow for automatic negotiation of the service context profile and according to user privacy and preferences. During the interaction between the client and service, the context is monitored on either the client or server machine or both of them. The response can be triggered on either ends, and not necessarily the same end where location was monitored.

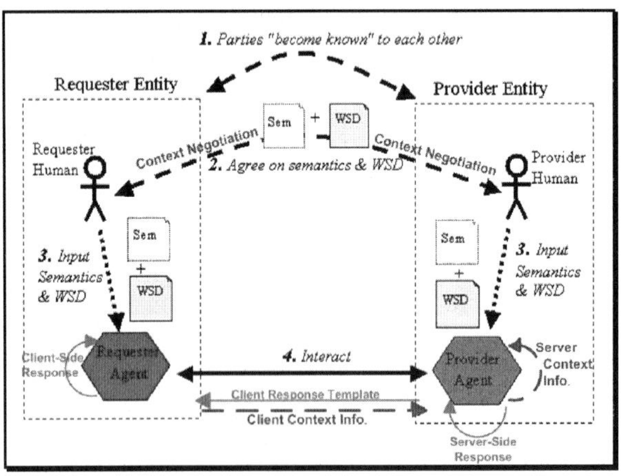

Fig. 7. The interaction with context-aware services (Figure modified from Booth et al. 2004)

4 OWL-S Extension

The Semantic Ontology for Web Services (OWL-S) version 1.2 is now being reviewed as one of DARPA agent markup languages. OWL-S aims at achieving the automatic discovery, invocation, composition and monitoring of web services. OWL-S also aims at providing for context awareness, one target that is yet to be realized through further development of the ontology [12]. In this section, we answer to the requirements of building a context aware infrastructure that relies on OWL-S described semantic web services. We propose an extension to the OWL-S standard to allow for encoding the context behavior of the web service as well as its client. The proposed extension is two folds: extending the profile ontology and extending the process model ontology.

In this section we list several attributes that are to be added to the OWL-S ontology of a context-aware web service to describe the entities and interactions involved in monitoring context and handling its changes. Having studied the OWL-S ontology, we believe that such extension is needed as the context model cannot be represented within the current ontology without losing focus and missing details. Perhaps one candidate for representing the context model within the current ontology is the Profile Attributes; which include a list of ServicePrameter where parameters can be defined and assigned values. However, the Profile attributes are intended for representing non-functional information that to be handled by humans, making its use for context representation inadequate. An alternative design is to extend the OWL-S ontology to include a different model; in addition to the profile model, the process model and the grounding model, that is the context model.

We choose to add attributes to the existing profile and process models of OWL-S. The profile model defines the service advertisement. Profile information is expected to be used in searching and assessing service in the registry. The context model of a service gives an insight on service behavior, for example the user – whether a human or a software agent- will prefer using a service that is bandwidth sensitive while a GPRS connection is in-use. We thus believe that the service context model is an important criterion while choosing a service to be used. As the profile model is used in search services, context model of a service has to be exposed in its profile. We introduce the "sensitive-to" attribute in the profile. To model how context information are to be exchanged among the involved entities, and how the response to context changes is to take place, the OWL-S process model is extended with the context block in addition to the already included preconditions, postconditions and effects. The context model is different from service effects in that it does not represent service impact on its environment, rather a model for interaction between the service provider and requester.

```
<!ENTITY soupa "http://pervasive.semanticweb.org/ont/dev/">
<!ENTITY con "http://localhost/ContextFramework/Components/">

<owl: ObjectProperty rdf:ID="monitors">
<rdfs:domain rdf:resource="# Profile"/>
<rdfs:range rdf:resource="&soupa;"/>
</owl: ObjectProperty >

<owl:ObjectProperty rdf:ID="uses">
<rdfs:domain rdf:resource="#Profile"/>
<rdfs:range rdf:resource="#List;"/>
</owl:ObjectProperty>

<owl:ObjectProperty rdf:ID="responds">
<rdfs:domain rdf:resource="#Profile"/>
<rdfs:range rdf:resource="#List;"/>
</owl:ObjectProperty>
```

Fig. 8. Part of the extension to the profile ontology

4.1 Profile Ontology Extension

The profile ontology is extended by three attributes that declare the type of context(s) the service monitors, in addition to two attributes, "uses" and "responds" that list the sensor and response components respectively, which the service can use to monitor and respond to the context type.

Figure 8 shows the extension to the OWL-S profile ontology. The three attributes are added as object data type attributes, as they range over object types; in case of "monitors" attribute, it ranges over context types, while the other attributes range over the list type, used to encapsulate list of alternative components.

4.2 Monitors Attribute

A profile attribute, the values of this attribute ranges over the context types declared in SOUPA context ontology or context types defined by the user as an extension to SOUPA. Context types are defined as a hierarchical OWL ontology. The value of this attribute is the URI of the context type the web service is to monitor and respond to.

4.3 Uses Attribute

As part of the profile ontology, this attribute lists the identifier of the sensor components the service could use to monitor its context.

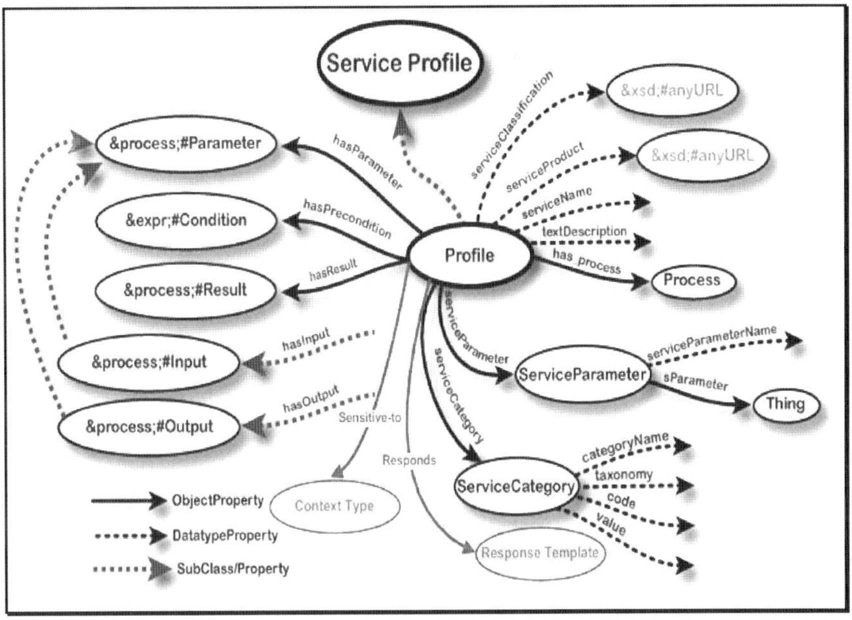

Fig. 9. Modified OWL-S profile ontology (Figure modified from [12]: Martin et al. 2003)

4.4 Responds Attribute

The Responds attribute is part of the service profile that includes the URI of the routine capable of handling the type of context at hand. The location of its components specifies where the routine is to run: client, server or both. The invocation of a web service can be regarded as running at an "external" location, however the location here refers to whether the client or the server is responsible for the invocation of the web service.

Figure 9 shows the added attributes within the original ontology. The added attributes come on the same level with attributes like "hasInput" and "hasOuput". We thus declare context types and components used by a service as an extension to the service signature in OWL-S.

4.5 Process Model Extension

The process model is extended by four attributes Monitors, uses, responds and Exposes. These attributes provide the needed semantic specification for the interaction between the service provider and requester while monitoring, processing and responding to a context type. We choose to follow the OWL-S convention of duplicating important attributes in the profile and process model ontology. Here we are duplicating the monitors, uses and responds attributes while exposes is unique to the process model.

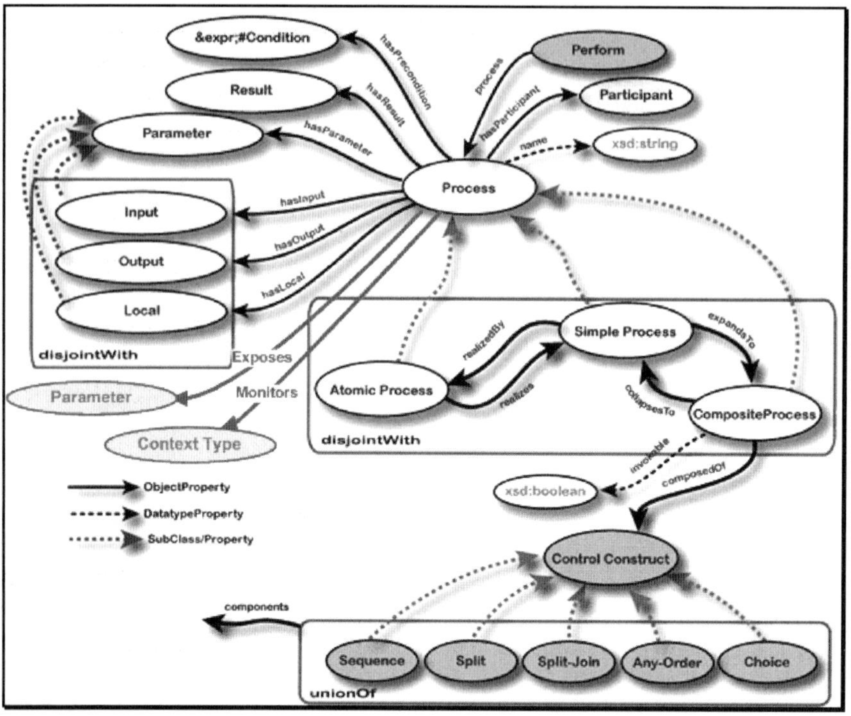

Fig. 10. OWL-S process model including the context block (Figure modified from [12]: Martin et al. 2003)

Figure 10 shows the extension to the process model relative to the original ontology. The exposes attribute is similar to the input and output attributes of the ontology, however it is not disjoint with them. It ranges over the variable name that is to be set by response components.

4.6 Exposes Attribute in the Process Model

The aim of this attribute is to allow the service response object to possibly change the state of some service variables or its control flow according to context change. If the response component responsible for processing the context information at hand is a plug-in separate from the web service, only exposed variable are accessible by the response object to affect the service's internal state and work flow. Changes in the values of exposed variables are published to the service using a setter method provided by the service developer.

Figure 11 shows the extended process model ontology. The "exposes" attribute is a data type attribute of the process model. Its value ranges over the string type, where the value is the name of the exposed variable.

```
<owl: ObjectProperty rdf:ID="monitors">
<rdfs:label>monitors</rdfs:label>
<rdfs:domain rdf:resource="#Profile"/>
<rdfs:range rdf:resource="&soupa;"/>
</owl: ObjectProperty >

<owl:ObjectProperty rdf:ID="uses">
<rdfs:label>uses</rdfs:label>
<rdfs:domain rdf:resource="#Profile"/>
<rdfs:range rdf:resource="&con;#List;"/>
</owl:ObjectProperty>

<owl:ObjectProperty rdf:ID="responds">
<rdfs:label>responds</rdfs:label>
<rdfs:domain rdf:resource="#Profile"/>
<rdfs:range rdf:resource="&con;#List;"/>
</owl:ObjectProperty>

<owl:DatatypeProperty rdf:ID="exposes">
<rdfs:label>exposes</rdfs:label>
<rdfs:domain rdf:resource="#Process"/>
<rdfs:range rdf:resource="&xsd;#string"/>
</owl:DatatypeProperty>
```

Fig. 11. The OWL-S extension to the process model

5 Results and Analysis: The File Storage Scenario

We adopt the common approach in the areas of pervasive systems, which is, we validate the proposed framework through putting it into action. We use the developed framework to build sample scenarios of a searchable online bookstore. This scenario is an example for retrieving resources based on user context. In general, user's environment or status is a factor that impacts what is a best format for retrieving a remote resource. In this example, according to user location, the resource (a file in this case) is retrieved in either the encrypted or the decrypted format. As the file includes critical business information, and the user can be accessing the file from a potentially hostile environment, the file is encrypted and the decrypt is invalidated shortly after the user closes it.

5.1 Scenario Context

In this scenario, a business woman is giving a presentation in a competing company's headquarters in the Far East. In this part of the scenario, she is able to work on updating her presentation from her hotel room and to store it on the server back in her company in Europe. While giving the presentation, the file is retrieved from the company server in encrypted format, the life time of the decrypt expires only a minute and half after giving the presentation, to sustain the data confidentiality.

> "She stores the presentation on the secure server at headquarters in Europe. In the hotel's seminar room where the sales pitch is to take place, she will be able to call down an encrypted version of the presentation and give it a post presentation decrypt life of 1.5 minutes." [30]

5.2 Scenario Formalization in the Framework

We implement a web service for file retrieval of confidential files. The web service is aware of the user location context. In the scenario, the client's location is detected to be outside of the home company headquarters. The client environment is thus considered insecure, being in headquarters of a competing business environment. The web service detects the status of the client, and responds through retrieving the file in encrypted format, in addition to setting the expiry of the decrypt to 1.5 min after the file use time. The location of the user is detected through a LocalRoutine sensor on the server side that processes the client request detecting her IP and thus location relative to the headquarters. The LocalRoutine response takes place on the server side as well, where the retrieved file is encrypted and the expiry time of the decrypt is set accordingly.

5.3 Semantic Model

The semantic encoding of the scenario is three folds. First: the service declaration to be monitoring the location context. Second: The service Profile and Finally the client profile.

Figure 12 shows the encoding of the sensor agreed upon in the negotiation, i.e. the sensor in the active profile. The sensor is a java routine that processes client request and infers his relative location to the company headquarters. The URI attribute of the sensor shows the location of the routine class file on the Context Repository. As the monitoring type is "Receiver Pull", the service request sensor reading from the framework's server instance. The sensor reading is thus obtained only when needed, that is, when a new request is received. The reading type is string, which is the encoding of the relative user location.

5.4 Interaction Model

Both the sensor and response components are of type local routine, and the needed executable files are located on the component repository. The sensor is queried for location readings before the file retrieval, through the Receiver-Pull mode. The sensor reason on the client location according to its IP and

```
<conFM:Sensor rdf:ID="LocationLocalRoutineSensor">
<Sensor:type>LocalRoutine</Sensor:type>
<Sensor:location>Server</Sensor:location>
<Sensor:uri>"http://ContextRepository/Component/Sensors/loactionLR.class"</Sensor:
uri>
<Sensor:monitoringType>ReceiverPull</Sensor:monitoringType>
<Sensor:readingType>String</Sensor:readingType>
<Sensor:condition>"!=InHeadquarters"</Sensor:condition>
</conFM:Sensor>
```

Fig. 12. The semantic description for location detection sensor

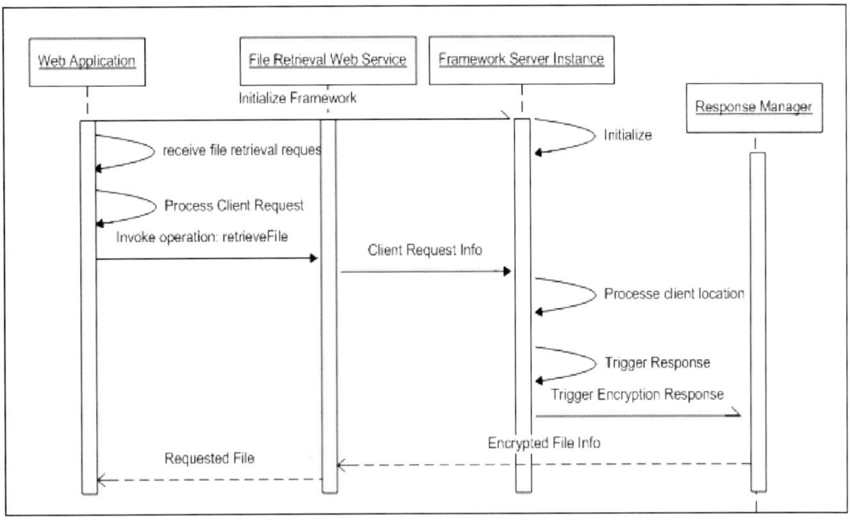

Fig. 13. Interaction diagram for the file retrieval scenario

represents the reading in string format, using the strings "InHeadquarters" and "OutHeadquarters". The response is of type local Routine and implemented in Java like the sensor. It uses an encryption library to encrypt the requested file and send it. The response takes as a parameter the path to the file to be encrypted.

Figure 13 shows the interaction model for the scenario. The service triggers the location sensor after receiving client request, the framework proceeds to trigger an encryption response if needed. The web service is then notified when the response is performed. This scenario shows a tight coupling between the processing of the framework and the web service; where by the activation of sensor is done by the service, in addition the response takes place on the server side.

To enforce the expiry of the file decrypt, a client side sensor is used. The client side sensor uses a WMI (Windows Management Instrumentation) script to check file status. When the file is closed, a timer is activated to invalidate

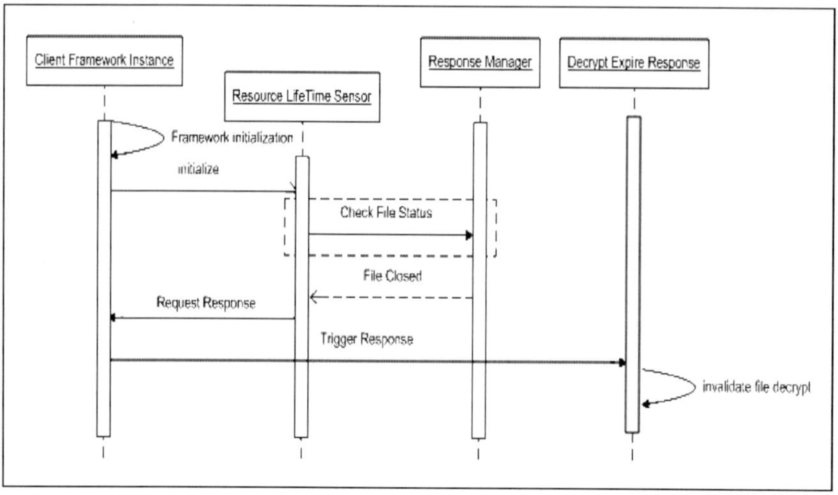

Fig. 14. The expiry of file decrypt sequence diagram

the file decryption. The UML sequence diagram for triggering the file expiry response is illustrated in Fig. 14.

6 Conclusion

In this work, we illustrated the importance of building applications that are context aware, and that are also adaptive to their ever changing context. We presented and elaborated on the concept so called "The Context Oriented Architecture" as a responsive architecture for building context aware and adaptive web services. Through utilizing the existing and powerful web services, and through expanding the famous OWL-S ontology language, we were able to specify an architecture that allows applications to sense their context and respond to it. Through depending only on the stable web services standards, the context oriented architecture provides an open and scalable infrastructure for adaptive applications. We fully implemented a prototype for this architecture. We extend the OWL-S API underlying the OWL-S Editor [12]. The API extension, the architecture prototype and the validation scenario are all built in Java, using the Web Tools Project Eclipse plug-ins. Several test scenarios and performance tests against the architecture prototype were conducted to measure and validate its completeness. We demonstrate a scenario that requires applications to react to ambient security requirements, and we show how applications can be built in compliance with the context oriented architecture.

References

1. Gu, T., Pung, H. K., Zhang D. Q. A service-oriented middleware for building context-aware services. Journal of Network and Computer Applications 1, 2005. 1–8.
2. Anand Ranganathan, Shiva Chetan, Roy H. Campbell. Mobile Polymorphic Applications in Ubiquitous Computing Environments. MobiQuitous 2004. 402–411.
3. Grimm, R. et al. Programming for pervasive computing environments. Technical Report UW-CSE-01-06-01, University of Washington, Department of Computer Science and Engineering, June 2001.
4. Dey, A. K., G. D. Abowd. A conceptual framework and a toolkit for supporting the rapid prototyping of context-aware applications. Human–Computer Interaction (HCI) Journal 16. 2–4.
5. Baldauf, M., S. Dustdar, F. Rosenberg. A survey on context aware systems. International Journal of Ad Hoc and Ubiquitous Computing 1(3), 2006.
6. Wang, X. H. et al. Ontology Based Context Modeling and Reasoning using OWL. Workshop on Context Modeling and Reasoning (CoMoRea). PerCom'04. 14 March 2004.
7. Chen, H., F. Perich, T. Finin, A. Joshi. SOUPA: Standard ontology for ubiquitous and pervasive applications. (Mobiquitous 2004), Boston, MA. 2004.
8. Elenius, D., Denker, G., Martin, D., Gilham, F., Khouri, J., Sadaati, S., Senanyake, OWL-S: Semantic Markup for Web Services. the OWL Services Coalition.
9. Chen, H., T. Finin, A. Joshi. An intelligent broker for context-aware systems. Adjunct Proceedings of Ubicomp 2003, Seattle, Washington, WA. 12–15 October 2003.
10. Keidl, M., A. Kemper. Towards context-aware adaptable web services. 13th International WWW Conference, New York, NY. May 19–21, 2004.
11. Robinson, P., Vukovic, M. SHOP2 and TLPlan for proactive service composition. UK-Russia Workshop on Proactive Computing, Nizhniy Novgorod, USSR. 2005.
12. Elenius, D., Denker, G., Martin, D., Gilham, F., Khouri, J., Sadaati, S., Senanyake, R.: The OWL-S editor: A development tool for Semantic Web services. Second European Semantic Web Conference, ESWC, Greece, May 29–June 1, 2005. 78–92.
13. The OWL-S 1.1 Release Technical Overview. The OWL Services Coalition. http://www.daml.org/services/owl-s/
14. Bartolini, C., Chris Preist, Harumi Kuno. Requirements for Automated Negotiation. Hewlett-Packard Labs. 2002.
15. Paolucci, M., T. Kawamura, T. R. Payne, K. Sycara. Importing the Semantic Web in UDDI. Proceedings of Web Services, E-business and Semantic Web Workshop. Toronto, Canada. 2002.
16. Elsafty, A., S. G. Aly, A. Sameh. The Context Oriented Architecture: Integrating Context into Semantic Web Services. The 1st International Workshop on Semantic Media Adaptation and Personalization, Athens, Greece, 2006.

Knowledge-Assisted Analysis of Video for Content-Adaptive Coding and Transmission

Vasileios Mezaris[1], Nikolaos Thomos[2], Nikolaos V. Boulgouris[3], and Ioannis Kompatsiaris[1]

[1] Informatics and Telematics Institute/Centre for Research and Technology Hellas, 1st Km Thermi-Panorama Road, Thessaloniki 57001, Greece
[2] Signal Processing Institute, ITS, Ecole Polytechnique Federale de Lausanne (EPFL), Lausanne, 1015, Switzerland
[3] Department of Electronic Engineering, Division of Engineering, King's College London, London WC2R 2LS, UK

Summary. In this chapter a knowledge-assisted, domain-specific video analysis framework is introduced and employed for content-adaptive video coding and transmission. Domain knowledge employed under the proposed framework considers both low-level features and spatial behavior of video content for the purpose of analysis, as well as domain-, application- and user-specific importance factors associated with each domain concept that guide content-adaptive coding and transmission. The analysis approach relies on a genetic algorithm for supporting efficient object localization and identification with the use of domain knowledge. The application of the genetic algorithm is preceded by the automatic generation of a set of atom-regions by means of segmentation and the subsequent extraction of the atom-region low-level descriptors. The output of the analysis process is used for the content-adaptive optimization of the coding and transmission of the video. Several methodologies for the coding and transmission of video over unreliable wireline and wireless channels are presented, utilizing advanced channel coding techniques for unequally protecting the objects of the video stream on the basis of the importance factors defined in the domain knowledge. Experimental results on a test set comprising Formula One and Tennis domain videos demonstrate the effectiveness of the proposed framework.

1 Introduction

Recent advances in computing technologies have made available vast amounts of digital video content, leading to ever-increasing flow of audiovisual information. This results in a growing demand for efficient video content management and exploitation, including coding and transmission. A key enabling factor for this is the acquisition of higher-level information about the meaning of the video content, which however is a non-trivial problem.

Initial attempts to understand video content were based on segmentation methods alone, i.e. signal-level processing methods aiming to partition a video

V. Mezaris et al.: *Knowledge-Assisted Analysis of Video for Content-Adaptive Coding and Transmission*, Studies in Computational Intelligence (SCI) **93**, 221–240 (2008)
www.springerlink.com

into (hopefully) meaningful elementary parts termed segments. Depending on the selected level of granularity of the partitioning, the term segmentation was used to describe a range of different processes: temporal segmentation, aiming to break down the video to scenes or shots [1]; spatial segmentation, addressing the problem of independently segmenting each video frame to arbitrarily shaped regions [2]; spatio-temporal segmentation, extending the previous case to the generation of temporal sequences of arbitrarily shaped spatial regions [3]; foreground/background separation, which can be seen as a special case of spatio-temporal segmentation [4]. In any case, the application of any segmentation method was followed by a feature extraction step, for compactly representing the resulting video segments in terms of their visual characteristics, e.g. color, texture, motion features, etc. [3].

The result of pure segmentation techniques, though conveying some semantics, such as the complexity of the key-frame or video, measured by the number of generated regions, the existence of moving objects in the shot (as a result of spatio-temporal segmentation or forground/background separation), or the visual characteristics of the video, is still far from revealing the complete semantic content of the video. To alleviate this problem, the introduction of prior knowledge to the segmentation procedure, leading to the development of domain-specific knowledge-assisted analysis techniques, has been proposed. The motivation behind introducing prior knowledge lies in the acknowledgement that extending existing techniques requires the introduction of a mechanism for mapping video segments and visual characteristics to objects and events; this is impossible without somehow defining the latter. Consequently, prior knowledge for a domain (e.g. F1 racing) typically includes the important objects that can be found in any given image or frame belonging to this domain (e.g. car, road, grass, etc. for the F1 racing domain), their characteristics (e.g. corresponding color models), any relations between them, etc. Given this knowledge, there exists the well-defined problem of deciding, for each part of visual information (whether that is a region or even a single pixel), if it represents any of the defined objects (and if so, which one) or none of them.

In order to represent the knowledge necessary for the task of video analysis and implement its introduction in the signal-level analysis process, a number of diverse approaches have been proposed. These include formal knowledge representation theories, semantic web technologies, Dynamic Belief networks, etc. to name a few. In particular, ontologies, being a formal specification of a shared conceptualization [5], provide by definition the formal framework required for representing the necessary knowledge. By making semantics explicit to machines, ontologies enable automatic inference support, thus allowing users, agents, and applications to communicate and negotiate over the meaning of information. In addition to techniques for representing explicit knowledge, machine learning techniques have proven useful for the problem at hand, serving as a robust methodology for discovering complex relationships and interdependencies between numerical image data and the perceptually higher-level concepts. Moreover, in many cases they elegantly handle problems of high dimensionality, which are common in visual information processing. Among the most commonly adopted machine learning techniques are Neural Networks (NNs), Hidden Markov Models (HMMs), Bayesian Networks (BNs), Support Vector Machines (SVMs) and Genetic Algorithms (GAs) [6,7]. Examples of applying the above techniques to the video analysis problem include [8], where semantic web technologies are used, and [9], where *a priori* knowledge representation models are used as a knowledge base that assists semantic-based classification and clustering. In [3], an object ontology coupled with a relevance feedback mechanism is introduced for the purpose of retrieval, in [10], semantic entities in the context of the MPEG-7 standard are used for knowledge-assisted video analysis and object detection, while in [11], associating low-level representations and high-level semantics is formulated as a probabilistic pattern recognition problem.

Once a semantic interpretation of the video is extracted, this can be used for introducing semantics-based functionalities to a number of video manipulation applications, including the semantic retrieval of video content from large multimedia databases and the semantics-based content-adaptive coding and transmission of video. In the application of retrieval of video content, the use of the knowledge-assisted analysis results can support the combination of visual-similarity-based queries with semantic-similarity-based ones, considerably improving the efficiency of access to visual content [12]. In the application of content-adaptive coding and transmission of video, the knowledge of the content semantics can guide the video encoding process to unequally assign error protection levels to different parts of the video based on their importance, so as to improve the perceptual quality of the encoded video when transmitted over unreliable channels.

In this chapter a knowledge-assisted, domain-specific video analysis framework is introduced and employed for content-adaptive video coding and transmission. The analysis framework uses a genetic algorithm to support efficient object localization and identification. An initial segmentation generates automatically a set of atom-regions and subsequently their low-level descriptors are extracted. Analysis is then performed using the necessary tools (genetic algorithm, domain knowledge) for relating high-level symbolic representations included in the ontology to visual features extracted from the signal domain. Additionally, the genetic algorithm decides how the atom-regions should be merged in order to form objects in compliance with the object models defined in the domain ontology. The output of this analysis process is then used for the optimization of the coding and transmission of the video. Specifically, several methodologies for the coding and transmission of video over unreliable wireline and wireless channels are presented, utilizing advanced channel coding techniques for protecting the objects. The optimal allocation of channel protection is also examined. The unequal amounts of channel protection are determined using either dynamic programming or lagrangian techniques. For more efficient data protection the system allows the

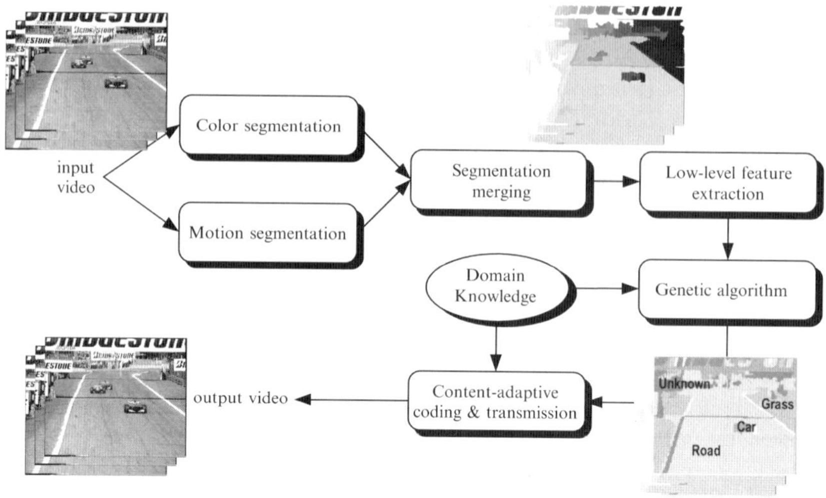

Fig. 1. Overview of the proposed domain-specific video analysis, coding and transmission framework

optimization of video delivery based on the video content and the significance of its constituent objects. The proposed framework is outlined in Fig. 1.

The remainder of the chapter is structured as follows: Sect. 2 considers domain ontology development and Sect. 3 contains a presentation of the segmentation and descriptor extraction algorithms and discusses the implementation of the genetic algorithm. Section 4 addresses the issues of coding and transmission exploiting the previously generated analysis results. Indicative results are presented in Sect. 5 and finally, conclusions are drawn in Sect. 6.

2 Domain Knowledge

The knowledge about the examined domain is encoded in the form of an ontology. The developed ontology includes the objects that need to be detected, their visual features and their spatiotemporal relations, as well as domain-, application- and user-specific importance factors associated with each domain concept. These descriptions provide the system with the required knowledge to find the optimal interpretation for each of the examined video scenes, i.e. the optimal set of mappings among the available atom-regions and the corresponding domain-specific semantic definitions, and to subsequently employ them for guiding efficient coding and transmission. To account for objects of no interest that may be present in a particular domain and for atom-regions that fail to comply with any of the object models included in the ontology, the unknown object concept is introduced; this concept is assigned the minimum of the domain-, application- and user-specific importance factors. In addition, support is provided for the definition of associations between low-level descriptions and the algorithms to be applied for their extraction. In the following, a brief description of the main classes is presented.

Class **Object** is the superclass of all objects to be detected during the analysis process: when the ontology is enriched with the domain specific information it is subclassed to the corresponding domain salient objects. Class **Object Interrelation Description** describes the objects spatiotemporal behavior, while **Low-Level Description** refers to the set of their representative low-level visual features. Since real-world objects tend to have multiple different instantiations, it follows that each object prototype instance can be associated with more than one spatial description and respectively multiple low-level representations. Different classes have been defined to account for the different types of low-level information (color, shape, motion, etc.) These are further subclassed to reflect the different ways to represent such a feature (e.g. color information could be represented by any of the color descriptors standardized by MPEG-7, and also by a wide range of non-standardized descriptors). The actual values that comprise the low-level descriptors (e.g. the DC value elements, color space, etc.) are under the **Low-Level Descriptor Parameter** class.

Class **Importance Factors** is the main class containing knowledge about the coding of the given domain object. It is subclassed to classes **Domain-specific Importance Factor**, **Application-specific Importance Factor** and **User-specific Importance Factor**, which define the importance factor values I_d, I_a and I_u. During coding, these are combined using an appropriate function $f(I_d, I_a, I_u)$ to drive the coding process.

Providing domain-specific spatiotemporal information proves to be particularly useful for the identification of specific objects, since it allows discrimination of objects with similar low-level characteristics as well as of objects whose low-level features alone are not adequate for their identification. The applied spatial relations consider two-dimensional, binary relations, defined between regions with connected boundaries. In the current implementation the included spatial relations are a subset of the eight topological relations resulting from the 9-intersection model as described in earlier works on spatial relations representation and reasoning [13, 14], enhanced by the four relative directional relations, i.e. right, left, above, below. The used low-level descriptors are the MPEG-7 Dominant Color [15] and Region Shape [16] descriptors, the motion norm of the averaged global motion-compensated block motion vectors for each region blocks and the ratio between a region's area and the square of its perimeter (compactness). The employed ontology for one of the domains considered in this work, the Formula One domain, is shown in Fig. 2.

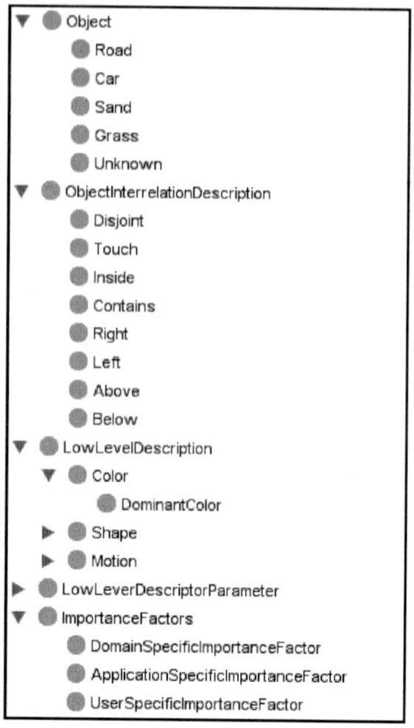

Fig. 2. Ontology for the Formula One domain

3 Knowledge-Assisted Video Analysis

3.1 Color and Motion Initial Segmentation

The color segmentation under the proposed approach serves the purpose of breaking down each video frame to connected, color homogeneous regions. This segmentation is subsequently combined with the result of motion segmentation, to generate a set of atom regions that are homogeneous in terms of both color and motion. These atom regions are the input to the genetic algorithm.

The employed color segmentation algorithm belongs to the K-means family. Clustering based on the K-Means algorithm, originally proposed by McQueen [17] is a widely used region segmentation method which, however, tends to produce unconnected regions. This is due to the classical K-Means algorithm ignoring spatial information about the intensity values in an image. In order to perform segmentation to connected regions, we use in this work a variant of the K-means algorithm, the K-Means-with-Connectivity-Constraint algorithm (KMCC). In this algorithm the spatial proximity of each region is also taken into account by defining a new center for the K-Means algorithm and by integrating the K-Means with a component labelling procedure. This ensures the connectivity of the generated regions, as discussed in [2]. The initialization of the KMCC algorithm is performed using the initialization procedure proposed in [2].

The following functions are defined to estimate the degree of matching in terms of low-level visual and spatial features respectively between an atom-region r_i and an object concept o_j.

- the interpretation function $\mathcal{I}_M^t(r_i, o_j)$, assuming that gene g_t associates region r_i with object o_j, to provide an estimation of the degree of matching between o_j and r_i. $\mathcal{I}_M^t(r_i, o_j)$ is calculated using the descriptor distance functions realized in the MPEG-7 XM and is subsequently normalized so that $\mathcal{I}_M^t(r_i, o_j)$ belongs to $[0, 1]$, with a value of 1 indicating a perfect match.
- the interpretation function $\mathcal{I}_R^t(r_i, o_j, r_k, o_l)$, which provides an estimation of the degree to which the spatial relation between atom-regions r_i and r_k satisfies the relation $\mathcal{R}$ defined in the ontology between objects o_j, o_l to which r_i and r_k are respectively mapped by gene g_t.

Since each individual represents the scene interpretation, the Fitness function has to consider the above defined low-level visual and spatial matching estimations for all atom-regions. As a consequence the employed Fitness function is defined as follows:

$$Fitness(g_t) = (\sum_{i}^{N_R} \mathcal{I}_M^t(r_i, o_m)) \prod_{i}^{N_R} \prod_{j \in S_i} \mathcal{I}_R^t(r_i, o_m, r_j, o_l) \tag{1}$$

where S_i denotes the set of neighboring atom-regions of r_i, since the spatial relations used have been defined only for regions with connected boundaries as mentioned in Sect. 2. It follows from the above definitions that the optimal solution is the one that maximizes the Fitness function. This process elegantly handles the merging of atom-regions: when the optimal or near-optimal solution is reached, any adjacent atom regions that have been estimated to belong to the same object, according to the generated solution, are assigned the same color (representing the detected object) in the segmentation mask and thus are merged without the need for any further processing. In our implementation, the following genetic operators were used: selection, crossover and mutation.

- Selection: a pair of chromosomes from the current generation are selected to serve as parents for the next generation. In the proposed framework, the Tournament Selection Operator [23], with replacement, is used.
- Crossover: two selected chromosomes serve as parents for the computation of two new offsprings. Uniform crossover with probability of 0.7 is used.
- Mutation: every gene of the processed offspring chromosome is likely to be mutated with probability of 0.008.

The iterative procedure of population evolution using the aforementioned operators continues until the diversity of the current generation is equal to or less than 0.001 or the number of generations exceeds 50.

4 Content-Adaptive Coding and Transmission

The availability of a systematic way for the semantic description of video sequences provides new means to deal with the subsequent delivery of video. In this section the reliable transmission of semantic objects is investigated. Various channel rate allocation algorithms are developed which ensure robust communication of objects over unreliable channels.

We assume that each object is compressed using embedded coding methods. This means that every object is represented using a scalable stream that can be decoded at arbitrary source rates depending on the required quality. The source coder aims to reduce the transmitted chunks of data. First it removes spatial redundancy by applying either Discrete Wavelet Transform (DWT) or Discrete Cosine Transform (DCT). Then block coding techniques or zero-trees are used in conjunction with entropy coding methods like arithmetic or run length coding method. The entropy coding improves the coding performance by exploiting the conditional probabilities. Unfortunately entropy coding results to more sensitive bitstreams to transmission errors which are encountered because of physical channel impairments and bottlenecks. Specifically, a single bit errors may cause synchronization loss between the encoder and the decoder. Thus, few errors can render large portions of the bitstream useless even if some portions are received error free.

For reliable communication advanced channel coding techniques like LDPC [24] and Turbo codes [25] should be used. The channel codes aim to enhance error resiliency by adding some redundancy to the bitstream. In theory, source and channel coding can be treated separately as stated by "Shannon" Principle. Although this is true for networks without time constraints and infinite codewords length, it does not hold for real networks. Therefore source and channel rates should be jointly determined for robust object communication. This leads to better resource exploitation and maximization of objects visual quality.

4.1 Basic Problem Formulation

In the following we try to find the optimal source and channel rate allocation. For this purpose several joint source and channel optimization algorithms appropriate for object based coding methods are proposed. The optimization algorithms seek for the optimal rate allocation policies which jointly maximize each video objects visual quality. Specifically, the algorithm first determines the channel protection of each object and right after it refines the allocated channel protection of each object. The refinement is necessary because of the progressive bitstreams. This enables our scheme to retrieve high quality objects at the receiver end.

Let denote with $D_i(\rho_i, n)$ the decoded quality of the ith object, where ρ_i is the source+channel bitrate that is devoted to the source and channel coding of the ith object, and n is Binary Symmetric Channel (BSC) bit error rate (BER) of the channel over which the video sequence is transmitted.

Based on the above, the overall distortion is the summation of independent objects distortion. Thus, it is given by the following formula

$$D = \sum_{i=0}^{N-1} f_i D_i(\rho_i, n) \tag{2}$$

where f_i represents the relative importance of the ith object, as defined in Sect. 2. Respectively, the cumulative bitrate for the coding and transmission of the video sequence is equal to

$$R = \sum_{i=0}^{N-1} \rho_i \tag{3}$$

Every objects rate is defined as

$$\rho_i = s_i + c_i \tag{4}$$

where s_i and c_i are the source and channel rates of the ith object. Therefore, the overall rate is expressed as

$$R = \sum_{i=0}^{N-1} (s_i + c_i) \tag{5}$$

The average decoded quality $D_i(\rho_i, n)$ is given by

$$D_i(\rho_i, n) = D_i \cdot (1 - P_i(s_i, c_i, n)) + P_i(s_i, c_i, n) \cdot \overline{D_i}$$

where D_i, $\overline{D_i}$ are respectively the decoded quality achieved when the source stream corresponding to the ith object is error less received and the average decoded quality of the ith object in case of error corruption. The probability $P_i(s_i, c_i, n)$ is the probability that the channel protection of an object is not enough to cope with transmission errors. In this case part of the stream is still decodable depending on the position where first error appears. Therefore, the total expected distortion in (2) can be expressed as

$$D = \sum_{i=0}^{N-1} f_i \cdot \{D_i \cdot (1 - P_i(s_i, c_i, n)) + P_i(s_i, c_i, n) \cdot \overline{D_i}\} \tag{6}$$

Using the above formulation, the allocation of bits of ontologies objects can be achieved using Lagrangian methods. Object-wise optimization of rate allocation is possible if the optimal source and channel rates for each block are known. In practice, this can be achieved using the techniques in [26], i.e. where the original problem is transformed to an unconstrained problem. The optimization minimizes an objective function F of the form

$$F = D + \lambda_L R \tag{7}$$

where D and R and are given by (6) and (5), respectively, and λ_L is a Lagrange multiplier. A similar optimization for the block-wise coding of images was presented in [27].

Provided that the channel conditions are known, i.e. the bit error rate n is known, the error probability $P_i(s_i, c_i, n)$, $i = 1, \ldots, N$ can be calculated for each object. However, the optimal determination of the channel rate for each object involves an iterative solution of (7) as λ_L converges to it's optimal value λ_L^*.

4.2 Channel Rate Optimization for Wireless Transmission Scenarios

The method discussed in the previous section provides sufficient protection for BSC channels with moderate to low BERs. Transmission over heterogeneous and wireless channels, however, is more difficult due to the fast altering channel conditions. For wireless transmission more sophisticated protection policies should be followed since conventional protection methods do not provide adequate protection. Product codes can be used instead as suggested in [28]. The product codes use two or more channel codes for improving bitstream resiliency. Specifically, they work with information arrays where the objects data is placed. They protect unequally the information by applying, in general, unequal amounts of channel protection in both the horizontal and vertical direction of the product code array. The utilized product coder [29] is illustrated in Fig. 4. In the horizontal direction of the product code array

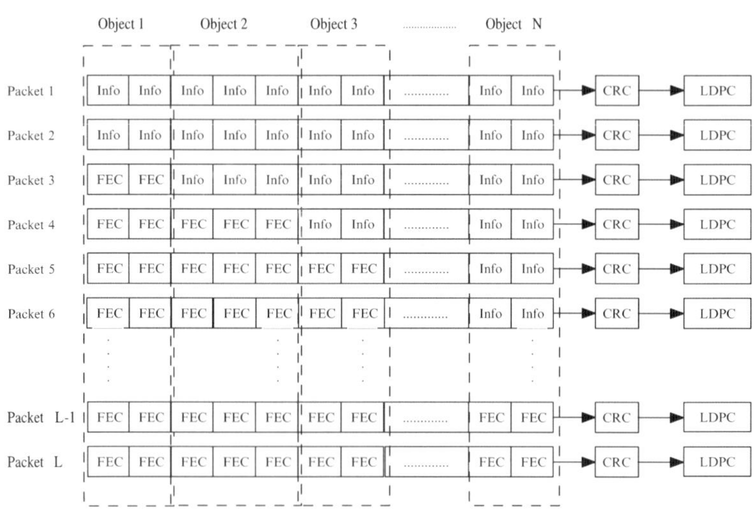

Fig. 4. Product codes based on LDPC and RS codes

it applies LDPC codes, while vertically the information is protected by Reed Solomon codes. The LDPC codes are used to cope with bit errors. Whenever LDPC codes can not recover the information of a row (the term packet is also used for every row), they sign the row as erased. The RS codes applied vertically are used as erasure correcting codes. They are able to recover all symbols from any set of rows larger or equal to the number of source symbols.

Since the objects are not of equal importance, the objects are placed into sub-arrays in the product code array and separately protected. The objects in the product code are ordered according to their relative importance. Specifically, the most significant object is first. Next the second most important object follows etc. Unequal amount of RS protection are allocated to each object according to its significance. Thus, the optimization algorithm seeks for the optimal RS protection of each object. Optimally unequal protection in both directions of the product code array should be provided. However, this is not trivial task. For this reason we equally protect all rows of the product code using LDPC codes with constant channel code rate. The code rate of the LDPC codes is determined by experimentation. For every channel condition a small set of LDPC code rates are appropriate since some code rates can not correct the bit errors while others overprotect the information.

The optimization object is the determination of optimal RS channel protection. The average decoded quality is given by

$$\overline{D} = d_0 \cdot P_d(1) + \sum_{k=2}^{N} d_{k-1} \cdot P_d(k) + (1 - P_d(N)) \cdot d_N \tag{8}$$

where d_k is the resulting distortion after the successful transmission of the first k objects. $P_d(k)$ is the probability difference between the event of losing the k_{th} object and the event of losing the $(k-1)_{th}$ object. However the distortion is not accurate metric for object based rate allocation since the semantic objects have in general different significance. It is obvious that in each object should be assigned a weighting factor w_i where $w_{k-1} \geq w_k$. This factor depends on the importance of the information carried by each object. Optimally, the weighting factor should be channel adaptive and determined for the expected channel condition. Thus, the averaged decoded quality of $\overline{D}$ when the significance of each object is taken into account is expressed as

$$\overline{D} = d_0 \cdot P_d(1) \cdot w_0 + \sum_{k=2}^{N} d_{k-1} \cdot P_d(k) \cdot w_k + (1 - P_d(N)) \cdot d_N \cdot w_N \tag{9}$$

Since the objects are placed in the product code array according to their relative importance, the most important objects should be protected by RS codes with lower code rates. This is reasonable since other allocation policies either overprotect less significant objects or does not provide enough protection. Due to the descending RS protection, i.e. for two arbitrary objects k and $k^{'}$, $k < k^{'}$, the k_{th} objects is always protected using at least as powerful

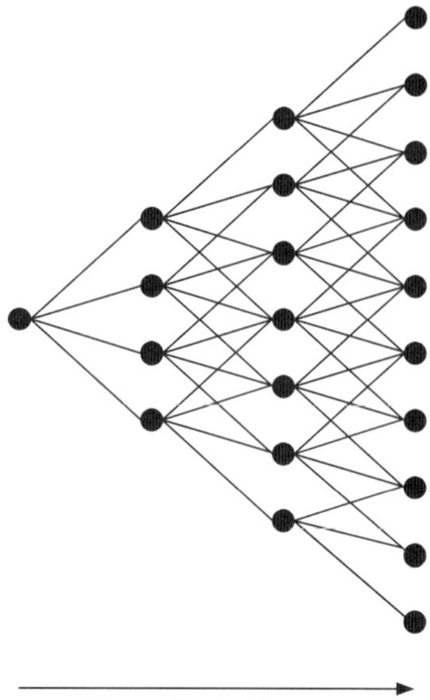

Fig. 5. Trellis diagram for RS allocation

codes as the codes used for the protection of the k'_{th} block. Therefore, the probability that all blocks are decodable is equal to the probability that the last (least protected) object is correctly decoded.

The minimization of the average distortion D can be achieved using dynamic programming based on the viterbi algorithm [30]. Trellis diagrams are used for devising a fast channel rate allocation algorithm. The trellis diagram corresponding to the minimization of (9), subject to a rate constraint, is shown in Fig. 5. Each branch in the trellis corresponds to the application of a specific RS code to an object. The algorithm first determines the RS protection of the more important object and then the respective protection of the less important objects. The nodes in the trellis represent the intermediate stages where decisions are made about the best RS allocation up to the s_{th} object protection. Paths merging in a single node correspond to allocations that yield not only the equal source rates but also equal transmission rates. Among the paths converging to a node, the path attaining the lower expected distortion is retained (survivor) while the rest are pruned. In the final stage, among the survivor paths, the one with the lowest overall expected distortion corresponds to the optimal RS allocation. The number of states in the trellis depends on the allowable RS protection levels.

The optimization algorithm finds the optimal RS protection of each object. The RS rate allocation can be further improved if the RS protection is reallocated within each object. The motivation behind this choice arises from the

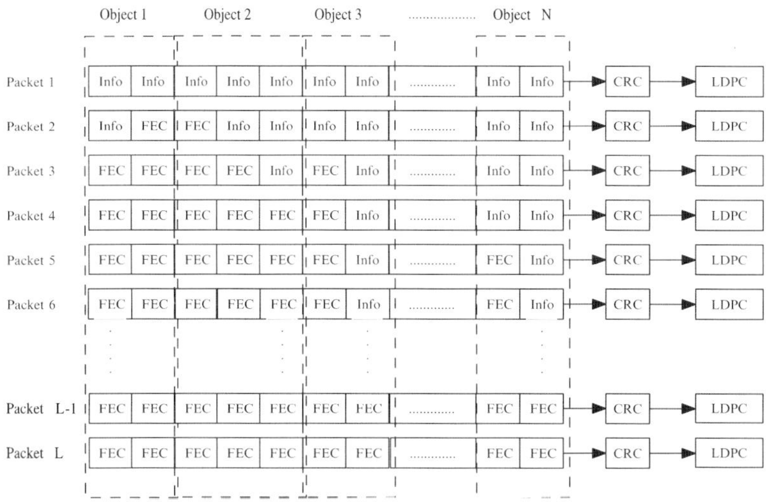

Fig. 6. Reorganized product codes based on LDPC and RS codes

fact that objects' bitstreams are progressive. Due to progressivity the first portion of each bitstream contains the most important information. Therefore, more RS symbols should be allocated to first portion of each object bitstream and fewer to the remaining portions. The modified channel protection in the product-code array is presented in Fig. 6. The reallocation procedure guarantees that the most important information is better protected. The reallocation can be performed either by dynamic programming or lagrangian techniques. For each object the optimization objective is the minimization of distortion subject to the rate determined by the inter-object rate allocation. Actually, the optimization is a succession of two independent optimization processes. Although the above procedure results to a sub-optimal allocation policy in practice yields to excellent allocations.

5 Experimental Results

The proposed approach was tested on a variety of Formula One and Tennis domain videos. Indicative analysis results for these two domains are shown in Fig. 7 and 8. As illustrated in these figures the analysis output is a segmentation mask outlining the semantic interpretation, i.e. a mask where different colors representing the objects defined in the ontology are assigned to each of the produced regions. The objects of interest included in each domain ontology are illustrated in Table 1. For all experimental domains, the low-level descriptors values included in the corresponding knowledge base were extracted from a training set of manually annotated frames.

The time required for performing the previously described tests was between 5 and 10 seconds per frame, excluding the process of motion

Input Images Segmentations Interpretations

Fig. 7. Sample frame analysis results for the Formula One domain

Input Images Segmentations Interpretations

Fig. 8. Sample frame analysis results for the Tennis domain

Table 1. Formula One and Tennis domain objects of interest

Domain	Concept
Formula One domain	Road
	Car
	Sand
	Grass
	Unknown
Tennis domain	Field
	Player
	Line
	Ball
	Wall
	Unknown

information extraction via block matching for which efficient and inexpensive hardware implementations exist [31]. More specifically, the time to perform pixel-level segmentation was about 2 s, while the time required by the genetic algorithm to reach an optimal solution varied depending on the number of atom-regions and the number of spatial relations. The extraction of the low-level and spatial descriptions is performed before the application of the genetic algorithm. In general, the proposed approach proved to produce satisfactory results as long as the initial color-based segmentation did not result in under-segmentation, i.e. in the generation of one or more atom-regions, each of which represented two (or more) of the objects depicted in the frame, or parts of such multiple objects.

The subsequent coding and transmission of the test set comprising the aforementioned Formula One and Tennis domain videos, using the techniques described previously, demonstrated the merits of the proposed novel content-aware approach. Specifically, objects designated as being of higher importance in our experiments (e.g. the Car and Player objects for the Formula One and Tennis domains, respectively) were consistently decoded at higher quality than other objects in the ontology that were designated as being of lower importance. This resulted in higher overall perceptual quality of the decoded video, after simulated transmission of it over unreliable channels, when the proposed content-aware coding approach was employed. The perceptual quality of the resulting video was subjectively judged in all cases by a pool of evaluators. This experimental outcome is consistent with our assumption that different spatio-temporal parts of the video stream have different importance for the viewer, according to their semantic content, and consequently contribute with different weights to the judging of the overall perceptual quality of the video, thus rendering meaningful the use of the semantic analysis result as a basis for the unequal protection of them at the encoding stage.

6 Conclusions

In this chapter, a knowledge-assisted domain-specific video analysis approach, which exploits the fuzzy inference capabilities of a genetic algorithm, is employed for supporting content-adaptive video coding and transmission. Domain knowledge includes both low-level features and spatial relations of video content for the purpose of analysis, as well as domain-, application- and user-specific importance factors associated with each domain concept to guide coding and transmission. The developed domain ontology provides a flexible conceptualization that allows the easy addition of new concepts, low-level and spatiotemporal descriptors, as well as updated importance factors, thus supporting different abstraction levels and flexible adaptation of the analysis and coding process to different domains, applications and users.

Acknowledgements

This work was supported by the European Commission under contracts FP6-001765 aceMedia and FP6-027685 MESH, and by the GSRT under project DELTIO.

References

1. U. Gargi, R. Kasturi, and S.H. Strayer, "Performance Characterization of Video-Shot-Change Detection Methods," *IEEE Transaction on Circuits and Systems for Video Technology*, vol. 10, no. 1, pp. 1–13, Feb 2000.
2. V. Mezaris, I. Kompatsiaris, and M.G. Strintzis, "Still Image Segmentation Tools for Object-based Multimedia Applications," *International Journal of Pattern Recognition and Artificial Intelligence*, vol. 18, no. 4, pp. 701–725, June 2004.
3. V. Mezaris, I. Kompatsiaris, N.V. Boulgouris, and M.G. Strintzis, "Real-Time Compressed-Domain Spatiotemporal Segmentation and Ontologies for Video Indexing and Retrieval," *IEEE Transaction on Circuits and Systems for Video Technology*, vol. 14, no. 5, pp. 606–621, May 2004.
4. S.-Y. Chien, Y.-W. Huang, B.-Y. Hsieh, S.-Y. Ma, and L.-G. Chen, "Fast Video Segmentation Algorithm with Shadow Cancellation, Global Motion Compensation, and Adaptive Threshold Techniques," *IEEE Transactions on Multimedia*, vol. 6, no. 5, pp. 732–748, Oct 2004.
5. T. Gruber, "A Translation Approach to Portable Ontology Specifications," *Knowledge Acquisition*, vol. 5, no. 2, pp. 199–220, 1993.
6. J. Assfalg, M. Berlini, A. Del Bimbo, W. Nunziat, and P. Pala, "Soccer Highlights Detection and Recognition using HMMs," in *IEEE International Conference on Multimedia and Expo (ICME)*, pp. 825–828, 2005.
7. L. Zhang, F.Z. Lin, and B. Zhang, "Support Vector Machine Learning for Image Retrieval," in *IEEE International Conference on Image Processing (ICIP)*, pp. 721–724, 2001.

8. J. Hunter, J. Drennan, and S. Little, "Realizing the Hydrogen Economy through Semantic Web Technologies," *IEEE Intelligent Systems Journal - Special Issue on eScience*, vol. 19, pp. 40–47, 2004.
9. A. Yoshitaka, S. Kishida, M. Hirakawa, and T. Ichikawa, "Knowledge-Assisted Content-Based Retrieval for Multimedia Databases," *IEEE Multimedia*, vol. 1, no. 4, pp. 12–21, Winter 1994.
10. G. Tsechpenakis, G. Akrivas, G. Andreou, G. Stamou, and S.D. Kollias, "Knowledge-Assisted Video Analysis and Object Detection," in *Proceedings of European Symposium on Intelligent Technologies, Hybrid Systems and their implementation on Smart Adaptive Systems (Eunite02)*, Algarve, Portugal, Sep. 2002.
11. M. Ramesh Naphade, I.V. Kozintsev, and T.S. Huang, "A Factor Graph Framework for Semantic Video Indexing," *IEEE Transaction on Circuits and Systems for Video Technology*, vol. 12, no. 1, pp. 40–52, Jan. 2002.
12. S. Dasiopoulou, C. Doulaverakis, V. Mezaris, I. Kompatsiaris, and M. G. Strintzis, "An Ontology-Based Framework for Semantic Image Analysis and Retrieval," *Semantic-based Visual Information Retrieval, Y.-J. Zhang (Ed.)*, 2007.
13. D.A. Randell, Z. Cui, and A.G. Cohn, "A Spatial Logic Based on Regions and Connection.," in *KR*, pp. 165–176, 1992.
14. M.J. Egenhofer and R.D. Franzosa, "Point Set Topological Relations.," *International Journal of Geographical Information Systems*, vol. 5, pp. 161–174, 1991.
15. B.S. Manjunath, J.-R. Ohm, V.V. Vasudevan, and A. Yamada, "Color and Texture Descriptors," *IEEE Transaction on Circuits and Systems for Video Technology, special issue on MPEG-7*, vol. 11, no. 6, pp. 703–715, June 2001.
16. M. Bober, "MPEG-7 visual shape descriptors," *IEEE Transaction on Circuits and Systems for Video Technology, special issue on MPEG-7*, vol. 11, no. 6, pp. 716–719, June 2001.
17. J. McQueen, "Some Methods for Classification and Analyis of Multivariate Observations," in *5th Berkely Symposium on Mathematical Statistics and Probability*, vol. 1, pp. 281–296, 1967.
18. T. Yu and Y. Zhang, "Retrieval of Video Clips using Global Motion Information," *Electronics Letters*, vol. 37, no. 14, pp. 893–895, July 2001.
19. F. Precioso, M. Barlaud, T. Blu, and M. Unser, "Smoothing B-Spline Active Contour for Fast and Robust Image and Video Segmentation.," in *ICIP (1)*, pp. 137–140, 2003.
20. MPEG-7 XM software, "http://www.lis.ei.tum.de/research/bv/topics/mmdb/e_mpeg7.html,".
21. M. Jacob, T. Blu, and M. Unser, "An Exact Method for Computing the Area Moments of Wavelet and Spline Curves.," *IEEE Transactions on Pattern Analalysis and Machine Intelligence*, vol. 23, no. 6, pp. 633–642, 2001.
22. M. Mitchell, *An introduction to Genetic Algorithms*, MIT, 1996.
23. D. Goldberg and K. Deb, *A comparative analysis of selection schemes used in genetic algorithms*, pp. 69–93, G. Rawlins, 1991.
24. R. Gallager, "Low-Density Parity-Check Codes.," *IRE Transactions on Information Theory*, pp. 21–28, Jan 1962.
25. C. Berrou and A. Glavieux, "Near Optimum Error Correcting Coding And Decoding: Turbo Codes," *IEEE Transaction on Communications*, vol. 44, no. 10, pp. 1261–1271, Oct. 1996.

26. Y. Shoham and A. Gersho, "Efficient Bit Allocation for an Arbitrary Set of Quantizers," *IEEE Transaction on Acoustics, Speech, Signal Processing*, vol. 36, pp. 1445–1453, Sep. 1988.
27. N.V. Boulgouris, N. Thomos, and M.G. Strintzis, "Transmission of Images Over Noisy Channels Using Error-Resilient Wavelet Coding and Forward Error Correction," *IEEE Transaction on Circuits and Systems for Video Technology*, vol. 13, no. 12, pp. 1170–1181, Dec. 2003.
28. N. Thomos, N.V. Boulgouris, and M.G. Strintzis, "Wireless Image Transmission Using Turbo Codes and Optimal Unequal Error Protection," *IEEE Transaction on Image Processing*, vol. 14, no. 11, pp. 1890–1901, Nov. 2005.
29. N. Thomos, N.V. Boulgouris, and M.G. Strintzis, "Product Code Optimization for Determinate State LDPC Decoding in Robust Image Transmission," *IEEE Transaction on Image Processing*, vol. 15, no. 8, pp. 2113–2119, Aug. 2006.
30. G.D. Forney, "The Viterbi Algorithm," *Proceedings of IEEE*, vol. 61, no. 3, pp. 268–278, Mar. 1973.
31. J.-C. Tuan, T.-S. Chang, and C.-W. Jen, "On the data reuse and memory bandwidth analysis for full-search block-matching VLSI architecture," *IEEE Transaction on Circuits and Systems for Video Technology*, vol. 12, no. 1, pp. 61–72, Jan. 2002.

Mobile Photography within a Social Context

Take my Snapshot and Get to Know my Friends with PhotoFOAF

Kris Luyten, Ruben Thysv, Kristof Thys, and Karin Coninx

Hasselt University, transnationale Universiteit Limburg, Expertise Centre for Digital Media, IBBT, Wetenschapspark 2, B3590 Diepenbeek, Belgium
Kris.Luyten@uhasselt.be, Kristof.Thys@uhasselt.be,
Karin.Coninx@uhasselt.be, Ruben.Thys@student.uhasselt.be

Summary. Mobile digital photography has become a ubiquitous service: nearly every mobile phone is equipped with a camera. Huge collections of pictures are created with these phones that are hard to structure in a usable way. Problems in saving, managing and exchanging digital pictures are becoming typical obstacles that need to be tackled in order to have a usable system. We present PhotoFOAF: a new approach to mobile imaging that exploits the social context of a picture as an important part of the metadata. We present an approach that allows users to insert social metadata in pictures while capturing them. The resulting collections of captured pictures can be managed, searched and exchanged more easily by taking benefit of the personal social relevance of a picture. PhotoFOAF implements an end-to-end process to accomplish this and it exploits, enhances and stimulates social relationships between users during all the different steps.

1 Introduction

Mobile digital photography is available to the majority of people nowadays. In combination with popular online photo management and sharing Web sites, this results in a huge and growing archive of photographs available on the Internet. Metadata that helps to structure this collection in a reasonable way is still lacking. Adding metadata to a photo is still a manual and time-consuming activity, certainly since it is not always possible to do so when taking the photo (e.g. when the user is mobile and there is insufficient time to annotate the photo). There is a large body of research that discusses automatic metadata annotation. Most of this work concentrates on the use of diverse types of sensors that enable the system to automatically extract metadata from the sensor readings. Well-known examples are the place the photo has been taken, the orientation and direction of the camera while taking the photo, the time, the user that took the photo,...

K. Luyten et al.: *Mobile Photography within a Social Context*, Studies in Computational Intelligence (SCI) **93**, 241–260 (2008)
www.springerlink.com

Next to annotating pictures, sharing pictures with friends usually takes a lot of time, and can be quite tedious. Furthermore, searching the collection of available photos of one person is hard but still feasible, but searching aggregated collections of different persons is practically impossible with current solutions. On top of this, linking photos to other data sources, such as including photos in a blog entry, could also benefit from a more powerful approach. This "connected data" approach is exactly one of the key benefits that the semantic Web offers [3]. Since there is a clear evolution toward very large distributed personal photo repositories new tools that help us to manage and link these photo's are required. Figure 1 shows the typical evolution that can be observed on a user's desktop after buying a new digital camera.

In contrast with other current approaches which focus on the inclusion of geo-spatial metadata to photos, PhotoFOAF supports *social context metadata.* PhotoFOAF offers a convenient way of capturing metadata, but it differs from other approaches because it allows to make use of a social network of

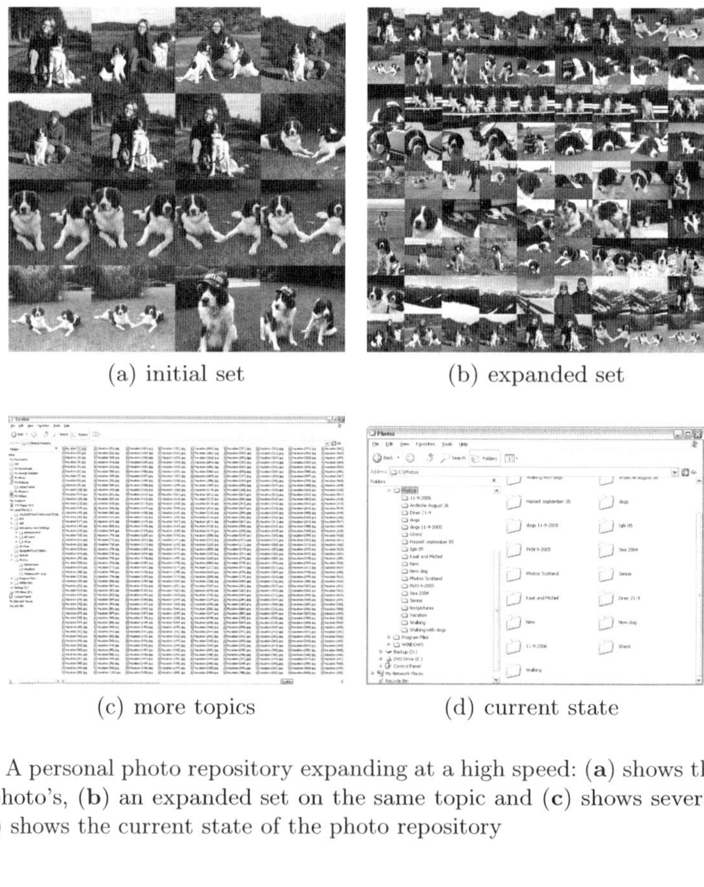

(a) initial set (b) expanded set

(c) more topics (d) current state

Fig. 1. A personal photo repository expanding at a high speed: (**a**) shows the initial set of photo's, (**b**) an expanded set on the same topic and (**c**) shows several topics and (**d**) shows the current state of the photo repository

Listing 1. Short extract from a FOAF profile

```
rdf:nodeID="me">
    <foaf:name>Kris Luyten</foaf:name>
    <foaf:givenname>Kris</foaf:givenname>
    <foaf:family_name>Luyten</foaf:family_name>
    <foaf:nick>Kris</foaf:nick>
    <foaf:mbox_sha1sum>
        489670ef790723bac81d5f72b65268ff2fb57959
    </foaf:mbox_sha1sum>
    <foaf:homepage rdf:resource="http://research.edm.uhasselt.be/kris"/>
    <foaf:workplaceHomepage rdf:resource="http://www.edm.uhasselt.be"/>
    <foaf:schoolHomepage rdf:resource="http://www.uhasselt.be"/>
    <foaf:interest dc:title="human-computer interaction" />
    <foaf:interest dc:title="photo sharing" />
    <foaf:knows>
        <foaf:nick>unzip</foaf:nick>
        <rdfs:seeAlso
        rdf:resource="http://photofoaf.edm.uhasselt.be/rthys/data/foaf" />
    </foaf:knows>
    <foaf:knows>
        <foaf:nick>OscarCookie</foaf:nick>
        <rdfs:seeAlso
        rdf:resource="http://photofoaf.edm.uhasselt.be/oscarcookie/data/foaf" />
    </foaf:knows>
</foaf:Person>
```

users that take photos or are depicted on photos. To accomplish this we use profiles based on the Friend-Of-A-Friend (FOAF) specification [5], a semantic Web profile that describes individual users and their relationships with other users. Listing 1 shows an extract from a FOAF profile. The FOAF profile describes the connections with other users positions the photos in the social network and indicates the *social relevance* of photos with respect to other users (in the social network). People appearing on photographs have a depicting relationship with the image. We use these depicting relationships to explore new human relationships between the depicted persons, often referred to as co-depiction (`http://rdfweb.org/2002/01/photo/`). This allows advanced possibilities such as querying photographs using the relationships between persons on the photo.

The remainder of this chapter is structured as follows: We start in the next section with the description of related work where we differentiate our approach with other similar initiatives. Next we describe the system in a case study using a concrete example. Section 4 explains the different processes in the PhotoFOAF system, where Sects. 5, 6 and 7 reveal specific aspects of the system. Finally, Sect. 9 discusses the future work and gives the conclusions.

2 Related Work

Online photo sharing services like Flickr and Zooomr have become very popular tools to store, share and search for photos. These Web sites make use of some type of folksonomy [15] to annotate photographs. The annotation of a photo in these tools is expressed as a list of tags from a folksonomy. In many cases adding the metadata can be done while uploading the pictures or afterwards when accessing the photos with a desktop computer. This results often in general and imprecise metadata [20] due to various reasons such as difficulties to remember the exact context when the photo was taken, lack of automatic metadata capturing tools and simply the burden of adding metadata afterwards.

Mobshare [21], a mobile phone picture sharing system, tries to simplify the exchange of pictures between mobile users. Blogging and visual representations of your photoalbum are simplified but metadata is neglected: time of capturing and a label that describes the picture are considered to be sufficient.

MMM2 [7, 12] tries to address the metadata capturing problem by using a complex algorithm. The algorithm uses the spatial, temporal and social context of the user and uses semi-automatic techniques. A list of possible subjects that can be used as metadata for the picture is generated. It is stated that mobile camera phones enable a new approach to annotating media by facilitating metadata capture at the time of image capture, facilitated by an automatized and collaborative metadata capturing technique [23]. However, MMM2 has less support then PhotoFOAF for searching photographs, blogging and building community. Social relationships that result from photographs are not exploited.

Möller et al. [17] emphasize the growing importance of semantics to support blogging and propose a framework that allows more powerful searches in published content [17]. They propose a structured schema of metadata that can be used including FOAF. Our approach only looks at the use of social networks to determine the social relevance of photo's with respect to a user and would fit nicely in their framework because of the usage of FOAF, an RDF-based format, for this purpose.

PhotoStuff [8, 9] presents a toolkit that allows users to markup multimedia using concepts from ontologies that are associated with regions of an image. As an example on their Web site (`http://www.mindswap.org/2003/PhotoStuff/`), they use FOAF as an ontology. PhotoStuff allows to select a region of a photo and add metadata to that region. This is a process that mostly happens online and is probably time consuming when a large set of photos has to be annotated. However, this tool, which allows to add more specific metadata afterwards, could be a complementary tool to PhotoFOAF.

Flipper is an experimental system, built around design themes for such systems with regard to their enhancement of social presence [6]. Field studies show that people share twice as many photos and have an increased sense of social presence when using the experimental system. The increase in photo sharing indicates the interest in combining social presence and photo sharing while still offering a light-weight and mobile tool to support photo sharing.

The previously described approaches use semi-automatic metadata capture or allow to add metadata afterwards using folksonomies. Our approach allows to add social metadata at the time of image capture based on the users' FOAF-profile and its location in the social network. Afterwards the social relationships resulting from the pictures and the relationships in the users' profile are combined to create a personalized adapted experience on the PhotoFOAF Web site.

3 Case Study

In this section we start with a scenario that involves taking and sharing pictures, to describe the workflow of the system. The optimizations PhotoFOAF offers are emphasized throughout this section. Figure 2 shows the different steps that are required to capture and share photos with friends without PhotoFOAF support:

Step 1 Steven decides to take a picture of Sarah and Kris in the park.

Step 2 At home, he uses specialized software to upload his pictures from his camera to his personal computer. He only finds time to add the name of the picture (Sarah_and_Kris_Park.jpg) as metadata.

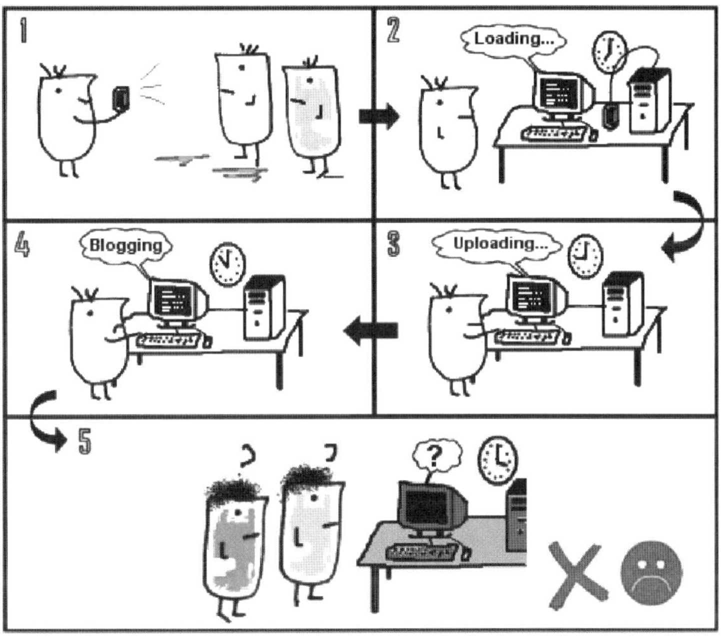

Fig. 2. The traditional way to capture and publish pictures

Step 3 Afterwards he uploads his pictures to his personal Web space, again using a different application which takes a few minutes.

Step 4 Steven posts a blog entry on his own Web space, and links to the pictures, using the URL from his own Web space.

Step 5 Monica and Emily, two friends of Sarah that are unacquainted with Steven, try to find the pictures of Sarah and Kris. However search terms "Sarah" or "Kris" do not return the required results. In this situation, Steven has to take the initiative to share the pictures with Monica and Emily, two people he might not even know. The situation described here requires a *push*-mechanism to share pictures with other people.

Figure 3 describes how the previous scenario is enhanced by using PhotoFOAF, with fewer steps:

Step 1 Steven takes pictures of Sarah and Kris in the park using his smartphone. Using the PhotoFOAF application, he can add "Sarah", "Kris" and "Going out with friends" as metadata to each picture easily right after making the photograph. The pictures together with their metadata are uploaded immediately when network connection is available. After the upload, the photographs are available according to the privacy rights on the PhotoFOAF Web site.

Fig. 3. New approach of taking and sharing pictures using PhotoFOAF

Step 2 Monica and Emily can find the pictures they want to see on the PhotoFOAF Web site, using the search term "Sarah Abrahams". Pictures are obtained using a *pull*-mechanism and Steven or even Sarah can blog easily, even on pictures they did not take themselves.

4 Using PhotoFOAF for Social Photo Management

PhotoFOAF supports an end-to-end process to capture, annotate, upload, manage and query pictures. The first stage is always the image capturing activity, for which PhotoFOAF provides client side software that manages capturing, annotating and uploading pictures. We label this the "acquisition" or "production" stage. The PhotoFOAF acquisition client runs on a mobile device. The second stage is the "consumption" stage where an end-user interacts with a Web server that manages the photo collection. PhotoFOAF has a clear division between both, meaning the mobile client responsible for photo capturing has no functionality for photo browsing. However, once the photo's are stored on a PhotoFOAF server they can be browsed by a regular browser on the mobile device provided the mobile device has network connectivity. The mobile acquisition client does make the transition from the producer to the consumer stage as smooth as possible, by providing an automatic upload tool that stores the photos together with the metadata on a PhotoFOAF server once a network connection becomes available.

4.1 Photo Production Phase

There are several important steps involved in capturing digital photos. To enhance the user experience afterward, capturing related metadata is also an important part. During the "production phase", the user makes use of a mobile device that includes a camera to capture photos. The mobile device allows us to provide additional software for adding metadata to the photos (Sect. 7 discusses this further into detail).

Several steps are involved for producing the photos:

1. Use mobile device to take a picture
2. When taking pictures, users get the opportunity to add metadata in a convenient way to their photographs:
   (a) Using the users' profile, the system suggests people and/or interests that can be added as metadata. A simple selection on the screen of the mobile device adds the metadata.
   (b) Additional metadata can be added if a picture is taken. We provide an intuitive and easy to use selection mechanism that lowers the threshold of doing this instantaneously. The system gives suggestions for appropriate metadata information based on a folksonomy that is stored on a central server. The folksonomy can also be used offline by the mobile

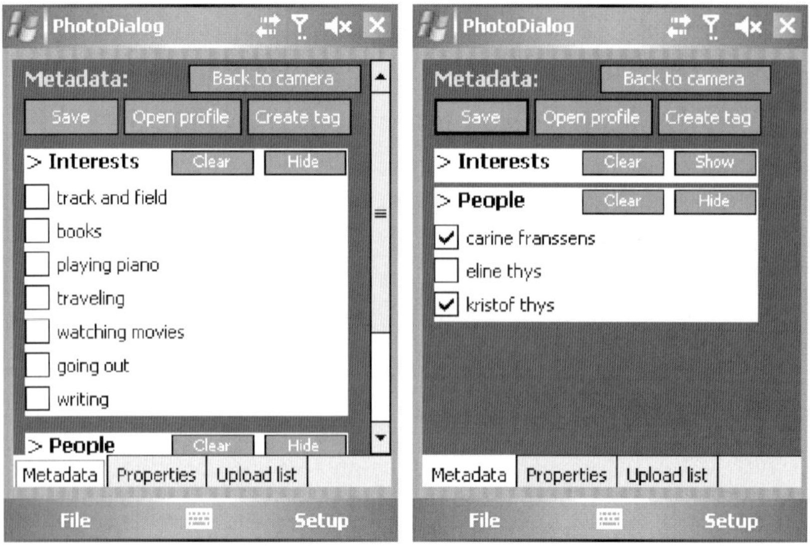

Fig. 4. Adding metadata to pictures with the mobile client

device. The user benefits from more accurate and complete metadata this way. The metadata gets updated automatically as people take more pictures and the folksonomy supported by the Web site is used by other people.

3. The user comes into reach of a wireless network: a connection will be automatically made and the photos and the related metadata are stored on the PhotoFOAF server.

Shneiderman et al. [22] clearly state users are most likely not motivated to annotate photos. A well designed interface to annotate the photos can make a big difference. For this reason we integrated the annotation of the photo in the production stage and provide an appropriate interface on the mobile device to support this task (Fig. 4).

4.2 Photo Consumption Phase

After the digital photos and their metadata are stored on a server, the user can exploit the social relationships to query the set of photos. This user-centric view of photo collections is a powerful approach toward photo management.

The following steps provide the reader with a more detailed overview of how the PhotoFOAF system helps to manage the photo collection:

1. The social relationships that can be derived from pictures e.g. persons `A` and `B4` appear on the same photograph (Fig. 5) are added to the FOAF-profile of `A` and `B4`, often referred to as co-depiction.

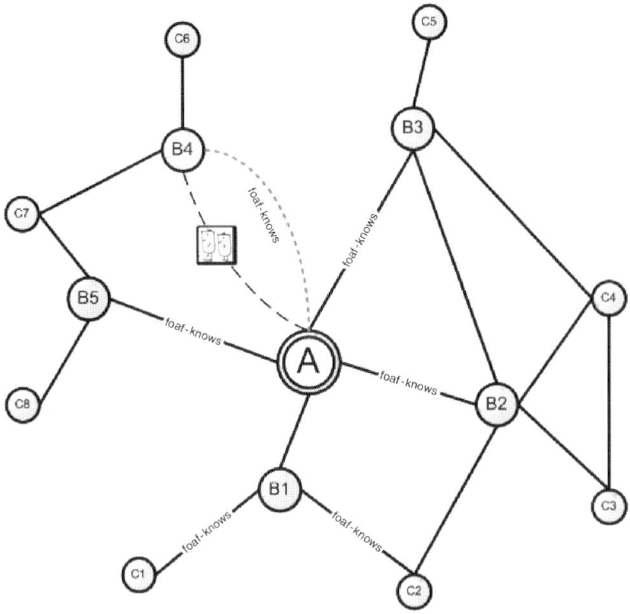

Fig. 5. Social network created by FOAF-knows relationships and expanded by taking pictures

2. Users can set privacy rights on all of their pictures using the social network available. E.g. in Fig. 5, user `A` can easily allow access to all `B`-users and disallow access to all `C`-users. Furthermore, people can give access to pictures based on the level of social relevance, e.g. the social relevance between `A` and `B2` is larger than the relevance between `A` and `B1` (because of the relation between `B2` and `B3`, where `B3` is also related to `A`). Calculation of social relevance is described into detail in Sect. 5.
3. Pictures are uploaded to the central server and relationships are stored in the database and in the picture itself. Central storage solves the problem of photographs that quickly become scattered over numerous different media, going from harddisk to DVD and flashdisk. Data transferal between these media can be hard and time consuming and not all media are reliable over time.
4. Users visiting the Web site get a personalized interface thanks to the relations in the FOAF-profile and the social relationships deducted from the pictures. E.g. in Fig. 5 user `B2` will also be presented with pictures of `B4`, a relationship derived from a combination of FOAF-knows and co-depiction.
5. Searching becomes a lot more convenient thanks to the directly added metadata combined with the afterwards added folksonomy metadata. Complex searches of pictures, that contain people with certain

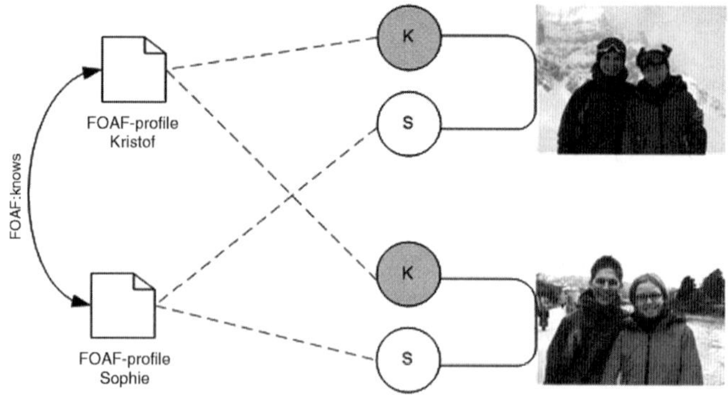

Fig. 6. Two persons appearing together in two different images as indicated by the related FOAF profiles

characteristics or even with certain social connections ("who she/he knows"), become easy from a user's perspective.

6. Blogging becomes easier thanks to the centralized storage of pictures, blogging on your own pictures, or pictures that depict you or your friends becomes straightforward.

Figure 6 demonstrates how co-depiction works: a photo has a set of related FOAF profiles of people that appear on the photo. In contrast with PhotoStuff [8, 9], PhotoFOAF does not allow to link regions inside a photo with metadata. It would be to cumbersome for a mobile user to do this type of metadata annotation on a mobile device.

5 Determining Social Relevance

PhotoFOAF identifies people by their Friend-Of-A-Friend (FOAF, [5]) profiles. This type of profile contains information about the social relations of the user next to traditional information that identifies various aspects of the user (such as name, address, hobbies, hair color...). Most people tend to take pictures of their friends, their activities and pictures of random objects related with their interests, which can all be described in detail in the FOAF-profile. This way the FOAF-profile gives a suggestion of the metadata that can be added to the picture. Listing 1 shows an example of a FOAF profile that is used by the PhotoFOAF system.

FOAF adheres a Resource Description Framework vocabulary (RDF, `http://www.w3.org/RDF/`) for describing people and social networks. Its RDF syntax allows us to easily query relations between different FOAF files and build a graph of related FOAF profiles. In this example we use the del.icio.us social bookmarking tags to specify the interests of a person: because of the fussiness of what human interests exactly are we use this folksonomy since it is a list of terms created by the end-users themselves. Although tags can have different names, sometimes they mean the same or are closely related with each other. Even if the tags do not exactly match (a problem stated in [20]), an algorithm can be used to calculate how closely two tags are related and use a threshold to determine whether or not the two tags indicate the same interest.

One approach to calculate whether two different tags have a similar meaning uses the del.icio.us social bookmarking service. This can be done as follows: suppose we have two tags a and b that have different names. A set U_a of x URLs that is associated with the tag a is retrieved, where x is the number of URLs in U_a. A second set U_b of URLs tagged with b is retrieved, also containing x elements. A simple approach is to count the number of elements in the intersection $U_a \bigcap U_b$. If the number of elements relative to both sets exceeds a threshold t (with $0 \leq t \leq 1$) we can assume tags a and b have a similar meaning. More granularity can be provided and the level of certainty (LoC) whether two tags share the same meaning can be measured by using the equation (assuming an union has no duplicate elements):

$$LoC = \frac{\#(U_a \bigcap U_b)}{\#(U_a \bigcup U_b)} \tag{1}$$

From this equation it is clear LoC has a value in the same range of t, more specific $0 \leq LoC \leq 1$. If LoC equals 1, we consider tags a and b as fully equivalent tags (they are synonyms, have exactly the same meaning and there are no ambiguities). If $LoC \leq t$, where t is experimentally defined or manually set by the user, we consider tags a and b are insufficiently related to draw any conclusions. If LoC equals 0, tags a and b have no relation at all. To summarize, two images I_1 and I_2 have the same context or they have a set of associated tags that are equivalent two-by-two. Equivalence between tags is found by using a threshold t and the level of certainty LoC for two tags, which gives us:

$$0 \leq LoC \leq t \rightarrow 0$$
$$t \leq LoC \leq 1 \rightarrow 1$$

In this case 1 indicates tags a and b are considered equivalent, 0 indicates tags a and b are not sufficiently related to be considered equivalent. The approach described here allows users to use their own tags (e.g. in different languages) and is still able to find whether these tags can be considered equivalent or not.

Since it is not the scope of this work to have complex matching algorithms, we did not explore this in further detail. One could imagine more complex algorithms that take more parameters into account. Since this matching process can be done without additional interaction and the results can be stored at the server, this does not influence the responsiveness of the system while being used by the end-user. Using FOAF a social network of related FOAF-profiles

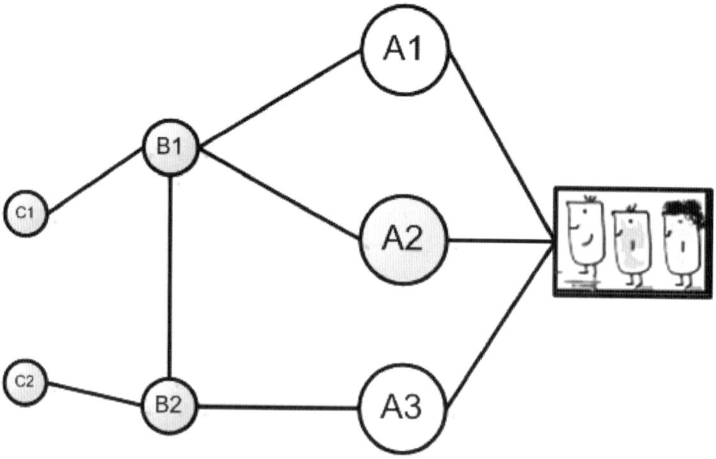

Fig. 7. Social relevance network

is built and this network is used as input for calculating social relevance, a measure to detect how interesting a photograph is to a PhotoFOAF user.

Social relevance is a measure that defines the social importance of a picture to a person. Formula (2) is an heuristic used to calculate the social relevance (SR) of an image, where $|p_i|$ is the length of unique path i, and n is the number of unique paths between the node and the picture.

$$SR = \begin{cases} 0 & \text{if number of edges} = 1; \\ \frac{\sum_{i=0}^{n} |p_i|}{n^2} & \text{if number of edges} > 1. \end{cases} \quad (2)$$

The following list shows the results of social relevance calculation for the graph shown in Fig. 7 (the smaller the result, the larger the social relevance):

$$\begin{aligned} \text{A1,A2,A3} &\rightarrow 0 \\ \text{B1} &\rightarrow \tfrac{7}{9} \\ \text{C1} &\rightarrow \tfrac{10}{9} \\ \text{B2} &\rightarrow \tfrac{8}{9} \\ \text{C2} &\rightarrow \tfrac{11}{9} \end{aligned}$$

Each person in the graph is mapped onto a social relevance value by formula 2. To persons `A1`, `A2` and `A3` the image is highly socially relevant, since these are the persons that appear on the image itself. The image is more socially relevant to `B1` in comparison to `B2`, since `B1` knows two persons on the picture in first degree, and `B2` only knows one.

PhotoFOAF exploits social awareness by using FOAF as the core of the system. FOAF profiles can be created from scratch with a simple user interface as shown in Fig. 8, or existing FOAF-profiles can be imported and used as user profile in PhotoFOAF. Furthermore, the created FOAF profile can

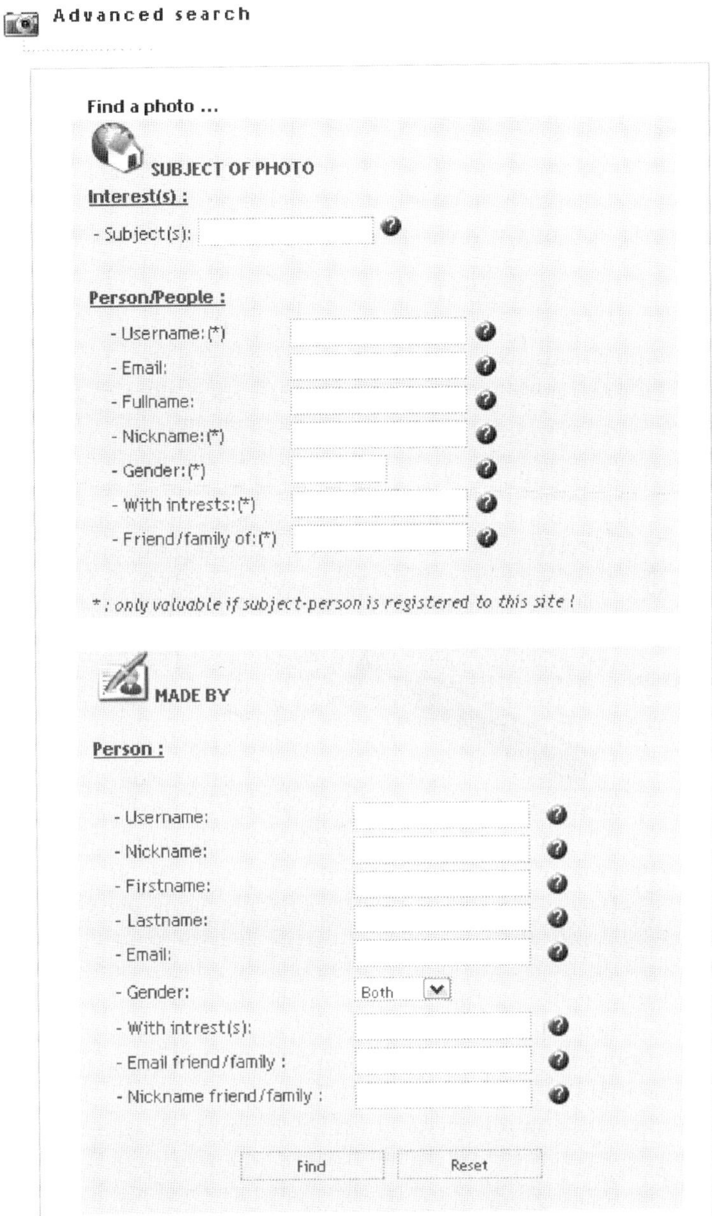

Fig. 8. Web-based form to enter a FOAF profile

be reused in other FOAF-compliant Web sites or applications. This way, we create a centralized user profile, that is kept up to date automatically. After registering (by creating or importing a FOAF profile) users can log in and make use of all the PhotoFOAF features. Afterwards, FOAF-profiles can be updated when new interests are discovered while taking pictures and new social relationships can be detected through co-depiction. This update can be done automatically (by inference in the FOAF graph), semi-automatically (after the users' approval) or manually through a form-based interface such as the one shown in Fig. 8.

As more people adopt FOAF by creating and publishing their own FOAF profile, a social network of related people is created, making it more interesting to use applications like PhotoFOAF. Unlike other approaches, e.g. [20], the persons that are co-depicted on a picture and the creator of the picture are identified unambiguously. Each FOAF profile is uniquely identified by a URI, but for performance reasons we parse the FOAF file, integrate it in the picture and add the information in a custom database to allow easy server-side querying. We are still evaluating the use of an RDF-based triple store to hold the information for querying, such as Jena [16], Redland (`http://librdf.org/`) or Semweb (`http://razor.occams.info/code/semweb/`).

6 Embedded Metadata

All pictures produced by PhotoFOAF contain social metadata described in RDF format. Metadata is added to a picture during the photo production stage, presented in Sect. 4. As discussed in Sect. 5 a collection of photographs results in a metadata graph, that is obtained by aggregating all RDF structures that are related with the pictures. RDF Query languages, like SparQL (http://www.w3.org/TR/rdf-sparql-query/) and RDQL (http://www.w3.org/Submission/RDQL/), can be used to query these graphs. Furthermore, RDF makes it possible to manually add and query new metadata types. Imagine for example adding a metadata type weather [18], making it possible to search all sunny or rainy pictures.

Adobe's Extensible Metadata Platform (XMP) shows it is feasible to add extensible metadata to media objects [1].

Because pictures contain metadata that refers to metadata contained in other documents (e.g. a FOAF profile "knows" another FOAF profile), complex queries are possible that consider the aggregated metadata of a collection of images. More specific, the linkage of a persons FOAF-profile to a picture results in a extended metadata graph. The primary metadata only consists of a link to the FOAF-profile itself. But the secondary metadata, such as the information in the FOAF-profile (e.g. hobbies and preferences) and the links to other FOAF-profiles with their own details, offers very powerful solutions. Finding a picture using several characteristics such as the owner of the image, people standing on the picture and their relations in the social network, shows the power of combining semantic Web techniques with digital image collections. Our technique makes it for example possible to find all pictures of two female friends that both like to play tennis.

7 System architecture

As shown in Fig. 9, the top-level of the PhotoFOAF system consists of two important parts: the acquisition client running on a smartphone (used in the photo production stage from Sect. 4) and a consumption client (used in the photo consumption stage, Sect. 4), which is Web-based. The smartphone application allows users to take pictures with the PDA. Immediately after taking a picture the user is presented a list of possible metadata tags (shown in Fig. 4), based on the user's FOAF-profile (shown in Listing 1). The metadata suggestion is divided into two parts, interests (based on *foaf:interest*) and people (based on *foaf:knows*). When acquaintances are selected (part on the right of Fig. 4), the link to their FOAF-profile is added to the picture, resulting in a large collection of secondary metadata, as discussed in 6. When taking multiple pictures within a certain time span, the adding of metadata is highly simplified by reusing previous selections. Afterwards, the pictures are uploaded to the PhotoFOAF server, and directly visible on the Web site, according to selected privacy rights. These rights can be based on the FOAF-profile, pictures can for example be accessed by your *foaf:knows* contacts only. Metadata is stored in the database as well as in the picture itself using the Exchangeable Image File Format (EXIF). PhotoFOAF stores *interests* and *people* metadata in the EXIF 'subject'-field, in order to make pictures with their metadata also usable in other applications. After uploading new pictures, the PhotoFOAF system detects new *foaf:knows* relationships based on co-depiction and adds these to the database and FOAF-profile.

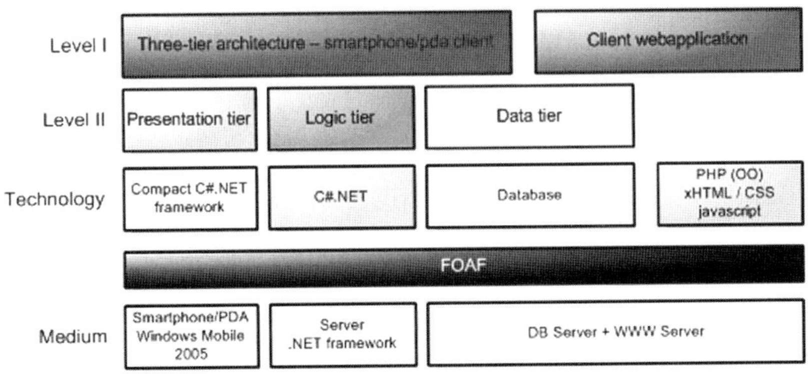

Fig. 9. System architecture PhotoFOAF

Fig. 10. A screenshot of the Web-based interface

The client Web-based application allows users to view photographs immediately after they have been uploaded. PhotoFOAF also integrates blogging facilities that allow to combine blog posts with pictures based on the social relevance. Bloggers get suggestions of pictures to include in their blog post. PhotoFOAF makes it easy to select your own pictures, pictures that depict you and pictures your friends took lately. Thanks to the power of FOAF and the social relevance calculations (Sect. 5) in combination with co-depiction, searching pictures with related people on it becomes straightforward. The client Web-based application also uses the social relevance calculation to adapt the interface according to the users' profile. E.g. socially relevant images and blogs are shown on the first page as shown in Fig. 10 (indicated with red rectangles).

8 Social Network Visualization Experiments

Recognizing the social relationships and navigating a social network is a reoccurring problem when using a social network to structure data on a Web site. There are several high quality visualizations of social networks available but they mainly focus on presenting the social network itself. We are interested in presenting the *social context* of a photo, thus combining the presentation of a set of related photographs with the presentation of a subpart of the social network that connects these photos. Such visualizations have proven to be of

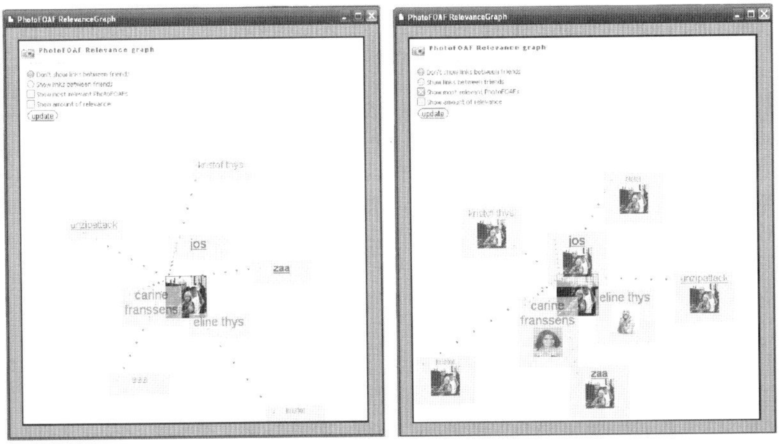

Fig. 11. Visualization of the local social context of a set of photographs

great importance for users to explore and find information in a social network and to understand the structure of the social network.

Figure 11 shows the social network visualized in combination with the relevant photo's embedded in the visualization. A node identifies a user, an edge between nodes a co-depiction relationship between users. A link between two nodes can be annotated with a set of photos that depict both persons identified by the nodes connected through the edge. The visualization is roughly inspired by Vizster [10], but simplified. In the future we hope to implement this feature with the prefuse toolkit [11] that supports powerful interactive visualization capabilities.

Considering the very large sets of pictures, the amount of items and relations that need to be visualized poses a problem. Consider a set of pictures taken during a biology field trip of typical class of a secondary school. Wit current day digital cameras,this would result in a large amount of picture of the field trip and almost every co-depiction is a meaningful relationship since these are students from the same class. One way to cope with the large amount of pictures being displayed to present the social context and navigate through it, is the use of a focus+context visualization strategy. For this purpose we need to identify which information exactly is required to give a comprehensible context to the information items (here pictures) we want to focus on. A focus+context visualization often uses a distorted graphical view on the data [14]. Well-known examples and variations of such visualizations are the magic lenses [4] and fisheye views [14]. Currently, we did not provide such a visualization for PhotoFOAF but consider this as the best option to cope with the large amount of pictures that should be displayed. We refer to the work of Schneiderman and Bederson for more information of appropriate visualizations of large amounts of pictures, e.g. [2] and [13].

9 Conclusions and Future Work

In this chapter, we presented PhotoFOAF, a system to easily capture, upload, manage, browse and share photographs. PhotoFOAF takes into account the social relationships of the users during all stages of usage. PhotoFOAF builds on FOAF-technology and uses this to provide a social context of a picture and to create a community for managing digital images. Instead of just linking pictures with FOAF profiles, we embed parts of the social metadata inside the picture in order to transport metadata and data as a whole. Semantic Web techniques make collections of pictures with social metadata easy to structure, query and manage the picture collection.

Adding metadata to pictures is always an extra burden however. A useful extension would be the automatic discovery of FOAF profiles that need to be linked to a picture while capturing it. With Bluetooth and related protocols becoming available on most mobile devices, we might be able to use this to gather all FOAF profiles of people that are in the vicinity. Unfortunately, there are still no sufficient standardized means to accomplish this. We also plan to investigate more appropriate visualization techniques on the PhotoFOAF mobile clients and the community Web site. The former needs an optimal interface to support on-the-spot tagging and FOAF selection, while the latter could benefit from techniques that present the different facets of picture collections (e.g. picture-persons relationships) in a smart way so these collections can be browsed according to one of these facets.

The fully operational prototype Web site is available at `http://photofoaf.edm.uhasselt.be` and is currently available for public usage and testing. We believe our approach can tackle problems that often occur with picture management [19]. While users currently prefer to have a large overview of pictures and prefer to order pictures by timestamp, PhotoFOAF leverages part of these problems by providing a social context to each picture in a collection. This context allows for a personal view on the collection, and view the collection from different angles. Complex queries based on the metadata can be composed with relative ease. For example; one could look for a new tennis partner and ask photoFOAF to show all pictures that contain 2 males and 2 females that play tennis together and from which you know at least one person. The rdfweb co-depiction example used in the introduction is a good example of using depicted people as a basis for browsing through a set of pictures; PhotoFOAF could even be used to present a photo collection ordered by co-depiction.

Acknowledgments

Part of the research at EDM is funded by EFRO (European Fund for Regional Development), the Flemish Government and the Flemish Interdisciplinary Institute for Broadband Technology (IBBT).

References

1. Adobe Systems. Adobe XMP for Creative Profesionals (white paper), 2006. `http://www.adobe.com/products/xmp/`.
2. Benjamin B. Bederson. PhotoMesa: a zoomable image browser using quantum treemaps and bubblemaps. In *UIST '01: Proceedings of the 14th annual ACM symposium on User interface software and technology*, pages 71–80. ACM, 2001.
3. Tim Berners-Lee. *Weaving the Web: The Original Design and Ultimate Destiny of the World Wide Web*. Harper San Francisco, 1999.
4. Eric A. Bier, Maureen C. Stone, Ken Pier, William Buxton, and Tony D. DeRose. Toolglass and magic lenses: the see-through interface. In *SIGGRAPH '93: Proceedings of the 20th annual conference on Computer graphics and interactive techniques*, pages 73–80. ACM, 1993.
5. Dan Brickley and Libby Miller. *FOAF Vocabulary Specification*. World Wide Web, 2005. `http://xmlns.com/foaf/0.1/`.
6. Scott Counts and Eric Fellheimer. Supporting social presence through lightweight photo sharing on and off the desktop. In *Proceedings of the SIGCHI conference on Human factors in computing systems*, pages 599–606. 2004.
7. Marc Davis, Nancy Van House, Jeffrey Towle, Simon King, Shane Ahern, Carrie Burgener, Dan Perkel, Megan Finn, Vijay Viswanathan, and Matthew Rothenberg. MMM2: mobile media metadata for media sharing. In *CHI '05: CHI '05 extended abstracts on Human factors in computing systems*, pages 1335–1338. ACM, 2005.
8. Christian Halaschek-Wiener, Jennifer Golbeck, Andrew Schain, Michael Grove, Bijan Parsia, and Jim Hendler. Photostuff – an image annotation tool for the semantic web. In *4th International Semantic Web Conference*, 2005.
9. Christian Halaschek-Wiener, Andrew Schain, Jennifer Golbeck, Michael Grove, Bijan Parsia, and Jim Hendler. A flexible approach for managing digital images on the semantic web. In *5th International Workshop on Knowledge Markup and Semantic Annotation*, 2005.
10. Jeffrey Heer and Danah Boyd. Vizster: Visualizing Online Social Networks. In *IEEE Symposium on Information Visualization (InfoVis 2005)*, pages 32–39. 2005.
11. Jeffrey Heer, Stuart K. Card, and James A. Landay. Prefuse: a toolkit for interactive information visualization. In Gerrit C. van der Veer and Carolyn Gale, editors, *Conference on Human Factors in Computing Systems, CHI 2005*, pages 421–430, 2005.
12. Nancy Van House, Marc Davis, Morgan Ames, Megan Finn, and Vijay Viswanathan. The uses of personal networked digital imaging: an empirical study of cameraphone photos and sharing. In *CHI '05 extended abstracts on Human Factors in Computing Systems*, 2005.
13. Jack Kustanowitz and Ben Shneiderman. Hierarchical Layouts for Photo Libraries. *IEEE MultiMedia*, 13(4):62–72, 2006.
14. Y. K. Leung and M. D. Apperley. A review and taxonomy of distortion-oriented presentation techniques. *ACM Trans. Comput.-Hum. Interact.*, 1(2):126–160, 1994.
15. Adam Mathes. Folksonomies: Cooperative Classification and Communication Through Metadata, 2004. `http://www.adammathes.com/academic/computer-mediated-communication/folksonomies.html`.

16. Brian McBride. Jena: Implementing the RDF Model and Syntax Specification. In *Semantic Web Workshop, WWW2001*, 2001.
17. Knud Möller, Uldis Bojars, and John G. Breslin. Using Semantics to Enhance the Blogging Experience. In York Sure and John Domingue, editors, *The Semantic Web: Research and Applications, 3rd European Semantic Web Conference*, pages 679–696, 2006.
18. Mor Naaman, Yee Jiun Song, Andreas Paepcke, and Hector Garcia-Molina. Automatically generating metadata for digital photographs with geographic coordinates. In *WWW Alt. '04: Proceedings of the 13th international World Wide Web conference on Alternate track papers & posters*, pages 244–245. ACM, 2004.
19. Kerry Rodden and Kenneth R. Wood. How do people manage their digital photographs? In *CHI '03: Proceedings of the SIGCHI conference on Human factors in computing systems*, pages 409–416. ACM, 2003.
20. Risto Sarvas. User-centric Metadata for Mobile Photos. In *Pervasive Image Capture and Sharing Workshop at Ubicomp 2005*, 2005.
21. Risto Sarvas, Mikko Viikari, Juha Pesonen, and Hanno Nevanlinna. Mobshare: controlled and immediate sharing of mobile images. In *MULTIMEDIA '04: Proceedings of the 12th annual ACM international conference on Multimedia*, pages 724–731. ACM, 2004.
22. Ben Shneiderman, Benjamin B. Bederson, and Steven M. Drucker. Find that photo!: interface strategies to annotate, browse, and share. *Communications of the ACM*, 49(4):69–71, 2006.
23. Anita Wilhelm, Yuri Takhteyev, Risto Sarvas, Nancy Van House, and Marc Davis. Photo annotation on a camera phone. In *CHI '04 extended abstracts on Human factors in computing systems*, 2004.

An Adaptive Semantic Framework to Support Multiple User Viewpoints over Multiple Databases

Stefan Poslad and Landong Zuo

Department of Electronic Engineering, Queen Mary University of London, London E1 4NS, UK
Stefan.poslad@elec.qmul.ac.uk, Landong.zuo@elec.qmul.ac.uk

Summary. A global domain Ontology is often proposed as a means to semantically interlink local heterogeneous database resources within a domain. The main advantage is that users need only understand one interface pertaining to the global ontology rather than to understand each of the local database resource interfaces. However, a global domain Ontology may contain considerable complexity, in order to cover increasing numbers of heterogeneous local data resources. Further, some users may not be familiar with the full global model in order to pose and refine their queries under the strict constraints of a single conceptualisation. The development and application of a flexible approach to derive more targeted views, oriented to different types of users and applications, is reported here. The key design challenge tackled here was to derive and constrain user views of the global view, in such a way that the quality of the global view model is maintained in each user view, and that each user view can be accurately mapped to the global view.

1 Introduction

Semantic models offer powerful benefits to mediate between and to reason about heterogeneous data sources, when data from multiple data resources needs to be combined. A single domain conceptualisation and semantics can be projected as a single information viewpoint over multiple information sources, targeted to answer core queries by domain expert type users. This has the advantage that it avoids the necessity of users needing to be familiar with the stored data schema and associated query interfaces. However, some users may not be familiar with the full semantic model in order to pose and refine their queries under the strict constraints of a single conceptualisation. The single conceptualisation may also not adequately capture the operational nuances, side effects and errors in answering data queries. The development and application of a more flexible approach to derive more targeted views, adapted for different types of users and applications, is reported in this chapter. This

S. Poslad and L. Zuo: *An Adaptive Semantic Framework to Support Multiple User Viewpoints over Multiple Databases*, Studies in Computational Intelligence (SCI) **93**, 261–284 (2008)
www.springerlink.com

research work has been undertaken as an extension to the Environmental Data Exchange Network for Inland Water or EDEN-IW project [14] that supports distributed information retrieval of environmental information.

1.1 Information Heterogeneity

At an abstract level, developing a semantic framework to support multiple user data views over a global data view conceptualisation is similar to supporting a global data view over local data views of data resources. This involves handling the different data heterogeneities that can be present between local data views and between the local views and the global view. Data heterogeneities will likely increase when supporting multiple user viewpoints, particularly when these are developed independently and support individual data modelling and presentation preferences.

Information heterogeneities can be classified into syntactic, terminology, semantic and pragmatic heterogeneity [9]. This chapter considers a subset of the classification and slightly extends the concept and introduces a four-type classification with specific definitions:

- *Syntactic heterogeneity*: Syntactic heterogeneity concerns all forms of heterogeneity that depend on the choice of the representation format including different language representations and logical structures. For example, query expression varies from language structure, query syntax and corresponding constraint relations.
- *Terminology heterogeneity*, encounters all forms of mismatches that are related to the process of naming the entities (e.g. individuals, classes, properties and relations) that occur in the domain Ontology covers all the naming differences according to linguistic representation such as synonyms and homonyms. The same named concept may have different meanings and be related differently to other concepts (homonyms) and different named concepts may have the same meaning (synonyms). Terminology heterogeneity also concerns other linguistic problem such as different abbreviations, spelling and multi-lingual support.
- *Semantic heterogeneity*: Semantic heterogeneity concerns mismatches in the content involving the meaningful interpretation of domain conceptualisation regarding different semantic expressions. Semantic heterogeneity indicates the interpretation of terminology meaning including any subsumption relations over data instances in a certain knowledge domain.
- *Semiotic or pragmatic heterogeneity* concerns the discrepancies that arise when different individuals and communities interpret the same Ontology in different ways, in different contexts. Semiotic heterogeneities can be further subdivided to support user view customisation along the dimensions of coverage, granularity and perspective [9]. Coverage identifies user interests as a portion of the domain knowledge. Granularity gives the level of a hierarchy for a user's understanding of the knowledge representation. Perspective indicates the beliefs or notions to convey the hypotheses, facts,

and assumptions that form the contents of a viewpoint, independent of how the beliefs are expressed [4].

The existence of such heterogeneities hampers the accessibility of information and information sharing. It can result in imprecise answers to user queries across multiple applications because of incompatibilities and information loss during the query translation, result aggregation and data harmonisation processes.

1.2 Outline of the Chapter

The remainder of this chapter is organised as follows: Sect. 2 reviews related work. Section 3 gives an overview of the method for viewpoint modelling and adaptation used. Section 4 describes the implementation. Section 5 presents the discussion and further work.

2 Related Work

This survey focuses on semantic based approaches to support multiple user viewpoints representations during user centred querying and results adaptation. First, the issue of using a global logical view to combine local data resource views and how this can be extended to support multiple user views over the global view is considered. Then some of the fundamental issues that arise when combining different semantic models that make different assumptions are considered.

2.1 Logical Data Views vs. User Views

In the layered view adaptation [2], [15], views are decomposed into layers so that a specific change of data schema and objects can be limited within a certain scope and the reusability of information system can be maximised. Sheth and Larson [21] have proposed a five-layer architecture for federated database systems. Two types of mapping approach are used to conduct schema translation between layers: explicit mapping which gives exact mapping relations between corresponding entities, and constrains rule mapping which specifies rules for how schema constraints are used during mapping. In Adnani et al. [2], a multi-layered functional data model is presented to support multiple data views in the Geospatial Information domain. Cross layer schema are derived using inheritance and class composition and mapped using equivalent and aggregation relations. The Object view [22] considers multiple classification problems, i.e. one single instance may belong to multiple information classes using a role mechanism. Ribière and Dieng-Kuntz [20] propose a multiple viewpoint solution to reconcile diverse developer interpretation processes upon the domain knowledge where a viewpoint is defined as different terminologies and

Table 1. Comparison of multiple viewpoint systems with respect to the types of information heterogeneities

Surveyed system	Terminologic	Syntax	Semantic	Semiotic	Coverage	Granularity	Perspective	Derivation approach for user views
Sheth [21]	√		√	√			√	*SQL view, term mapping*
Adnani [2]				√		√	√	*Instance category*
Ribière [20]	√			√	√	√		*Term mapping, instance category*
Jung [15]	√	√		√	√		√	*Instance category, concept composition*
Benchikha [7]		√	√	√	√	√		*SQL view, instance category, role*
Calvanese [10]	√		√	√	√		√	*SQL view, instance category*

instance category relations within the domain knowledge. DIF (Design Information Framework) [15] supports a translated, collaborated and integrated multiple user viewpoints via a consistent and explicit representation of metadata and data type information. Basic types such as attribute, entity, time and act are defined as basic units in concept layer. The basic units are used to build high level concepts of function, goal and profile in the viewpoint layer representing the different intensions and acts of project groups. Benchikha and Boufaida [7] extend the single integrated database schema at multiple levels: role, view and viewpoint in order to improve the representation flexibility and access interoperability.

Table 1 classifies the related work in terms of semiotic, terminological, syntactic or semantic with respect to information heterogeneities as defined earlier. In addition, the surveyed approaches are analysed in terms of their support for viewpoint adaptation at different semiotic levels of representations with respect to coverage, granularity and perspective. Some common drawbacks of surveyed systems given in Fig. 1 are:

- Semantic model development tends not to focus on adapting queries and their results to user views but focuses more on projecting a single domain view.
- There is a lack of support for a comprehensive set of data heterogeneities.
- There is a lack of support for user view models that combine coverage, granularity and coverage.

There are a lack of explicit well-defined processes to adapt information retrieval to user view and to support changing views and domain models.

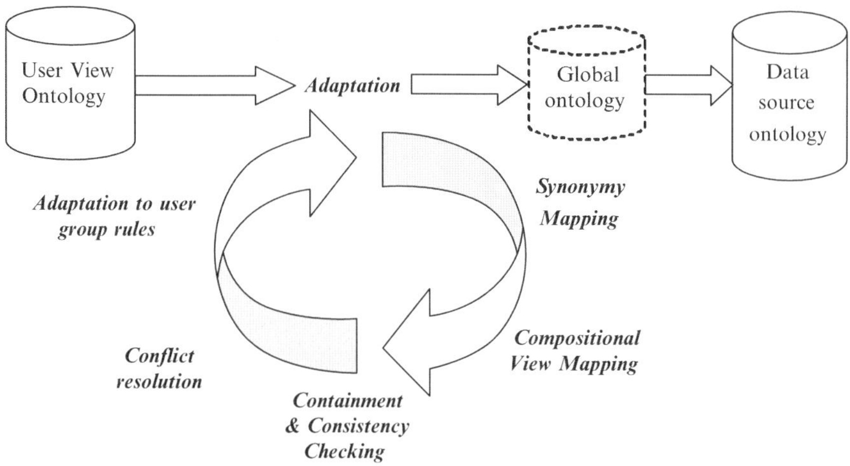

Fig. 1. Viewpoint adaptation using ontology mediation

2.2 Integrating Semantics, Rules, Logic and Databases

There are some fundamental issues when dealing with Semantic Web and Database Integration that have not been explicitly raised during approaches that focus on data integration. This is mainly because the approaches discussed so far haven't used the semantic model to attempt to reason about the relational model schema such as what can be said about queries that return no results but rather reason about derived semantic conceptualisations of the relational model schema. The main challenge here is that the database relational models operate under a closed world assumption whereas the Semantic Web operates under an open world assumption. Reasoning under an open world assumption can infer information about a closed world model that conflicts with it or causes the data integrity of the closed world model to be reduced. Reasoning using Semantic Web models that involves rules as constraints, are often needed in practice, but there is still a lack of agreement about whether any single way to interlink rule-based models, logic models and conceptual models is more beneficial than any other way. As a result there is as yet still no standard way to interlink these models in a system. This challenge and some projects that have attempted to address this issue are now discussed in more detail.

So called Strong Ontology models based on Description Logic or DL are briefly reviewed here in order to lead to problems of combining open world and closed world semantic models. A DL-based information system comprises two components, the TBox and the ABox. The TBox introduces the terminology, i.e. the vocabulary of an application domain, while ABox contains assertions about named individuals in terms of this vocabulary. The ABox of Ontology model can be used to represent the instance data in a relational database and their unary and binary relations. The semantics of relations amongst concept, property and individual are imposed in TBox, which does not exist in the relational data model.

An important semantic distinction between Ontology and database is the "open-world" and "close-world" assumption, i.e. ABox of Ontology indicates one subset of the information model satisfying the TBox, it may be incomplete as more assertions can be inserted at any time, whereas a database is often assumed to contain a complete data model. As a consequence, absence of information in a database is interpreted as negative information, while absence of information in an ABox only indicates lack of knowledge [3]. Inconsistencies can arise when system conducting information reasoning within a knowledge model.

A relational view over a database indicates a designated query to retrieve a data instance according to the schema, whereas an ontological viewpoint contains more content involving different representations of conceptual structures and relations upon the domain knowledge. Since each view over database can be derived from original database schema via relational operations of projection, selection, join and rename in a straightforward way, see virtual table [16], this ensures the consistency between two data models during the process of derivation. However, an ontological viewpoint may contain open information about domain knowledge, where representation confliction may exists in terms of different types of information heterogeneities.

Data retrieval via a conceptual viewpoint based upon an Ontology model can be reduced to SQL queries over relational view if no further information inferences are involved. Hence a database of tuple-sets is considered to be a closed subset of ABox assertions in the knowledge domain. Thereafter well-established relational view approaches for database can be adopted here to support data queries posed on different viewpoints.

Reasoning is an important feature in a description logic framework and is used to support information inference. Logical relational schema data integration assumes that each source is basically a database, i.e. a logical theory with a single model, such an assumption is not made in Ontology integration, where a local Ontology is an arbitrary logical theory, and hence can have multiple models [10].

Damasio et al. [11] consider closed-world reasoning in which negation-as-failure is the only negation mechanism supported. They then propose two major extensions to the semantics to better support open world reasoning: answer set semantics and well-founded semantics with explicit negation. These can be used to support two forms of negation, weak and strong. Weak negation is similar to the mechanism of non-monotonic negation-as-failure, and strong negation allows the user to express negative knowledge and is monotonic. The combination of these two forms of negation allow the distinction between open and closed predicates, is illustrated in the paper but practical computational versions of their model are not given. Pan and Heflin [18] present, DLDB, a knowledge base system that extends a relational database management system with additional capabilities to store and query DAML+OIL inference. The most significant aspect of theory approach is the use of a description logic FaCT reasoner to pre-compute the subsumption hierarchy in order to flatten it to be stored in relational database issues. However, they do not consider closed world vs. open world semantic issues.

In addition, since the early 1990s there has been much work that preceded the uptake of the semantic web. Description logic based approaches that have looked at extending database models to support logic based reasoning about database data, i.e. deductive databases [12]. Perhaps the most well-known of these is based upon Datalog but there are many others [8]. Datalog aims to separate out facts that relate to a closed world in an extensional database part from inference rules that can derive other data from facts in an open world in an intensional database part. It extends relational models but without negation and recursion support in the inference. Patel-Schneider and Horrocks [19] consider Datalog in relation to classical logics such as First-Order Logic and Description Logics, and their use as underlying formalisms for the Semantic Web. They argue however, that although they are similar, they have important differences at more expressive language levels and that after considering some of these differences, they argue that, although some of the characteristics of Datalog have their utility, the open environment of the Semantic Web is better served by standard description logics.

De Bruijn et al. [13] have undertaken a recent survey of the attempts by the Semantic Web community to combine classical, first-order logic and various description logics, with rule languages rooted in logic programming such as SWRL (a Semantic Web Rule Language Combining OWL and RuleML), dl-programs, and DL + log and highlight that they differ significantly in the way Ontologies combine with (non-monotonic) rules bases. However, each of these approaches overcomes the differences between the first-order and rules paradigms (open vs. closed domain, non-unique vs. unique names, open vs. closed world) in different ways and vary with respect to the ease of implementation and availability of reasoning techniques. There is as yet no clear recommendation for combining logic and rules. Ng [17] also considers the issues of combined Open and Closed world and Rules and queries in a common model using two use cases from Industry. He has outlined the necessity of a notion of negation-as-failure within these use cases and proposes an extension of OWL that supports two additional operators to support this and has provided an implementation approach using only open-world type query answering services.

3 A Multiple Viewpoint Semantic Framework

The adaptation of retrieved information to produce customised information viewpoints is akin to matching service provider service descriptions or capabilities to user preferences. To facilitate this matching, a shared conceptualisation between the information provider and the information user is useful. A key design issue is how many combinations of user preference and provider capabilities are available in the matching process, how the user preferences and provider capabilities are created and imported and how a domain independent model that supports customised user viewpoints of the data can be constructed. An information viewpoint represents a given representation for some reality of interest, among the set of possible representations [22]. It reflects the understanding, classification and operations upon the domain knowledge pertaining to a particular usage. Users can pose query on viewpoint without the necessity of understanding the whole domain conceptualisation. A viewpoint representation may contain conceptual information such as the user terminology, conceptual structures and logic rules representing information interests at a specified level of abstraction.

In the EDEN-IW project [14], the individual distributed heterogeneous databases are virtually integrated at global level using a common Ontology. The global conceptualisation is represented in concepts, relations and constraints with respect to TBox terminologies in the DL model, whilst the real instance ABox set is still stored in the underlying databases. By this means, no standard reasoning model can be applied here to support viewpoint projection and semantic adaptation to user preference as no instance data is available in either the common Ontology or the viewpoint Ontology.

The research work presented in this chapter has developed a computational approach to reason the semantic adaptation under such circumstances to support information retrieval using user terminology, conceptual structure and logic rule. The viewpoint project is conducted with finite logic operations with semantic mapping to resolve different types of information heterogeneities with consistent query transformation with respect to the instance domain in databases.

Generally, (Information) service providers publicly make available their descriptions and capabilities with respect to a finite set of dimensions that they think the typical users understand. But providers are often unable to understand or model the variability and range of usages at database service design time. Hence user viewpoints vary even if the provider capabilities are fixed. To support generic adaptation, customisation along the dimensions of coverage, granularity and perspective is chosen [9]. Coverage actually identifies user interests as a portion of the domain knowledge. Granularity gives the level of a hierarchy for a user's understanding of the knowledge representation. Perspective indicates the beliefs or notions to convey the hypotheses, facts, and assumptions that form the content of a viewpoint, independent of how the beliefs are expressed [4].

To support and computationally constrain the options for domain specific adaptation, users are stereotyped in the IW domain to scientist, aggregator and policy maker. Adaptation becomes a matching process to orientate the queried data to the user preferences. We anticipate that this matching is complex. It most likely involves reasoning because of the various heterogeneities and the variety of mappings that must be handled. To some extent, the heterogeneities that are supported increase in complexity in the framework developed in this chapter as multiple abstractions and their conceptualisations associated with the different user viewpoints must also be handled.

A general framework has been developed to manage query transformation and conceptual adaptation across multiple viewpoints based upon the Semantic Web and Description Logic models. This framework supports the following properties:

- Multiple viewpoint vocabularies, categorisations and user stereotyping or groups
- User preference and usage descriptions
- Viewpoint generation
- User queries answered in relation to an associated user view

Potentially, there are a large number of arbitrary user views that can be projected from the global schema and these would lead to an infinite number of mappings between the user view and global view. In order to constrain the user view to global view mappings, only a limited number of stereotype user or group user views are supported that individual users of these groups are allowed to adapt in a finite set of ways. Individual user views containing user preferences are considered as a further projection from the stereotype viewpoints.

User queries posed with respect to a particular viewpoint need to be aligned to the common domain conceptualisation. Conversely, the results of user queries need to be aligned to the user presentation preferences associated with the viewpoint. The approach given here focuses on viewpoint development, viewpoint management, query transformation and result adaptation in order to enhance the usability of the information system.

3.1 Viewpoint Conceptualisation

The semantics of the multiple viewpoint framework (MVF) is given as $A = \{V, G, M\}$, where $V = \{v_1, v_2 \dots v_n\}$ is a set of viewpoints where each element $v_i, 0 < i < n$, indicates a unique conceptual viewpoint. G is a common conceptual representation within a single domain, where atomic terminologies are defined in the conceptualisation as primary building blocks and further decomposition of terminologies is not allowed in G. $M = \{\overrightarrow{m}\}$ is the union of semantic mapping relations. $\overrightarrow{m}$ is a vector of semantic mapping, $\overrightarrow{m} = \{m_1, m_2 \dots m_n\}$, where each element $m_i, 0 < i < n$, indicates a set of mapping from v_i to G, $q^{Vi} \rightarrow q^G$, indicating a query over viewpoint v_i is equivalent to a semantically corresponding query over G.

A viewpoint model v_i conforms to a conceptual interpretation I_v over a knowledge domain Δ, denoted to be Δ_i^{IV}, whereas G gives another interpretation over knowledge domain Δ, denoted as Δ^{IG}. Since the viewpoint v_i is a model of G, it is supposed that interpretation of v_i over Δ must be a subset of interpretation G regarding their coverage over the same domain.

The viewpoint v_i is a valid model of G, if v_i that satisfies all constraints of G via conceptualisation mapping m_i: the instance set of v_i and its semantic correspondence in G via mapping relation m_i should satisfy the constraints in model G.

Thereafter, the DL-based theories are adopted in a multiple viewpoint system to support the semantic mapping and queries posed on multiple viewpoint system can also be formalised in terms of conjunctive type queries.

3.2 Semantic Mapping

Semantic mapping is a core technique used in the MVF. A view-based approach enables explicit semantics to be used to evaluate a corresponding query and to retrieve the corresponding results from a collection of databases. The approach mainly focuses on resolving terminology heterogeneity, semantic heterogeneity and semiotic heterogeneity between stereotyped viewpoint representation and common conceptualisation. The mapping is specified as the enumerated functions to map the corresponding entities and relations, describing the equivalent information sets across Ontologies. The semantic equivalence between queries is validated by domain experts. The semantic mapping falls into four categories regarding different types of heterogeneities that are handled: terminology mapping, value mapping, conceptual mapping and context mapping.

The terminology mapping resolves all synonym and homonym relations across user viewpoint. An independent multi-lingual glossary is developed by the domain experts indicating possible terms and their mapping in relation to their particular usage in stereotyped user viewpoints.

The value mapping encounters the expression of terms and instantiated values in relation to different coding formats or unit conventions. In such case, an intermediate ontology is introduced to support the indirect mapping between terms and values. For example, the semantics of measurement units and their compounds are explicitly defined in the Unit Ontology, where conversion operations between different units are able to be computed.

The conceptual mapping aims at overcoming the issue of semantic heterogeneities. The approach introduces a user-oriented, independent, viewpoint of the domain knowledge that is referred to as a TBox Ontology $V = (T_v, R_v, C_v)$, where viewpoint V is constituted with terminology set T_v, a unary or binary relation set R_v and constraints set C_v that is associated with T_v and R_v. The Semantic meaning of the viewpoint is indexed via an interpretation I of the ontological terms T_v and R_v into a non-empty domain knowledge set Δ^I, where $T_v^I \subseteq \Delta^I$, and $R_v^I \subseteq \Delta^I \times \Delta^I$. The mapping information has to be developed to connect semantic correspondences between an end-user viewpoint V and a common Ontology model $G = (T_g, R_g, C_g)$. If G is regarded as a primary viewpoint V_g of conceptual terms in domain Δ, a viewpoint V defines a compositional terminology and a relation set of domain knowledge with additional constraints and rule specification upon V_g that is similar to the GAV model used in a conventional data schema integration approach. The indexes of terminology T_{vi} can be expressed in a conjunctive logic form with equivalent relations such as $T_{vi}^{\Delta^1} = (disjunc_{i1} \wedge disjunc_{i2} \ldots disjunc_{in})^{\Delta}$, that specifies the interpretation I of terminology T_{vi} regarding domain Δ as an *exact view* of the right-hand logic expression over Δ, where each element of the conjunctive expression $disjunc_i, 0 < i < n$ is a disjunction of terms T_g, R_g with value or existential quantification. The following conditions must hold:

$$\forall t | t \in T_{Vi}^{\Delta^1} \rightarrow t \in (disjunc_{i1} \wedge disjunc_{i2} \ldots disjunc_{in})$$
$$\forall t | t \in (disjunc_{i1} \wedge disjunc_{i2} \ldots disjunc_{in})^{\Delta} \rightarrow t \in T_{Vi}^{\Delta}$$

In addition, equivalent relations for an exact-view mapping can also be defined: $R_{Vi}(a,b)^{\Delta} = R_g(a',b')^{\Delta}$ iff, $R_{vi}.a^{\Delta} = R_{gj}.a'^{\Delta} \wedge R_{vi}.b^{\Delta} = R_{gj}.b'^{\Delta}$, where R_{Vi} is a relation in V_i and a,b are concepts associated with it, R_{gj} is a relation in G with associated concept a' and b'. This specifies that an exact-view mapping between two conceptual relations R_{Vi} and R_g with respect to a viewpoint interpretation domain Δ can be established if and only if there are exact-view mappings between their corresponding properties over a viewpoint representation.

The context mapping consists of a sequence of designated concepts and relations in terms of the common conceptual model G. The mapping relations indicate the interpretation of viewpoint concepts or relations regarding a particular user context. The mapping is specified in numerated rule functions that are processed in an external rule-based reasoning engine, see Sect. 3.4.

3.3 The Process to Project a Viewpoint

Translation of a query corresponding to a particular viewpoint to the common Ontology uses a set of mapping rules called rule unfolding to substitute query concepts and any constrains with the corresponding ones in the common Ontology. The conceptualisation may be inconsistent because the mapping information was developed independent of any conceptual constraints upon the viewpoint constructs. When a viewpoint query is aligned to a common Ontology, constraints in the viewpoint need to be translated into the representation of the common Ontology, where conflicts may occur. Conflicts in the viewpoint conceptualisation are either removed or reconciled according to a common model. To this extent, the viewpoint is considered as a more restricted conceptualisation on the basis of their semantic correspondence to the common Ontology.

The framework supports multiple user viewpoints of a domain using the sub processes given in Fig. 1. These seek to find equivalences between the user view and the local data views via the common global view of the data. The modelling process is conducted in an iterative cycle of sub-processes: synonym mapping, compositional mapping, consistency checks, conflict resolution, and viewpoint adaptation. This is discussed in more detail below.

The aim is to eliminate all possible semantic ambiguities and conceptual inconsistencies between the common Ontology and the viewpoint Ontology. For the case of unsolved consistencies, the system will simply report to the user about the alternatives and allow the user to guide the system to make a choice to resolve the ambiguity. Translation of a query corresponding to a particular viewpoint to the common Ontology uses a set of mapping rules called rule unfolding to substitute query concepts and any constrains with the corresponding ones in the common Ontology.

Synonym mapping focuses on conceptual alignment between Ontologies via consistent interpretation functions and semantic mappings, i.e. it seeks to find the corresponding expressions in the global Ontology that have the equivalent semantics and that subsequently result in a non-empty instance with respect to possible interpretation in the knowledge domain. Terminology heterogeneity reflects different naming conventions amongst user groups. In EDEN-IW system, terminology heterogeneity also involves multi-lingual term usage. An independent terminology glossary has been developed on the basis of contents analyses of standard terminology sources such as EARTh and IOW thesauri [23] The main task is to build synonym equivalents amongst different term sources in different languages and involves comparing the meaning of different terms with their explicit definitions.

Compositional View Mapping is used to find any indirect correspondences between user views and data resource views via the global view. For example, a compositional mapping is the reverse derivation of a *monthly-report* in EDEN-IW viewpoint of an aggregator that is created by combining observations of determinands or even determinand groups and averaging these over time and space.

Consistency Check: A new constraint and rule may introduce unsatisfiability and inconsistency into a multiple viewpoint system. In such a case, the containment or equivalence and consistency of concepts and their relations are checked. An end-user is prompted with the conflict and the rule is tagged to be ignored.

Conflict Resolution: After consistency checking, any conceptual conflict and violation detected between the viewpoint and the common Ontology is resolved before any viewpoint query is evaluated because this avoids re-evaluation of the conceptualisation for the whole domain. The conceptual conflict and violation may appear in the following forms:

- Incompatible data types
- Contradictory hierarchical structures
- Cardinality constraints
- Other Conceptualisation conflicts

3.4 Modelling Individual Users

Group viewpoints can be further tailored according to the individual user's preferences for the conceptualisation and presentation, as specified in an individual profile. Each individual profile is associated with a certain user group so that a user preferences can be interpreted correctly. The connection between a user profile and associated group viewpoint is derived from key terminology concepts in the user profile and the group viewpoint regarding their synonym relations.

The representation of group viewpoints can be further tailored according to the particular demands of individual users, where preferences for conceptual representation are specified in an individual model separate from the group viewpoint. Each individual profile is associated with a certain user group, such that user preference can be interpreted correctly. The connection between a user profile and an associated viewpoint is derived from key terminology concepts in the user profile and group viewpoint regarding synonym relations.

Individual user profile contains preference descriptions for individual users consisting of query preference, access permission, presentation preference, and user identification. A user is required to provide a concept set that relates to the view of the domain knowledge in terms of key concepts such as time, territorial area and inland water quality parameters, e.g. a concern about status of Nutrient quality in UK River B during time period C. Presentation preferences may also be specified in the individual user viewpoint to indicate the manner by which the retrieved results are presented in the user interface. User preferences for the presentation style, language support and possible standards for result classification can be specified.

User identification contains personal information for the individual user including identification of a user group or business role in the environmental organisation. Only pre-defined roles are recognised. The role identification contributes to the evaluation of appropriate rules in the viewpoint model. For example, a user profile for a French Policy Maker may contain the following information, see Table 2.

A reasoning process is carried out in the light of terminology similarity regarding the synonym relations given in an external glossary. The output of the reasoning is to identify a predefined group viewpoint that an individual user can be associated with.

Table 2. User profile for a French policy maker

Attribute name	Value
Business role	Policy maker
Associated viewpoint	Policy maker's viewpoint
Access permission	France territory
Preferred language	French
Applied classification standard	EU

Group user viewpoints may be further restricted by explicit rules that have been applied to the domain knowledge to specify the processing strategy and user preferences for information retrieval under certain circumstances. The explicit rules are specified in logic formulae. Explicit rules are specified in associated relations as dynamic constraints complementary to the object properties. A user is allowed to specify their own perspective upon the domain knowledge by introducing a set of logic rules in the viewpoint Ontology. The interpretation of the viewpoint conceptualisation may further be adapted in compliance with the individual user's roles that are defined explicitly in a set of utilisation rules in addition to the viewpoint model.

The rules are of the form of an implication between an antecedent (body) and consequent (head). A rule has the form, *antecedent* $\Rightarrow$ *consequent*. The intended meaning can be read as: whenever the conditions specified in the antecedent holds, then the conditions specified in the consequent must also hold. Rule-based knowledge is considered as an addition to a KR system of TBox (intensional data) and ABox (extensional data) such as OWL. Some types of logical inferencing are not allowed, for example, the conjunction of two concepts implying a new class relationship.

Deployment of such rules in a viewpoint can further tailor the presentation of the IR results and improve the information usability with respect to the requirements for targeted users. Role-specific underlying knowledge is explicitly defined using rule specification that is shareable and reusable across different user groups. Role-specific underlying knowledge is a set of supplementary expression of operational conventions and utility functions during the process of information analysis regarding particular information usages. User-specified terminologies are expanded regarding their query context and underlying knowledge in an explicit manner using rules such that the query expression can be adapted into an appropriate viewpoint for further processing. For example, the status of water quality may imply a classification of the average measurement for different types of aggregated observation in a specified time period according to certain criteria. The classification standard may vary according to the concrete roles of the querier.

Rules are developed manually with the aid of domain experts and stored away from user profile in a separate base. Rule conditions are verified against the user profile content during the process of query answering and result adjustment. Other specifications such as preferred natural language and applied classification standards are also provided as optional information regarding the availability of user roles.

3.5 View-Based Query Answering and Result Adjustment

User queries are evaluated over the common Ontology. The process of query evaluation needs to reason about the containment relations, see Fig. 2. The process starts when a query is constructed in a user's terminology and associated with a selected viewpoint model, where synonyms and multi-lingual

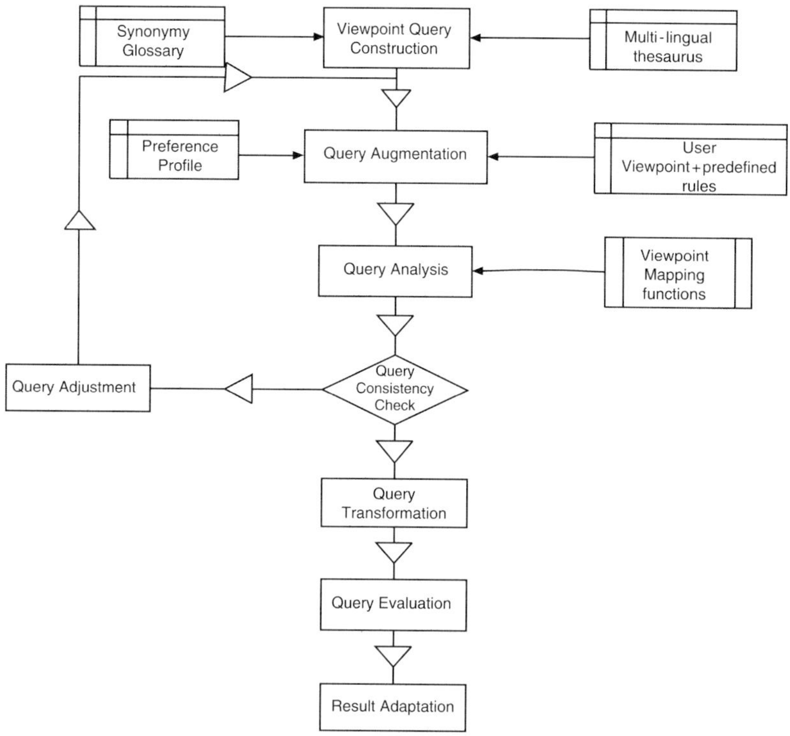

Fig. 2. The process of query answering and result adjustment

terms are translated according to the multi-lingual glossary. The query is further expanded using the underlying knowledge and default values are defined in the user profile and role-specified rules, such that, a user query is fully expressed and interpreted according to the conceptualisation of the associated viewpoint. Thereafter, the semantics of a user query is analysed and mapped into an equivalent or contained query expression in terms of a common Ontology with mapping relations computed via TBox reasoning. The transformed query is validated against viewpoint model and common Ontology to ensure conceptual consistency. When a query returns results from the database systems, the results need to be checked if they satisfy the viewpoint conceptual constraints. Any necessary adjustments are carried out in compliance with the viewpoint conceptualisation, user preference and role-specified rules.

User queries posed on viewpoint system need to be evaluated over the common Ontology in order to get the result sets from the underlying data sources. The process of query evaluation needs to reason about the containment relations between result sets over the common Ontology with respect to the initial viewpoint. The maximum-contained answers are computed semantically in order to find a proper expression over the Ontology that can be executed further in distributed databases.

The process of query answering using views is divided into sub-processes and performed in order as described in Fig. 2. The process starts when a query is constructed in user's terminologies and associated with a selected viewpoint model, where synonymy and multi-lingual terms are translated according to the synonymy glossary and multi-lingual thesaurus that are developed and maintained independently and commonly used within the knowledge domain.

The query is further expanded using the underlying knowledge and default values defined in the user profile and role-specified rules, such that, a user query is fully expressed and interpreted according to the conceptualisation of the associated viewpoint. Thereafter, the semantics of a user query is analysed and mapped into equivalent or contained query expression in terms of a common Ontology with mapping relations computed via TBox reasoning. The transformed query is validated against viewpoint model and common Ontology to ensure conceptual consistency, so that theoretically evaluation of such a query would produce non-empty results set over multiple viewpoint system. When a query returns results from the database systems, the results need to be checked to consider if they satisfy the viewpoint conceptual constraints. Any necessary adjustments are carried out in compliance with the viewpoint conceptualisation, user preference and role-specified rules.

Regarding the conjunctive query Q over a viewpoint $Q(\overrightarrow{x}) = conj_1(\overrightarrow{x}, \overrightarrow{y}_1) \vee conj_2(\overrightarrow{x}, \overrightarrow{y}_2) \ldots \vee conj_1(\overrightarrow{x}, \overrightarrow{y}_n)$, each variable of $\overrightarrow{x}$ and $\overrightarrow{y}$ would have predefined mapping functions to their semantic correspondences in the common Ontology that each $\overrightarrow{x}_i \approx q(g)$ and $\overrightarrow{y}_i \approx q'(g)$, where $q(g)$ and $q'(g)$ denote a query expression over the common Ontology as described earlier. The query $Q(\overrightarrow{x})$ can be reformulated into an expression of containment query over common Ontology consisting of terminology substitutions by attribute and concept unfolding. Thus, if the reformulated query has a valid interpretation over common Ontology, it can be proved that the original query would have a valid interpretation over the same domain. However the reasoning approach considers only terminology subsumptions, the actual conceptual constraints in viewpoint are ignored in the phase.

To solve this situation, retrieved results are validated against mismatched conceptual constraints in the post-processing phase. As a retrieved instance set corresponds to a valid interpretation of a transformed query over common Ontology, so all constraints of common Ontology must be satisfied. However, constraints of viewpoint conceptualisation are not enforced in the common Ontology during information retrieval. Suppose viewpoint constraints are only further restrictions, because conceptual conflicts have been filtered out in design phase. The post-processing approach mainly focuses on solving hierarchy-based constraint and functional dependencies throughout the viewpoint conceptualisation, whereas other mature logic-based reasoning is conducted in a well-developed logic algorithm that is embedded in a third-party reasoning engine. The following rules are presumed to conduct hierarchical and functional constraints:

- IS-A relations for generalisation and instance sets of child-class are combined to generate a new instance set.
- IS-PART relations for aggregation means that each instance is aligned with relevant instance according to aggregation relation to form a new instance.
- Functional Dependencies are processed in compliance with specific operations such as average, maximum, etc.

3.6 Applying Preference and Rules in Query Answering

Explicit role-specific rules contribute to the representation adaptation between viewpoint conceptualisation and user preference that are applied in the process of terminology expanding and result adjustment. For example an EU Policy maker may want to ask: "What is the status of the water quality of the river Thames in 1980?" In order to solve this, the following relevant rules would be taken into account for pre-processing and post-processing of query evaluation.

$$\begin{aligned} &\mathrm{River}(?\mathrm{r}) \wedge Country(?n) \wedge \mathrm{isLocatedIn}(?\mathrm{r}, ?\mathrm{n}) \\ &\Rightarrow \mathrm{appliedQualityStandard}(?\mathrm{n}) \end{aligned}$$

The above rule specifies that the country name for a given river associated with the query determines the classification standards for water quality. If the UK standard for water quality is applied, then Nutrient Grading can be derived from NatriteGrading and PhosphateGrading classification. Thereafter, the following rules can be applied to further interpret the grading of Nitrite and Phosphate according to the UK standards.

$$\begin{aligned} &appliedQualityS\tan\ dard(?n) \wedge Country(?n) \wedge (equal(?n, \text{“}uk\text{”}) \Rightarrow \\ &(NatriteGrading(?x) \wedge PhosphateGrading(?x) \Rightarrow \\ &NutrientsGrading(?x)) \end{aligned}$$

The following rules can further affect the meaningful interpretation of *NatriteGrading* and *PhosphateGrading* in context of UK standards.

$$\begin{aligned} &LessThan(averageValue(?t, ?t-3, ?c, ?x), 5) \wedge observe(?c, ?x) \\ &\wedge totalNitrate(?c) \wedge catchment(?x) \wedge inUnit(?c, mg/l) \\ &\Rightarrow NatriteGrading(1) \wedge appliedQualityS\tan dard(UK) \end{aligned}$$

$$\begin{aligned} &LessThan(averageValue(?t, ?t-3, ?c, ?x), 0.02) \wedge observe(?c, ?x) \\ &\wedge totalPhosphate(?c) \wedge catchment(?x) \wedge inUnit(?c, mg/l) \\ &\Rightarrow PhosphateGrading(1) \wedge appliedQualityS\tan dard(UK) \end{aligned}$$

According to UK General Quality Assessment standard, totalNitrate can be further defined in a viewpoint Ontology:

$$Observation(?x) \wedge hasDeter\min and(?x, Nitrate)$$
$$\wedge\, hasMedium(?x, water) \wedge isAnalyse(water, totalMedium))$$
$$\Rightarrow totalNitrate(?x) \wedge appliedQualityS\tan dard(UK)$$

The implicit facts associated with a user query are put into a knowledge model during the process of query expanding. The satisfied rule in an associated rule-based processing engine is fired. The process runs in an iterative manner until all effected rules have been fired.

3.7 User-Centre Query Answering

The process of query answering and result adjustment is given in the architecture given in Fig. 3, and consists of three processes. The pre-answering process involves terminology translation and query reformulation. Query reformulation involves possible decomposition of a query into sub-queries. In the Answering Process, the reformulated user query is further validated and optimised using logic-based query optimisation to remove redundant sub-goal expressions for the reformulated query. In the Post-answering Process, the results set returned from the answering process needs to be processed in

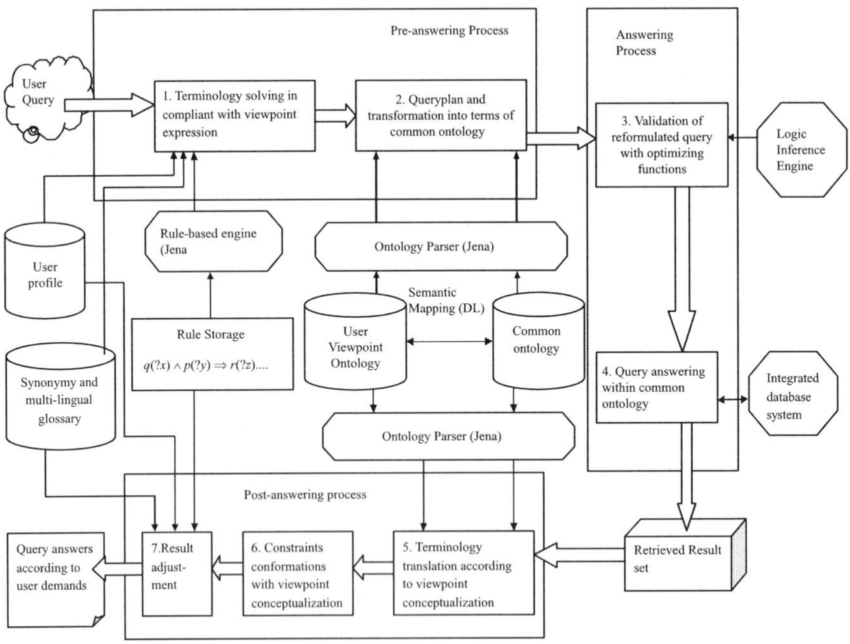

Fig. 3. Architecture of the adaptive viewpoint system

compliance with user specified representation of the conceptual model and terminologies defined in the individual user profile. The result of a successful evaluation is further adjusted according to the user preference and user roles.

Pre-Answering Process

The pre-answering phase contains two sub-processes for terminology resolution and query reformulation. In the former process, a user is guided to construct a query in their familiar terms that are then translated to viewpoint terminology via synonym and multi-lingual mappings defined in the glossary. The user query is constructed in a SQL-like syntax in RDF structure. User preference and usage rules are also imported to resolve relevant terminologies into an appropriate level in compliance with semantic mapping.

An individual user profile is associated with a specific group viewpoint, where user terms can be mapped appropriately using synonym relations. Identification of a business role of a user is linked to rule specifications in rule storage, where the underlying knowledge of the specific business roles is explicitly specified. The role specific rules mainly resolve the homonym problems in which the same term is interpreted differently according to business role. The rule specifications are loaded into rule engine in Jena, which is configured in a Forward-Chaining model. The terms in a query expression are validated against rule conditions. If any rule trigger satisfies the corresponding terms in a query expression, it is expanded and replaced using the appropriate literals defined in rules. The condition validation and rule firing is conducted iteratively. The expanded and replaced query is checked against rule conditions repeatedly until no new satisfactory rule is found. The output of the terminology resolution is forwarded to the next sub-process of query reformulation.

Query reformulation is conducted in two steps: sub-query planning and mapping unfolding. As the viewpoint conceptualisation is expressed in a hierarchy structure, the production of rule-based terminology resolution is not sufficient to be mapped into a common Ontology terms in a straightforward manner. A reasoning approach is required to cover the distance and decompose query into sub-queries. The reasoning reduces the query granularity into an appropriate level where all terms can be directly mapped to a common Ontology. The term unfolding uses a GAV like query-rewriting approach regarding the soundness and completeness of views for the semantic mapping of the viewpoint terminology to expressions in the common Ontology. The relevant terms in sub-queries are substituted by its semantic correspondences.

Answering Process

The reformulated user query is further validated and optimised using logic-based query optimising functions to remove the redundant sub-goal expression for reformulated query. The optimised query is forwarded to a virtual database for result answering when all constraints of common Ontology are satisfied. The virtual database is realised in EDEN-IW as an integrated database system with the EGV Ontology. The process of information retrieval from underlying data source has been described in [14] The result set of query in RDF structure is returned to the post-answering process for result adaptation.

Post-Answering Process

The results set returned from the answering process needs to be processed in compliance with user specified representation of the conceptual model and terminologies defined in the individual user profile. The results set in the instance table is reformulated into viewpoint structures by going through an adaptation path where conceptual operations have been defined in relevant semantic mappings. The reformulated results must reflect an ABox set of viewpoint conceptualisation in its OWL representation. The transformed ABox set assertion is evaluated with a TBox definition and constraints. Any part of the unsatisfied results is filtered out. The constraints integrity of viewpoint conceptualisation is guaranteed. The result of a successful evaluation is further adjusted according to the user preference and user roles. The adjustment operations including synonym translation, rule-specified terms resolution and result presentation.

4 Implementation and Validation

The MVF is implemented as an extension to multi-lateral Ontology in EDEN-IW system as described in Fig. 4. The dotted white boxes indicate the extensions to the EDEN-IW for MVF support. The MVF Ontology model is written in OWL-DL or Ontology Web Language-Description Logic, a subset language of OWL [6]. The Ontology model is edited using the Protégé Ontology editor. The OWL Ontology is parsed and loaded into internal programming structures for operational processing using Jena, a Java-based Ontology parser with a limited inference capability for reasoning about OWL-Lite. More complicated logic inferencing is processed in an external inference engine, Pellet [1] in which, Ontology validation and subsumption reasoning is reduced to a satisfiability problem of description logic SHIOQ. The DIG [5] interface of Jena supports the ability to plug-in any standard logic engine, such as Pellet. The multiple viewpoint system is validated by using a user-focused evaluation by posing user queries for specified stereotype viewpoints and with respect to particular individual user profiles.

Viewpoint conceptualisation and aggregation concepts are added as an extension to the EDEN-IW project system to support a multiple viewpoint conceptualisation for retrieving environmental information. The multiple viewpoint system is implemented as a Java-based application consisting of two sub-systems, one for viewpoint adaptation and management, the other

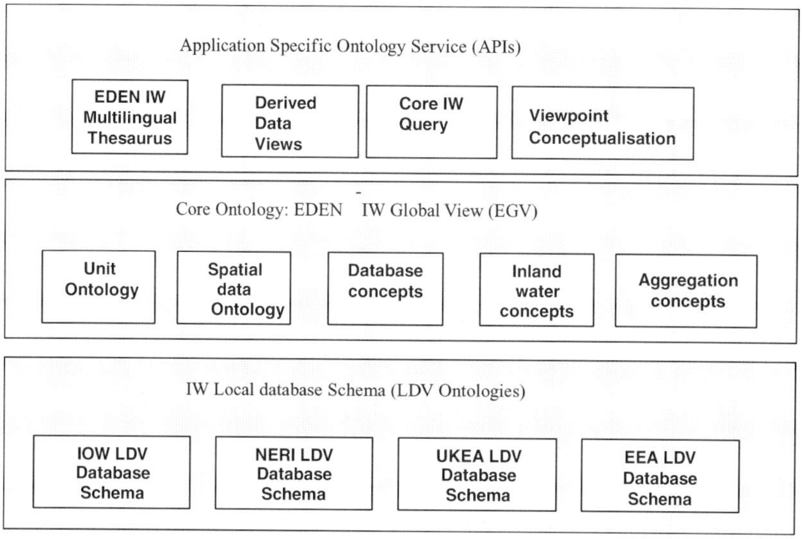

Fig. 4. Multi-lateral ontology used by the EDEN-IW system

for query processing and query result adjustment. A general conceptualisation framework consists of four key components: Ontology, Ontology parser, rule-base reasoner, and logic inference engine.

The multiple viewpoint system is validated by posing user queries at specified viewpoints with respect to a particular user profile and role specification. For example, the following information has been taken into account to answer a query in policy maker's viewpoint such as "What's the quality status of river Thames in 1980?":

- User role: UK environmental policy maker
- Applied Standard: UK General Quality Assessment
- Associated viewpoint: policy maker's viewpoint

In a traditional IR system, answering such a question is not possible, because the query specifies information that is too vague to be retrieved from the database directly. In a multiple viewpoint system, the original query is able to be expanded and adapted using semantic analysis according to a viewpoint conceptualisation. In this case, sub-queries would be generated for UK standards with respect to Nutrients and Total Phosphorus with restricted time and geographical scope. The retrieved result from an integrated database system is further adjusted according to an average function and grading standards for corresponding parameters. Test cases have been made for each viewpoint to validate its flexibility to handle queries with different conceptualisation and different user roles.

Table 3 shows that high-level queries posed on a group viewpoint with the underlying conceptualisation are successfully handled via a multiple viewpoint

Table 3. Validation of multi-viewpoint system for IR using test cases

Group viewpoint	Query expression	Traditional IR	Multi-viewpoint system for IR
Scientist	*Concentration of Nitrite in river Y at time T?*	*Needs to be transformed into SQL*	√
Legislator and aggregator	*Monthly mean concentration of total Nitrite in basin X of river Y?*	*Requires expert*	√
Policy maker	*What's the most polluted area in river X at time Y?*	*Requires expert*	√

framework in conjunction with the conceptual IR system. The answering of such queries in a traditional IR system requires more human support for information processing and conceptual interpretation. The multiple-viewpoint framework automates the processes via an explicit definition of domain conceptualisation. The ability to adapt information to multiple viewpoints provides more flexible usage.

5 Conclusions and Further Work

Research and application of semantic model to support database, user and application interoperability is challenging and goes far beyond creating a global view semantic domain model and sharing it. An IR system often can't be reengineered to use a single global semantic model throughout. Instead a global semantic model is used as a mediating model to interoperate with multiple heterogeneous database models, user and application data models. A framework is needed that can handle both simple direct and complex indirect and composite data mappings, and that uses explicitly orchestrated mapping processes to handle data queries posed by different users and applications to combinations of database resources.

A general framework is presented in the chapter to support dynamic representation of multiple viewpoints and to fulfil the customised IR requirement regarding different user stereotypes. The usability of an IR system is improved by supporting dynamic interoperability and IR result adaptation between viewpoint representation and the common conceptual model. A DL-based semantic approach is utilised to support terminology, semantic and semiotic mapping and a query answering process on the basis of identified information heterogeneities. Such explicit mapping processes are facilitated using logic reasoning in order to perform consistency and containment checks for data mappings.

The reasoning is performed on semantic conceptualisations of the stored data and user models rather directly on the stored, relational, data models and user models. Reasoning about the actual stored relational data model itself is advantageous as it can help to explain how queries work, how to compare queries and why queries fail. However, operations, conceptualisations and constraints between the DL and other data models such as relations models differ, e.g. the use of an open world vs. closed world assumption. There is as yet no standard way to combine these. This needs further investigation.

References

1. Pellet, an open-source Java based OWL DL reasoner, Home page, retrieved from http://www.mindswap.org/2003/pellet/index.shtml on 2006-01.
2. Adnani, M.E., Ygtongnon, K. and Benslimane, D., A multiple layered functional data model to support multiple representations and interoperability of GIS: Application to urban management systems. In *Proceedings of the 9th ACM international symposium on Advances in geographic information systems* (Atlanta, Georgia, USA), 2001, 70–75.
3. Baader, F. and Nutt, W., *Basic Description Logics.* Cambridge University Press, Cambridge, 2003.
4. Baldwin, D., Applying multiple views to information systems: A Preliminary framework. *ACM SIGMIS Database, 24* (4), 15–30.
5. Bechhofer, S., The DIG description logic interface: DIG/1.0, University of Manchester, Manchester, UK, http://dl-web.man.ac.uk/dig/2002/10/interface.pdf on 2006-01.
6. Bechhofer, S., Harmelen, F.v., Hendler, J., Horrocks, I., McGuinness, D.L., Patel-Schneider, P.F., Stein, L.A. and Olin, F.W., OWL Web ontology language reference, W3C recommendation, 10 February 2004, http://www.w3.org/TR/owl-ref/#Sublanguages on 2006-01.
7. Benchikha, F. and Boufaida, M., Roles, views and viewpoints in object-oriented databases. *Workshop at European Conference on Object-Oriented Programming (ECOOP 2005)*, Glasgow, UK, 2005.
8. Bertino, E., Catania, B. and Zarri, G.P., *Intelligent Database Systems.* Addison-Wesley, Boston, MA, 2001, 170–178.
9. Bouquet, P., Ehrig, M., Euzenat, J., Franconi, E., Hitzler, P., Krötzsch, M., Serafini, L., Stamou, G., Sure, Y. and Tessaris, S., Knowledge Web project deliverable version 2.2.1, specification of a common framework for characterizing alignment, Home page, retrieved from http://knowledgeweb.semanticweb.org on 2005-02.
10. Calvanese, D., Giacomo, G.D. and Lenzerini, M., A Framework for ontology integration. In *Proceedings of the 2001 Int. Semantic Web Working Symposium*, 2001, 303–316.
11. Damasio, C.V., Analyti, A., Antoniou, G. and Wagner, G., Supporting open and closed world reasoning on the Web. In *Proceedings of Principles and Practice of Semantic Web Reasoning (PPSWR06)* LNCS, Springer, Berlin Heidelberg New York, 2005, 21–36.

12. Das, S., *Deductive Databases and Logic Programming.* Addison-Wesley, Wokingham, 1992.
13. De Bruijn, J., Eiter, T., Polleres, A. and Tompits H., On representational issues about combinations of classical theories with nonmonotonic rules. DERI Technical Report, retrieved from http://www.deri.at/digital-library/browse/technical-reports/ on 2006-03.
14. Haastrup, P. and Wurtz, J. (eds.) *Environmental Data Exchange Network for Inland Water. ISBN 978-0-444-52973-2.* Elsevier, New York, 2007.
15. Jung, E.-C., Sato, K., Chen, Y., He, X., MacTavish, T. and Cracchiolo, D., DIF knowledge management system: bridging viewpoints for interactive system design. In *Proceeding 11th Human Computer Interaction International Conference* (Las Vegas, Nevada USA), 2005.
16. Michael Stonebraker, Paul Brown and Martine Herbach, Interoperability, distributed applications and distributed databases: The virtual table interface. *IEEE Data Eng. Bull.*, *21* (3), 25–33.
17. Ng., G., Open vs. closed world, rules vs. queries: use cases from industry. In *OWL experiences and directions workshop* (Galway, Ireland), 2005.
18. Pan, Z. and Heflin, J., DLDB: Extending relational databases to support semantic Web queries. Technical Report LU-CSE-04-006, Dept. of Computer Science and Engineering, Lehigh University, http://www.cse.lehigh.edu/~heflin/pubs/psss03-poster.pdf on 2006-01.
19. Patel-Schneider, P.F. and Horrocks I., A Comparison of two modelling paradigms in the semantic Web. In *Proceedings of the Fifteenth International World Wide Web Conference (WWW 2006)*, ACM, 2006, 3–12.
20. Ribière, M. and Dieng-Kuntz, R., A viewpoint model for cooperative building of an ontology. In *Proceeding of 10th International Conference on Conceptual Structures (ICCS 2002)*, (Borovets, Bulgaria), Springer, Berlin Heidelberg New York, 2002, 220–234.
21. Sheth, A. and Larson, J., Federated database systems for managing distributed, heterogeneous, and autonomous databases. *ACM Computing Surveys*, *22* (3), 183–230.
22. Spaccapietra, S., Parent, C. and Vangenot, C., GIS databases: From multiscale to MultiRepresentation. In *Proceedings 4th InternationalSymposium, SARA-2000* (Horseshoe Bay, Texas, USA), 2000, 57–70.
23. Stjernholm, M., Preux, D., Sortkjaer, O. and Zuo, L., DILIVERABLE 17, Structured list integration into IW distributed semantic hybrid agents application, The Environmental Data Exchange Network for Inland Water (EDEN-IW) project, retrieved from http://www.eden-iw.org on 2006-05.

User Preferences for Access to Textual Information: Model, Tools and Experiments

Thibault Roy and Stéphane Ferrari

GREYC-CNRS UMR 6072, Computer Science Laboratory, University of Caen, F14032 Caen Cedex, France Thibault.Roy@info.unicaen.fr, Stephane.Ferrari@info.unicaen.fr

Summary. There are more and more documents produced and exchanged on both public and private networks. At the same time, the tools proposed to access their content do not fully satisfy the users. Most of them do not really take into account the user's point of view or knowledge. The aim of the works we describe in this chapter is to fill this gap between users and the collections of documents they browse through. Thus, we propose a user-centered model of lexical knowledge as well as related graphical interfaces. The application of this model for access to textual information is realised by the *ProxiDocs* platform. This tool provides users with interactive maps and hypertexts improved with mark-up directly related to their own choices and preferences.

1 Introduction

There are more and more documents produced and exchanged on both public and private networks. At the same time, the tools proposed to access their content do not fully satisfy the users. Most of them do not really take into account the user's point of view or knowledge. The aim of the works we describe in this chapter is to fill this gap between users and the collections of documents they browse through. Thus, we propose a user-centered model of lexical knowledge as well as related graphical interfaces. The application of this model for access to textual information is realised by the *ProxiDocs* platform. This tool provides users with interactive maps and hypertexts improved with mark-up directly related to their own choices and preferences.

In Sect. 2, we present the motivation of our research, and existing works related to the representation of users' point of view and to textual data visualisation. We position our approach between these two kinds of works: viewpoints representation and visual methods for access to information. Section 3 gives an overview of our propositions. First, we detail the model main principles and illustrate their use through an example of structured resources. Then, we present the related interactive tools, one developed for building the resources,

T. Roy and S. Ferrari: *User Preferences for Access to Textual Information: Model, Tools and Experiments*, Studies in Computational Intelligence (SCI) **93**, 285–305 (2008)
www.springerlink.com

the other one for using them to access textual information. Section 4 presents an experiment in information retrieval, which is the standard use of the model and tools. The context is to find information about European decisions. In Sect. 5, we illustrate the flexibility of these model and tools. We present a second experiment in Natural Language Processing (NLP) domain. The context is the observation of conceptual metaphors in a domain-specific corpus. To conclude, we briefly discuss our results and point out the main perspectives of our works.

2 Motivations

2.1 Textual Information and Users' Point of View

A task we realised every day is information retrieval on the Web. In order to illustrate the users' satisfaction in such a task, Véronis made an experiment on search engines with many users [30]. They had to formulate many requests of different topics on six classical search engines such as Google, Yahoo, MSN Search, etc. A grade of relevance between 0 for a bad result and 5 for a good one is then given by users for each search. As we can see on Fig. 1, scores are not very good.

Well-known search engines such as Google or Yahoo have the best scores but lower than the average one. This experiment reveals the dissatisfaction of users for a classical task of information retrieval on the Web. In such experiment, the users' point of view is represented by the search keywords and nothing else.

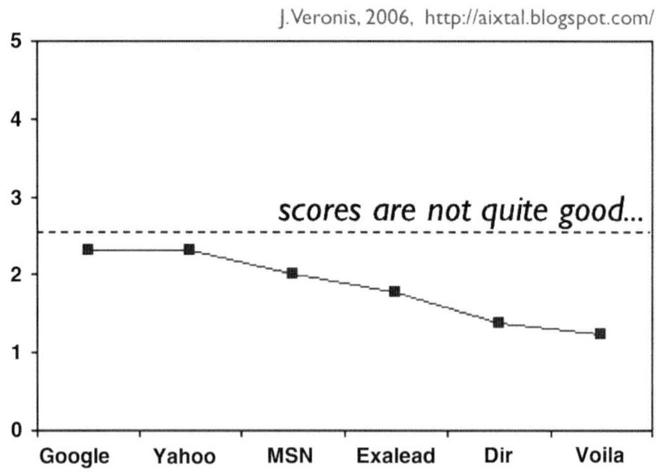

Fig. 1. Results of a six search engines evaluation

Not considering the users' point of view on the task is one reason of such a dissatisfaction. Some works are dedicated to representations of users' preferences and points of view. In [20], the authors present a review of existing methods for viewpoints representation in a database. They also describe a new formal model in order to symbolise a user viewpoint and methods to project such viewpoints on data for a question–answering system. The authors of [29] propose to select and filter relevant concepts for a user in an ontology. Such "personal" concepts are then used in order to regroup users in communities. In [19], the lexical database WordNet[1] is used to extend and adapt users' requests in image retrieval on the Web.

Briefly, all of these works describe models and representations of users' point of view with personal filters and selections in databases and in ontologies.

2.2 Visual and Interactive Tools for Access to Textual Information

Considering such an interest of visual methods for accessing information in sets of textual documents, a lot of works have been realized. In [27], the author suggests experimenting with users the main methods of visualization and investigation of corpora of texts. The proposed methods were metric representations in 2 or 3-D spaces, trees, graphs, etc. The results of this experiment show that users prefer a metric representation of the corpus on a plan.

In the field of information retrieval, Salton [25] suggests using such graphical representations by positioning on the perimeter of a circle the pages returned by a search engine. In this way, the proximity on the circle between pages indicates possible similarities of contents between the pages. In the same domain, Hearst [11] proposes to answer a request on a search engine with a set of rectangles, each one corresponding to a document considered relevant by the system. In these rectangles, every line corresponds to a keyword of the request and every column is coloured according to the frequency of the keyword within the segment of the document which is linked to it.

Other techniques were proposed in tasks of electronic management of a set of documents. Some of these techniques present the set of documents as hierarchies in two or three dimensions, such as *Cone Trees* [22] or *Hyperbolic Trees* [14] (cf. Fig. 2). To reach information in long documents, Jacquemin and Jardino [12] describe the *3D-XV* tool which proposes a three dimensional interface based on a thematic segmentation.

For a few years, some tools of textual analysis have used a visualisation technique called *cartography*. Like a roadmap revealing cities and roads connecting them, a map of a set of textual documents displays the nearness and the links between textual entities. Since 2001, the two cartographic metasearch engines *KartOO* [5] and *MapStan* [28] have been available on the Web (Fig. 3).[2] These two tools return maps representing the sites proposed in

[1] `http://wordnet.princeton.edu`.
[2] `http://www.kartoo.com` and `http://search.social-computing.com`.

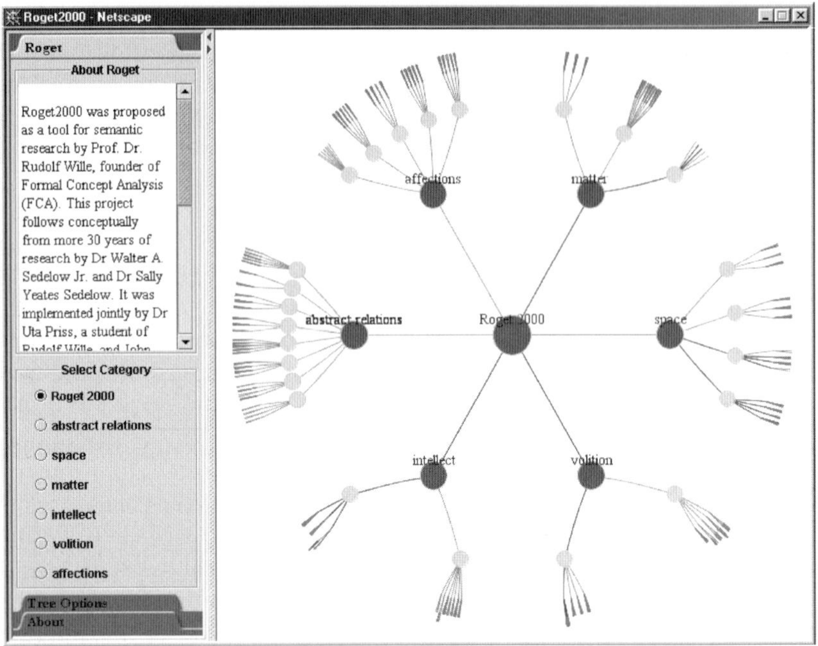

Fig. 2. Interface of *Hyperbolic Trees* showing links between concepts

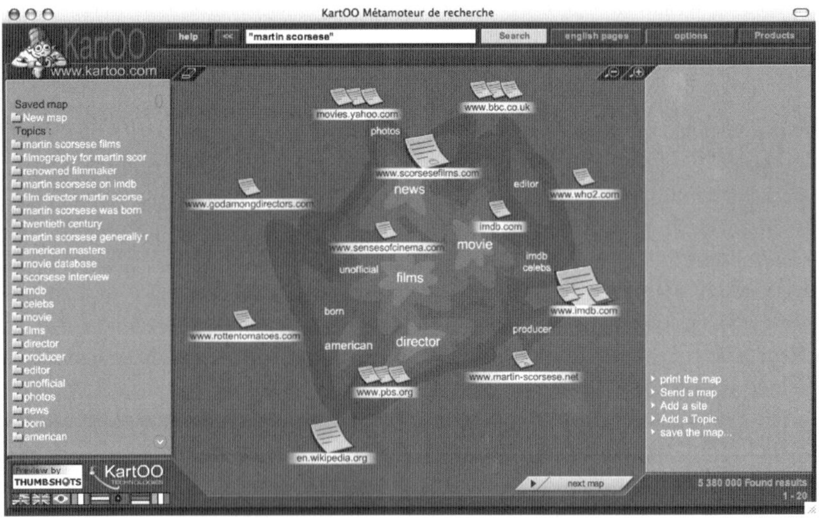

Fig. 3. Interface of *KartOO* showing a Web information retrieval with keywords "Martin Scorsese"

answer to a user's request. Theses systems position the sites estimated similar at the same place on the maps. It is also possible to distinguish the main categories of information proposed in answer to a user's request.

Graphical tools presented in this section have two different main goals. In [18], the authors also pointed that two main interactive steps must be taken into account to reach information. The first one consists in providing users with help for browsing through a collection. The second one concerns visual representations of a specific document.

2.3 Our Approach

The two previous approaches motivate our own works. It seems necessary to take the users' point of view into account to increase their satisfaction by returning textual information which is relevant according to their own tasks. It also seems necessary to use graphical interactive tools and visual representations in order to browse through collections of texts and to navigate in long textual documents.

Therefore, we propose both to provide the users with graphical interactive representations and to take the users' point of view into account, merging these two approaches. Moreover, we propose the users to structure their own knowledge rather than accessing to filtered pre-built ontologies. Our hypothesis is by this means it becomes easier for the graphical tools to directly reflect the users' viewpoint.

Section 3 describes the model and the tools we developed in order to implement our propositions.

3 Models and Tools

3.1 LUCIA: A Model for Representing User's Knowledge on Domains

Main Principles

The LUCIA model, proposed by Perlerin [16], is a differential one, inspired by Rastier's works on Interpretative Semantics [21]. The basic hypothesis is the following: when describing things we want to talk about, in order to set their semiotic value, differentiating them from things for which they could be mistaken is enough.

According to this hypothesis, a lexical unit is described with semantic features. These semantic features are relevant only in specific contexts. The notion of *isotopy*, introduced by Greimas in [10], characterises these contexts. An isotopy is the recurrence of one semantic feature in a linguistic unit, like a sentence, a text or even a set of texts.

Furthermore, in this model, the user, as the core position, describes the domains of his choice, according to his own point of view and with his own words. Domains descriptions are not supposed to be exhaustive, but they reflect the user's opinion and vocabulary.

The principle for knowledge representation is structuring and describing lexical items (i.e. words and compounds) according to two main criteria:

- Bringing together similar lexical items
- Describing local differences between close items

Such a representation is called a *device*. The user can define it for each domain of interest. A device is a set of *tables* bringing together lexical units of a same semantic category, according to the user's point of view. In each table, the user has to make explicit differences between lexical units with couples of *attributes* and *values*. A table can be linked to a specific line of another table in order to represent semantic associations between the lexical units of the two tables. All the units of the second table inherit of the *attributes* and related *values* describing the row it is linked to. In the following, an example illustrates these notions.

Examples of LUCIA Devices

This section illustrates the use of the model for a device representing knowledge about cinema. Let us consider the following lexical items, translation of the ones observed in a French corpus: *actor*, *director*, *cameraman*, *montage specialist*, *minor actor*, *soundman*, *filmmaker*, *Jean-Pierre Jeunet*, *Steven Spielberg*, *Georges Lucas*, *Alfred Hitchcock*, *John Woo*, etc.

With these lexical units, it is possible to build a first set of LUCIA tables in order to bring them together. Table 1 shows an example of such a first step. Using the model in such an incremental approach is recommended, with step-by-step enrichments. The differentiation between close lexical items, i.e. items in a same table, can be realised in a second time, by defining and using attributes and values. Here, for instance, two attributes can characterise the **Staff** table items: *Professional*, with values *Yes* vs. *No*, and *Job*, with values *Performer* vs. *Director* vs. *Technician*. Another point of view can be reflected in the **Director** table, using an attribute *Nationality* with values *American*

Table 1. Bringing similar words together

Staff
Actor, director, cameraman, montage specialist, minor actor, soundman, filmmaker

Director
Jean-Pierre Jeunet, Steven Spielberg, Georges Lucas, Alfred Hitchcock, John Woo

Table 2. Differentiating similar words

Staff	*Professional*	*Job*
Actor	*Yes*	*Performer*
Director, filmmaker	*Yes*	*Director*
Cameraman, soundman, montage specialist	*Yes*	*Technician*
Minor actor	*No*	*Performer*
–	*No*	*Director*
–	*No*	*Technician*

Director	*Nationality*
Steven Spielberg, Georges Lucas	*American*
Jean-Pierre Jeunet	*French*
Alfred Hitchcock	*English*
John Woo	*Chinese*

vs. *French* vs. *English* vs. *Chinese*. Such choices result in the device shown in Table 2.

Cells can be blank in LUCIA tables, when the user finds no relevant lexical unit described by the combination of attributes and values on the same line (e.g. the last two lines of Staff table in Table 2). Finally, the user can specify *inheritance links* showing that the lexicon of a whole table is related to a specific line of another one. In the example, the **Staff** table can be linked to the line *Professional*: *Yes* and *Job*: *Director* of the **Staff** table. This means that each lexical unit of the **Director** table inherits of the attributes and values from the linked line. These links are used in further analysis.

3.2 User-Centred Tools

VisualLuciaBuilder: Building LUCIA Devices

VisualLuciaBuilder is an interactive tool for building LUCIA devices. It allows a user for the step-by-step creation and revision of devices through a graphical interface. This GUI (see Fig. 4) contains three distinct zones.

- Zone 1 contains one or many lists of lexical units selected by the user. They can be automatically built in interaction with a corpus. The user can add, modify or delete lexical units.
- Zone 2 represents one or many lists of attributes and values of attributes as defined by the user.
- Zone 3 is the area where the user "draws" his LUCIA devices. He can create and name new tables, drags and drops lexical units from zone 1 into the tables, attributes and values from zone 2, etc. He can also associate a colour to each table and device.

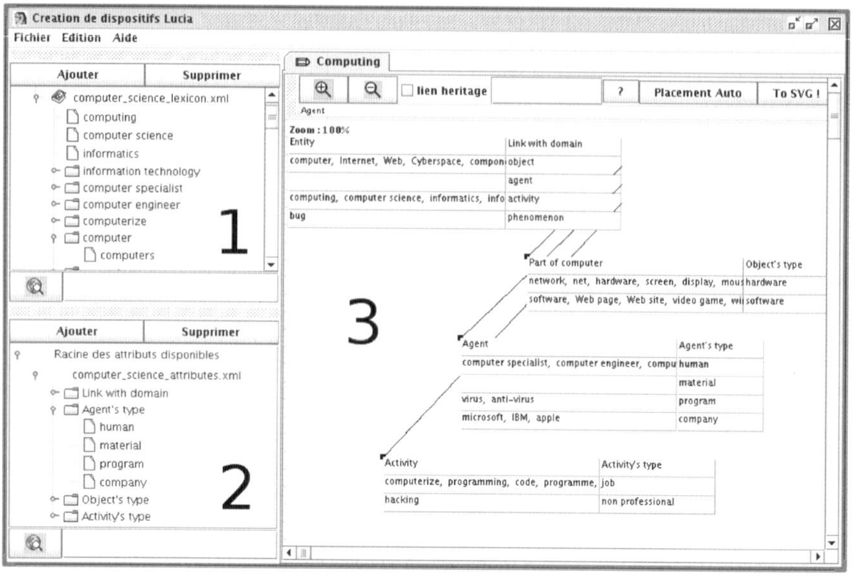

Fig. 4. *VisualLuciaBuilder*'s Interface

The tool allows SVG (Scalable Vector Graphics) exports of the devices. SVG is a text-based graphics language of the W3C[3] The lexical representation are stored in an XML format for further use (revision or application).

ProxiDocs: Projecting LUCIA Devices in a Corpus

The *ProxiDocs* tool, [23], builds global representations form LUCIA devices and a collection of texts. It returns maps built from the distribution of the lexicon of the LUCIA devices in the corpus. Maps reveal proximities and links between texts or between sets of texts. Other graphical representations of a corpus can also be returned by the tool, such as the "cloud" of lexical units presented in the following section.

In order to build maps of a set of texts, *ProxiDocs* realized different processes.

Figure 5 sums up such processes.

In a first stage, the tool how many lexical units from each device appear in each text of the set. A list of graphical forms is associated with each lexical unit (for example, the graphical form "politics" is associated to the lexical unit "politic"). It returns a result proportional to the size of the text. A list of numbers is thus joined to each text, an N dimensional vector, in which N is the number of devices specified by the user.

The next stage consists in a projection of the N dimensional vectors into a 2 or 3 dimensional space we can visualise. Many methods are then proposed

[3] Specifications available: `http://www.w3.org/TR/SVG`.

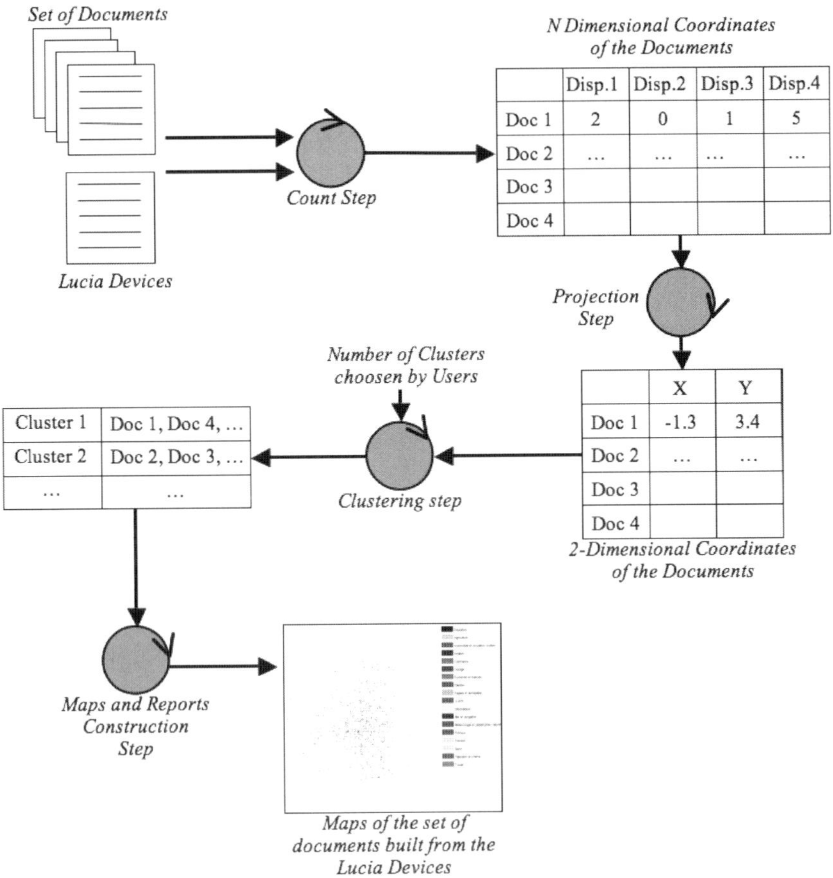

Fig. 5. Processes realised by the *ProxiDocs* tool in order to build maps of a set of documents according to Lucia devices

such as the Principal Components Analysis (PCA) method [3] or the Sammon method [26]. Each text can then be represented by a point on a map. Proximity between different points informs the user that there are some domain similarities between the related documents.

In order to emphasize such proximities, a clustering method is applied. We propose in *ProxiDocs* the Ascendant Hierarchical Clustering (AHC) [3] method or the KMeans [15] method. Maps representing groups of texts can be built from the clusters. Analyses reports are also returned to the user, with information about most frequent lexical units, attributes and values, etc. All maps and texts are interactive, linked to each other and to the source documents, providing the user with a helpful tool for accessing the textual information of the corpus.

Like *VisualLuciaBuilder*, *ProxiDocs* is developed in Java and open-source. All maps and texts are interactive, using SVG and HTML formats. These are linked together and to the source documents, providing the user with a helpful tool for accessing the textual information of a collection.

Examples of graphical outputs built with ProxiDocs are shown in the two following sections. The first one relates an experiment directly addressing access to textual information for Web users. The second one is dedicated to a NLP experiment realised in a research environnement.

The tools presented in this section and the LUCIA devices used in the two following experiments are available on the Web at: `http://www.info.unicaen.fr/~troy/lucia`.

4 Experiment 1: Accessing Information

4.1 Context and Materials

The first experiment concerns information retrieval and documents scanning on the Web. The objective is to perform a search for information on the Web in a broad context: the "European decisions". This search is realized with regards to the domains interesting the user.

The domains representing the user's point of view are **agriculture**, **pollution**, **road safety**, **space**, **sport** and **computer science**. These six domains are represented by LUCIA devices built using the *VisualLuciaBuilder* tool. The devices contain from 3 to 5 tables and from 30 to 60 lexical units. Some common attributes are used to structure the devices, such as the attribute *Role in the domain* with the values *Object* vs. *Agent* vs. *Phenomenon*, and the attribute *Evaluation* with the values *Good* vs. *Bad*.

Figure 6 presents one of the devices used during this experiment. It represents the computer science domain. Four tables are used: the main table named *Entity* and the three other tables respectively named *Part of Computer*, *Agent* and *Activity*. Each of these three tables is linked to the main table. For instance, the *Part of Computer* table is linked to the first line of the *Entity* table. Therefore, all its lexical units inherit of the *Link with domain* attribute with the *Object* value.

In order to constitute the collection of texts, the key words "European decision" have been searched using the Yahoo engine[4] for texts in English. The returned first 150 links were automatically collected. The textual part of these documents, which were in three formats, HTML, PDF and DOC, were automatically isolated in order to constitute a corpus of text documents, each one between 1,000 and 50,000 tokens.

ProxiDocs is used in order to project the devices in the corpus, building both "clouds" of lexical units and maps of texts, discussed below.

[4] `http://www.yahoo.com`.

Entity	Link with domain
computer, Internet, Web, Cyberspace, etc.	object
-	agent
computing, computer science, informatics, etc.	activity
bug	phenomenon

Activity	Activity's type
computerize, programming, code, programme, etc.	job
hacking, cracking	non professional

Agent	Agent's type
computer specialist, computer engineer, etc.	human
robot	material
virus, anti-virus	program
Microsoft, Apple, IBM, etc.	company

Part of computer	Object's type
network, screen, display, mouse, keyboard, etc.	hardware
Wep page, Web site, video game, windows, etc.	software

Fig. 6. Computer Science device used during the experiment

Fig. 7. Cloud showing frequent words

4.2 Results and Discussion

Figure 7 is called a "cloud" of lexical units. Such clouds have been introduced on the Web site *TagCloud*[5] to give a global view on blogs. A cloud reveals which lexical units from the selected devices have been found in the documents of the corpus. They are sorted out in alphabetical order and their size is proportional to their number of occurrences in the corpus. Here, lexical units from the computer science domain are particulary found, with the words

[5] `http://www.tagcloud.com`.

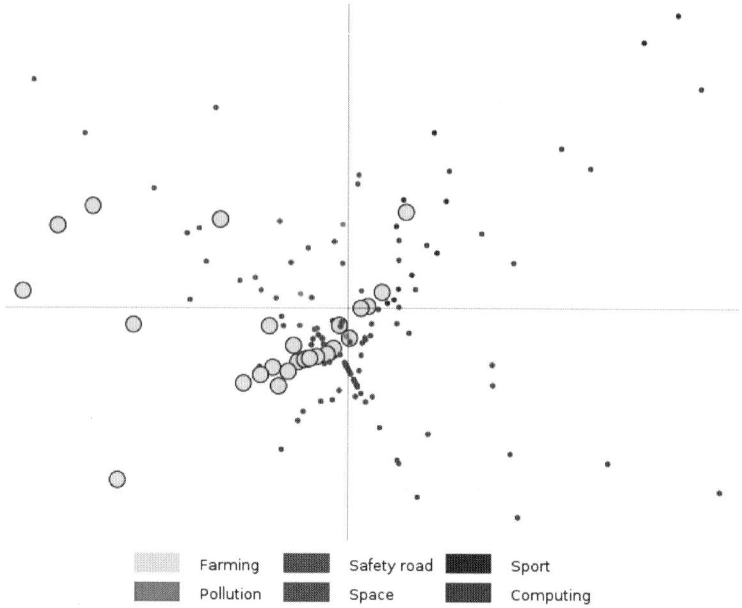

Fig. 8. Web Pages Map of the analysed set

programme, *network*, *Microsoft*, *software*, etc. Some words from the pollution domain and from the agriculture domain are also emphasised.

Such clouds constitute a first corpus analysis which can help the user accessing textual information by simply bringing frequent terms to the fore, according to his own lexicon.

Figure 8 reveals proximities between Web pages of the set according to the users' devices. Each point or disc on the map represents a Web page. Its color is the one of the most often represented device in the document. Each point or disc is a link to the represented document. The map is interactive, when users put their mouse on a device name on the caption (at the bottom of the map), documents mainly dealing of this device are emphasized. On Fig. 8, pages mainly about the farming domain are brought to the fore: documents mainly about this domain are represented by discs. Such interactions enable user to have a concrete idea on domains present or not and on links between domains in the documents of the set.

Figure 9 reveals proximities between documents according to the user's devices. Each disc on the map represents a cluster. Its size is proportional to the number of documents contained in the cluster. Its colour is the one of the most often represented device in the cluster and its label contains the five most frequent lexical units. The map itself is interactive, each disc is also a "hypertext" link to a description of the cluster. The description shows, sorted by frequency, the lexical units, the attributes and values found in the cluster, etc.

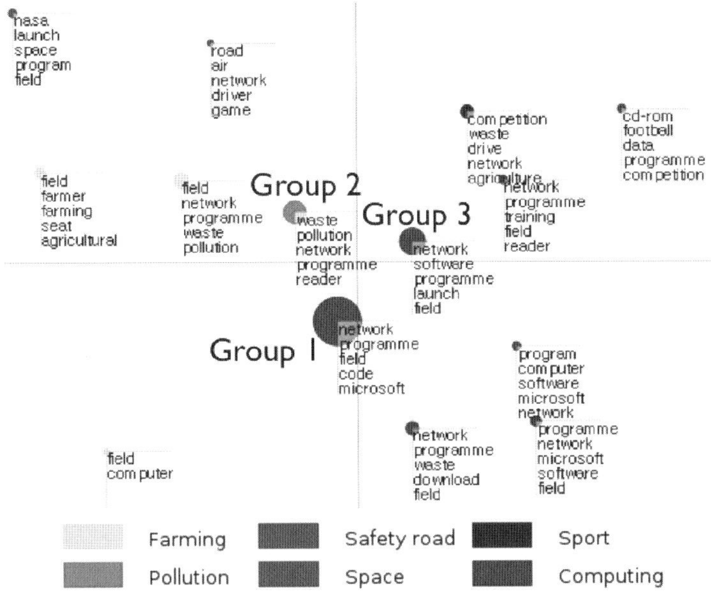

Fig. 9. Map of clusters

The map, as well as the previous cloud, reveals that the computer science domain is particularly represented in the corpus. The largest disc (manually annotated *group 1* on Fig. 9) has the colour of this domain. But an analysis of this cluster shows that the documents are related to many themes (health, politics, broadcasting of the information, etc). The computer science domain is not really the main theme. It is rather the notion of vector of communication which is often mentioned in this corpus, whatever the theme of the documents.

The attributes and values frequently repeated in the documents of group 1 are *Object type* with values *hardware* and *software* and *Activity type* with value *job*. They highlight that the documents of this group are mostly dealing with objects and jobs in computer science.

Group 2 is mainly about the pollution domain. Here, an analysis of the cluster shows documents really dealing with problems related to the pollution, and more particularly with European decisions on the sustainable development, where the couples (attribute: value) *State: gas* and *Evaluation: bad* are the most frequent. These two groups illustrate two different interpretations of proximities and maps.

Like Group 1, Group 3 is mainly about the computer science domain. Contrary to this first group, it really addresses computer science and more specifically problems between Microsoft and the European Union. Here, the attributes and values *Object type: hardware and software* and *Evaluation: bad* are the most frequent, which illustrates the main topic of this group of Web pages.

The graphical outputs presented in this section provide the user a personalised help for accessing textual information, reflecting the way his own knowledge on domains he describes is related to the documents in a collection. It is the main objective of the model and tools developed. Section 5 presents a completely different kind of experiment to show the flexibility and the adaptability of these model and tools.

5 Experiment 2: Conceptual Metaphors

In this second experiment, the objective is a corpus-oriented study of the way the lexicon related to conceptual metaphors is used. A possible application of such a study in NLP is a help for text interpretation or semantic analysis when conventional metaphorical meanings are suspected. This work has been realised under a project called *IsoMeta*, which stands for *isotopy* and *metaphor*. It is not an isolated experiment, for the whole project *IsoMeta* involves a set of experiments in an incremental approach. The first part, now completed, consisted in adapting the LUCIA model for lexical representation in order to characterise the main properties of metaphorical meanings, presented in 5.1. The second part, Sect. 5.2, is a study of what could be called *metaphoricity* of texts in a domain-specific corpus.

5.1 Constraints on the Model for Metaphor Characterisation

This work is based on the existence of recurrent metaphoric systems in a domain-specific corpus. It is closely related to conceptual metaphors as introduced by Lakoff and Johnson [13], more specifically ones with a common target domain, which is the theme of the corpus. Previous works have already shown different conceptual metaphors in a corpus of articles about Stock Market, extracted from the French newschapter *Le Monde*: "the *meteorology* of the Stock Market", "the *health* of Economics", "the *war* in finance", etc. The first part of the *IsoMeta* project focussed on how the LUCIA model for lexical representation could help to describe a specific metaphorical meaning.

Rather than changing the core of the LUCIA model, a protocol for building the lexical representations has been defined, with constraints taking the main properties of metaphors into account. The first property is the existence, for a conceptual metaphor, of a source domain and a target domain. In [7], Fass proposed a classification of the different approaches to metaphor, discriminating between the comparison point of view and the novelty point of view. The two last properties reflect these two points of view. The second property is the existence of an underlying analogy between the source and the target of a metaphor, which is the comparison point of view. The third and last property is the possible transfer of a new piece of meaning from the source, then considered as a vehicle, to the target, which is here the novelty point of view.

The hypothesis on metaphors studied in the *IsoMeta* project can not be detailed in this chapter. See previous works for specific information, e.g. [2,17]. See also [8] for further works on metaphors, tropes and rhetoric.

Source and Target Domains

Conceptual metaphors involve a source domain and a target domain. Thus, a first constraint consists in building a LUCIA device for the source domain and another one for the target domain. For instance, to study the "meteorology of the Stock Market", a device describing the lexicon related to *meteorology* must be built, and another one for the *Stock Market* lexicon. But conceptual metaphors only use semantic domains, and when they are used in language, the resulting figure is not necessarily a metaphor. It can be a conventional one, *lexicalised*, and no longer perceived as a metaphor.

For instance, in our corpus, the French word "baromètre" (*barometer*) is commonly used to talk about the Stock Exchange. It can be considered as a lexicalisation, and "baromètre" becomes a word of the *Stock Market* lexicon. In this case, using the LUCIA model, the word is simply considered as polysemous, and can be described in both devices, for each of its meanings. For the purpose of this study, describing the meaning related to the conventional metaphor is forbidden, the word must not appear in the target device. The goal is here to use the model to "rebuild" the metaphorical meaning, not to literally code it as an ad hoc resource. The other constraints must help this "rebuilding".

Analogy

The analogy between the source and the target of a metaphor is usually a clue in NLP for semantic analysis. In the LUCIA model, the constraint reflecting this analogy is a set of common attributes shared by the source and target devices. For instance, the couple (attribute: value) (*tool*: *prevision*) can be used to describe *barometer* in the source domain. The same couple can also be used in a description from the target device, e.g. for *computer simulation*. Thus, this shared couple reflects the underlying analogy between the two domains, and allow to rebuild the conventional metaphorical meaning of *barometer* in a sentence like:

The Dow Jones is a Stock Exchange barometer

Furthermore, describing related words in the same device allows an interpretation of variations:

The Dow Jones, for instance, the thermometer of Wall Street, which had fallen from 508 points...

The Dow Jones is the New York Stock Exchange mercury

In these two examples, the same attribute *tool* with another value e.g. *measuring device* can explain the nuance of meaning.

The analogy underlying a metaphorical meaning is hard to find again using resources such as Lucia devices. Indeed, this kind of semantic representation is rather dedicated to a surface description than to deep ones. Thus, when used to describe a lexical entry of the source domain of a metaphor, the user must be aware of both the metaphorical meaning and the usual one. It is then possible to propose couples (attribute: value) that are sufficient for an interpretation of the metaphorical meaning, and compatible with the usual one. But compared to the complexity of resources used in approaches to metaphor or analogy such as e.g. in [6, 7, 9], our representation does not contain enough information to justify the existence of an analogy or a metaphor between the source and the target.

In our approach, the existence of the relation between the source and the target is presupposed, as in recent works on the matter [1, 4]. The shared attributes and values may only reflect this relation. In the experiment, their main purpose is to help interpreting the metaphorical meanings, not to find them.

Novelty

Somehow the novelty property consists in using the metaphor to bring something new in the target domain. For instance, in:

The storm has now reached the Stock Markets.

Storm not only denotes agitation, it also differs from other words referring to the same kind of turbulences: *wind*, *breeze*, *tornado*, etc. Therefore, the strength of the phenomenon, which is mainly what characterises this particular word compared to the other ones, is also the piece of new information it brings to the target. A *storm* in a financial place is not only agitation, it is a strong, violent one. A specific attribute *strength* with the value *high* is enough to help interpreting the novelty part of the metaphorical meaning in the previous example.

The novelty property can be rendered if the corresponding specific attributes are well identified as being "transferable" from the source domain to the target domain. Our hypothesis is that they belong to the same class as the shared attributes. They can become shared when more domains are described. Thus, in the semantic representation, it is not necessary to distinguish these attributes from the shared ones used for the analogy constraint. Therefore, the constraints for analogy and novelty can finally be viewed as a unique one: a set of "sharable" attributes must exist for the description of the source and the target domain, clearly identified as transferable to reflect metaphorical meanings.

5.2 Map and Texts "Metaphoricity"

In the second part of the *IsoMeta* project, the previous protocol is used to study multiple conceptual metaphors in the same domain-specific corpus. A LUCIA device is built for each domain, the three source domains, *meteorology*, *war* and *health*, as well as one unique target domain *Stock Market*. Words from the three source domains can be used both for metaphorical and literal meanings in this corpus. Usually, NLP approaches to metaphor focus on locally disambiguating such polysemy. Our hypothesis is the language of the whole text may be viewed as more or less metaphorical. This can be compared to NLP methods used to determine the main language of a text: the whole text is then viewed as monolingual, even if other languages can be used locally. In our hypothesis, we consider the degree of *metaphoricity* of a whole text as a general tendency, even if local exceptions can exist.

Therefore, experiment 2 consists in using the *ProxiDocs* tools to classify texts from the lexical resources related to conceptual metaphors. Results are detailed in [24]. Figure 10 shows the most relevant one. After the analysis of

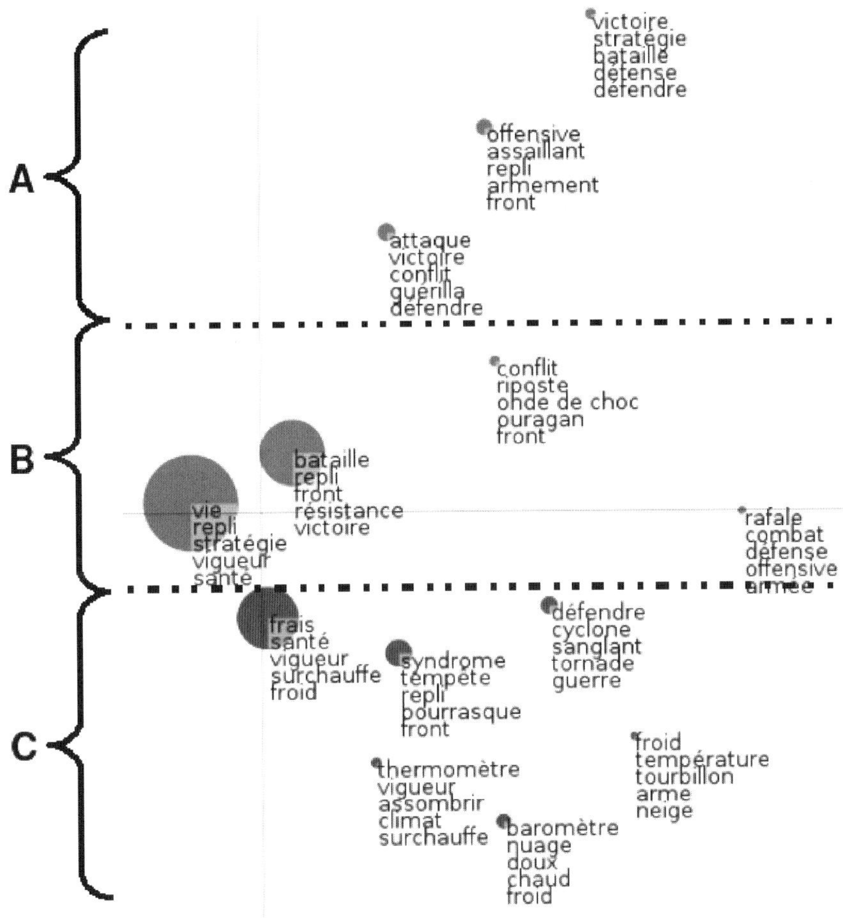

Fig. 10. Cartography reflecting the "metaphoricity" of texts

the map three zones can be drawn. Zone **A** contains texts' clusters in which mostly literal meanings are used, e.g. in:

Pour se déplacer (...), des officiers de la guérilla utilisent les motos récupérées pendant les attaques. (*For their movements, the guerrilla war officers used the motorbikes found in the assaults.*) Le Monde, 13/04/1987

The *war* lexicon is not metaphorical. Zone **B** contains mostly conventional metaphors, e.g. in:

En neuf mois, six firmes sur les trente-trois OPA ont été l'objet de véritables batailles boursières. (*In nine months, 6 firms out of the 33 takeover bids were subjected to real financial battles.*) Le Monde, 26/09/1988

In which the phrase "bataille boursière" is a common one. Zone **C** contains rare and more varied metaphors, e.g. in:

Porteur du terrible virus de la défiance, il se propage à la vitesse de l'éclair et les tentatives désespérées de réanimation (...) sont inopérantes. (*Carrying the dreadful virus of distrust, it spreads in a flash and the desperate attempts of reanimation are vain.*) Le Monde, 30/10/1987

Un petit vent frisquet a soufflé, ces derniers jours rue Vivienne, qui, sans crier gare, s'est soudain éclipsé à la dernière minute pour laisser la place à une brise nettement plus chaude. (*A gentle chilly wind was blowing other the last days on the French Stock Market, which, without warning, suddenly disappeared at the last minute to make place for a noticeably warmer breeze.*) Le Monde, 15/05/1989

The map reveals what can be called the "metaphoricity" of texts, from a degree 0 at the top of the map to the highest degree at its bottom. This is an interesting result for our study on the use of conceptual metaphors in collections of text. But in this chapter, our aim is merely to illustrate the high flexibility of the model and tools we used. A user may add his own rules, as the protocol defined for building devices, in order to fulfill his own task involving semantic access to a collection of texts.

6 Conclusion

In this chapter, we presented a centred-user approach for accessing textual information. Previous works focussed on how to take user's preferences into account for such tasks, when other works merely studied the interaction between users and documents, proposing specific visual representations and graphical tools. Our aim is to combine these two aspects into a generic approach. Founded on a model for lexical representation, a set of interactive tools have been developed to help users to specify their own point of view on a domain and using this knowledge to browse through a collection of texts.

We presented two different experiments in order to illustrate how to use such model and tools as well as to point their high flexibility. The first experiment consisted in providing help for a traditional task of access to textual information. The second one concerned a study of the use of conceptual metaphors in specific domains. It obviously showed that a user can easily appropriate the model and adapt it to a task far from its original purpose.

This result raises interesting questions we hope to answer in future works: What is the role of the graphical tools in the process of appropriation? Can models and tools be both flexible and not diverted?, etc. For the time being, our perspectives mostly concern the evaluation of the model in a well-defined task with a large number of users. Our current works focus on how to characterise the contribution of the user's point of view in tasks of access to textual information.

Acknowledgements

First, we want to thank for their comments and advice both the reviewers of this book and the ones of the First Semantic Media Adaptation and Personalization Workshop (SMAP'06).

We are also very grateful to Mrs. Dominique Goron, Mr. Yves Lepage and Mr. Pierre Beust of the University of Caen (France) for their help in the realization and the presentation of this chapter.

References

1. A. Alonge and M. Castelli. Encoding information on metaphoric expressions in wordnet-like resources. In Mark Lee John Barnden, Sheila Glasbey and Alan Wallington, editors, *Proceeding of the ACL 2003 Workshop on the Lexicon and Figurative Language*, pages 10–17, 2003.
2. P. Beust, S. Ferrari, and V. Perlerin. Nlp model and tools for detecting and interpreting metaphors in domain-specific corpora. In Dawn Archer, Paul Rayson, Andrew Wilson, and Tony McEnery, editors, *Proceedings of the Corpus Linguistics 2003 conference*, volume 16 of *UCREL technical papers*, pages 114–123, Lancaster, U.K., 2003.
3. J.M. Bouroche and G. Saporta. *L'analyse des données.* Collection Que sais-je? Presses Universitaires de France, Paris, 1980.
4. K. Chibout, A. Vilnat, and X. Briffault. Sémantique du lexique verbal: un modèle en arborescence avec les graphes conceptuels. *TAL*, 42(3):691–727, 2001.
5. W. Chung, H. Chen, and J.F. Numaker. Business intelligence explorer: A knowledge map framework for discovering business intelligence on the web. In *HICSS '03: Proceedings of the 36th Annual Hawaii International Conference on System Sciences (HICSS'03) – Track1*, page 10.2, Washington, DC, USA, 2003. IEEE Computer Society.
6. B. Falkenhainer, K.D. Forbus, and D. Gentner. The structure-mapping engine: Algorithm and examples. *Artificial Intelligence*, 41(1):1–63, November 1989.
7. D. Fass. *Processing metaphor and metonymy.* Ablex, Greenwich, Connecticut, 1997.
8. S. Ferrari. Rhétorique et compréhension. In Gérard Sabah, editor, *Compréhension des langues et interaction*, chapter 7, pages 195–224. Lavoisier, Paris, 2006.

9. D. Gentner. Structure-mapping: A theoretical framework for analogy. *Cognitive Science*, 7:155–170, 1983.
10. A.J. Greimas. *Sémantique Structurale.* Larousse, 1966.
11. M.A. Hearst. Tilebars: Visualization of term distribution information in full text information access. In *CHI '95: Proceedings of the SIGCHI conference on Human factors in computing systems*, pages 59–66. ACM Press/Addison-Wesley, USA, 1995.
12. C. Jacquemin and M. Jardino. Une interface 3D multi-échelle pour la visualisation et la navigation dans de grands documents XML. In *IHM '02: Proceedings of the 14th French-speaking conference on Human–computer interaction (Conférence Francophone sur l'Interaction Homme–Machine)*, pages 263–266. Poitiers, France, ACM Press, USA, 2002.
13. G. Lakoff and M. Johnson. *Metaphors we live by.* University of Chicago Press, Chicago, USA, 1980.
14. J. Lamping. A focus+context technique based on hyperbolic geometry for visualizing large hierarchies. In *CHI '95: Proceedings of the SIGCHI conference on Human factors in computing systems*, pages 401–408. ACM Press/Addison-Wesley, USA, 1995.
15. J. MacQueen. Some methods for classification and analysis of multivariate observations. In *Proceedings of the fifth Berkeley symposium on mathematical statistics and probability*, volume 1, pages 281–297. University of California Press, Berkeley, USA, 1967.
16. V. Perlerin. *Sémantique Légère pour le document.* PhD thesis, University of Caen – Basse-Normandie, 2004.
17. V. Perlerin, P. Beust, and S. Ferrari. Computer-assisted interpretation in domain-specific corpora: the case of the metaphor. In *Proceedings of NODALIDA'03, the 14th Nordic Conference on Computational Linguistics*, University of Iceland, Reykjavík, Iceland, 2003.
18. V. Perlerin and S. Ferrari. Modèle sémantique et interactions pour l'analyse de documents. In *Proceedings of the 7th French speacking International Conference on Electronic Document (Approches Sémantique du Document Electronique, Colloque International sur le Document Électronique CIDE 7)*, pages 231–251. 22–25 June 2004, La Rochelle, France, 2004.
19. A. Popescu, G. Grefenstette, and P.-A. Moellic. Using semantic commonsense ressources in image retrieval. In P. Mylonas, M. Wallace, and M. Angelelides, editors, *Proceedings of the 1st International Workshop on Semantic Media Adaptation and Personalization*, pages 31–36. 4–5 December 2006, Athens, Greece, IEEE Computer Science Society, 2006.
20. S. Poslad and L. Zuo. A dynamic semantic framework to support multiple user viewpoints during information retrieval. In Phivos Mylonas, Manolis Wallace, and Marios Angelelides, editors, *Proceedings of the 1st International Workshop on Semantic Media Adaptation and Personalization*, pages 103–108. 4–5 December 2006, Athens, Greece, IEEE Computer Science Society, 2006.
21. F. Rastier. *Sémantique Interprétative.* Presses Universitaires de France, Paris, 1987.
22. G. Robertson, J. Mackinlay, and S. Card. Cone trees: Animated 3d visualizations of hierarchical information. In *CHI '91: Proceedings of the SIGCHI conference on Human factors in computing systems*, pages 189–194, New York, NY, USA, 1991. ACM Press, USA.

23. T. Roy and P. Beust. Un outil de cartographie et de catégorisation thématique de corpus. In G. Purnelle, C. Fairon, and A. Dister, editors, *Proceedings of the 7th International Conference on the Statistical Analysis of Textual Data*, volume 2, pages 978–987. Presses Universitaires de Louvain, 2004.
24. T. Roy, S. Ferrari, and P. Beust. étude de métaphores conceptuelles à l'aide de vues globales et temporelles sur corpus. In P. Mertens, C. Fairon, A. Dister, and P. Watrin, editors, *Verbum ex machina – Proccedings of TALN'06, the 13th conference Natural Languages Processing*, volume 1, pages 580–589. Presses Universitaires de Louvain, Louvain-la-Neuve, Belgium, 2006.
25. G. Salton. *Automatic Text Processing: The Transformation, Analysis, and Retrieval of Information by Computer.* Addison-Wesley, 1989.
26. J. Sammon. A nonlinear mapping for data structure analysis. *IEEE Transactions on computers*, C-18-5:401–409, 1969.
27. B. Shneiderman. The eyes have it: a task by data type taxonomy for information visualization. In *VL '96: Proceedings of the 1996 IEEE Symposium on Visual Languages*, pages 336–343, Washington, DC, USA, 1996. IEEE Computer Society.
28. E. Spinat. Pourquoi intégrer des outils de cartographie au sein des systèmes d'information de l'entreprise? In *Actes du Colloque Cartographie de l'Information: De la visualisation à la prise de décision dans la veille et le management de la connaissance*, 2002.
29. D. Vallet, I. Cantador, M. Fernandez, and Pablo Castells. A Multi-Purpose Ontology-Based Approach for Personalized Content Filtering and Retrieval. In P. Mylonas, M. Wallace, and M. Angelelides, editors, *Proceedings of the 1st International Workshop on Semantic Media Adaptation and Personalization*, pages 19–24. 4–5 December 2006, Athens, Greece, IEEE Computer Science Society, December 2006.
30. J. Véronis. A comparative study of six search engines. Author's blog: http://aixtal.blogspot.com/2006/03/search-and-winner-is.html, March 2006.

Modeling the Reusable Content of Adaptive Web-Based Applications Using an Ontology

Mária Bieliková and Michal Moravčík

Institute of Informatics and Software Engineering, Faculty of Informatics and Information Technologies, Slovak University of Technology in Bratislava, Ilkovičova 3, 842 16 Bratislava, Slovakia bielik@fiit.stuba.sk

Summary. Personalization becomes more common in Web-based applications. More and more adaptive Web-based applications that adapt information presentation (the content or navigation) employing goals and other characteristics of a user or context are developed. By increasing the number of existing adaptive applications the need for more effective creating and reusing the content among adaptive applications rises. In this chapter we present an approach to creating the content using ontological representation that is suitable for reusing. The ontology includes a domain model and domain dependent part of a user model, both specified on several levels of abstraction. We support the content transformation between several applications and its cross-system use that enables better use of best characteristics of each application without the need of manually creation of the content, which already exists. We evaluated the method using the domain ontology of teaching programming by examples. We claim that the content and structure of the adaptive application represented by the ontology is suitable for sharing and reusing when defining new applications.

1 Introduction

Growths of available information presented by current Web-based information systems requires an assistance to a user in the task of finding relevant information. So an intelligent support of navigating in large information spaces by Web-based applications is a key requirement today. One approach to achieve this is adapting information presentation (its layout or the content itself) or navigation to each user or a context of the presentation. A need for adaptive Web-based applications was recognized a decade ago, and from that time new adaptive Web-based applications are being still developed.

With respect to the increasing usage of adaptive Web-based applications the need for effective authoring and content delivering becomes increasingly important. However, the content of current adaptive applications is generally not designed for cross-system usage. We can improve situation by providing means for developing reusable models of adaptive applications together with tools for creating instances of the models in various domains.

M. Bieliková and M. Moravčík: *Modeling the Reusable Content of Adaptive Web-Based Applications*, Studies in Computational Intelligence (SCI) **93**, 307–327 (2008)
www.springerlink.com

The base idea of modeling is to deliver general, simple, consistent, portable, reusable representation of the modeled subsystem. These characteristics are in the goals of several existing adaptive hypermedia reference models. By enabling a transformation of the content among adaptive applications the best features of individual applications can be employed. For example existing authoring systems can be reused as authoring tools not only for adaptive applications they were developed for but also for other applications which provide just content presentation. This principle is known as a metaphor *authoring once, delivering many* [16]. That is, an author should only to create a piece of information (e.g., educational material) once. This information should be easily re-usable in many other content presentation systems. This allows for the cost-effective creation of information as a user can use for authoring just one application.

Our aim is to support a reuse the content (domain) model of adaptive Web-based applications employing an ontology as the knowledge structuring approach used extensively for the Semantic Web applications. The goals and possibilities of the Semantic Web designate the advantage of using the ontology as a domain model of adaptive applications. Several approaches expressing particular models of adaptive applications by means of the Semantic Web technologies exist (e.g., [1, 13, 15]). Nevertheless, the lack of effective reuse of adaptive applications models is still in course.

We present a method for modeling content of adaptive applications using ontology, exporting the content into intermediate format and importing it into (possible more than one) adaptive application. This approach enables also reuse of existing content by its transformation into the ontology. For evaluation of the method we selected the CAF format (Common Adaptation Format, [9]) and adaptive Web-based application AHA! [10]. Moreover, we transformed the content of another adaptive Web-based system ALEA [4] into our modeling environment, so enable sharing the content between several applications. We present proposed core ontology of the content used throughout the method together with examples from domain of teaching programming.

The chapter is organized as follows. In Sect. 2 we present related work, i.e. existing approaches to adaptive application content modeling with stress on the Semantic Web technologies usage. We devote special attention to Web-based application content representation approaches and its transformation between adaptive applications. Section 3 contains description of proposed method for domain modeling, importing existing domain model to the ontology representation and delivering the ontology content into existing adaptive applications. In Sect. 4, we present ontology models of the content used for domain and user modeling. This section is followed by discussing meta-model of the content that aims at delivery the content into the adaptive system with defined concept sequences (Sect. 5). Section 6 includes discussion on proposed method evaluation in the domain of learning programming. The chapter concludes with conclusions and topis for future work.

2 Related Work

Content-oriented adaptive applications modeling can be seen from two points of view: modeling the architecture of an application (models of generic Web-based applications can be also considered) or modeling the content presented by the application (represented using a domain model). Architecture of adaptive Web-based applications can be expressed using methods for modeling generic Web-based applications (e.g., OOHDM [17], WebML [6], Hera [13]) even though the specification of adaptive behavior is not accordingly handled by all of these methods. Common architecture of adaptive hypermedia systems reflects reference models such as AHAM [20], Munich reference model [14] or LAOS [7], which define layers of the adaptive application by separating data models (e.g., domain, user, context, goals) from an adaptation engine. Reference models are used as a base for new adaptive applications.

Not all layers of adaptive application model are adequate to be shared and/or transferred between applications. The common for the applications is the content with definitions of its usage in a user model (expressed in domain dependent part of the user model).

2.1 Content Representation

Modeling the content of a content oriented Web-based application and its efficient representation is as well important as modeling the application itself. For content modeling it is important to analyze to what extent is a particular representation flexible for different domains together with the possibility of reasoning directed to decisions on properties of the information space (e.g., consistency). We do not consider proprietary formats as they almost totally prevent the sharing and reuse of the domain model.

Existing approaches to representing the content of a Web-based application include mainly approaches using a relational database or an XML based language. XML offers powerful enough expressiveness. The performance of this solution is limited by the performance of the used file system (it is effective for domain models with few instances and rich structure of concept's characteristics). Reusability and sharing is better than with the database approach, thanks to the platform independence of XML. Using XML has also the advantage that it can be used directly in the Web environment. However, XML as a meta-language defines only general syntax without formally defined semantics, which leads to difficulties when reasoning is required. Moreover, everyone can invent his own names for tags; somebody stores attributes as tags; somebody uses the attributes of tags defined by XML syntax.

Both above mentioned approaches offer only a way of describing characteristics of domain model concepts and do not offer any added value from the content modeling perspective. Ontology-based approach offers a way of moving content modeling from the low-level describing of domain concept characteristics to a higher-level with additional possibilities (reasoning).

According to the most cited definition of ontology in the Semantic Web community, ontology is an explicit specification of the conceptualization of a domain [12]. The term ontology includes a whole range of models that show varied semantic richness. We represent the ontology by RDF/OWL formalisms (Resource Description Framework; Web Ontology Language). An approach based on RDF and its extension OWL takes the previously mentioned XML representation (syntax) and eliminates its disadvantage by defining a vocabulary for describing properties and classes. OWL serves as a common language for automated reasoning about the content for the vision of the Semantic Web.

The advantages leading to using ontologies for content modeling come from the fundamentals of this formalism. Ontologies provide a common understanding of the domain to facilitate reuse and harmonization of different terminologies. They support reasoning, which is considered as an important contribution of the ontology-based models. Although there exist several approaches where ontology is used as a base for models representation [15], usually specialized ontology for particular case is developed only. We use the ontology as a universal format for representation of models and define a method for their reusability.

2.2 Content Transformation

Creating a schema of the domain and instances of it is serious bottleneck of content oriented Web-based applications. Using ontology for representation of the domain increases the probability that domain concepts together with their characteristics will be shared among a range of applications of the same domain (especially on the Web, where most ontologies are currently represented using OWL).

Reusing the content can be realized using commonly accepted model for the content of adaptive applications in particular domain, or mapping the content among adaptive applications. A commonly accepted domain model is ideal solution. Since we agree that building common vocabularies is important and useful (we remark the role of standards), considering a large distributed information space (e.g., the Web) we need to make a compromise between enabling diversity and looking for mappings between various models. The idea of commonly accepted domain ontology is simply impossible to reach in such diverse and distributed environment as the Web.

On the other hand, designing converters for each pair of applications is rather ineffective and demanding approach. A compromise lays in defining an intermediate format of the information content for adaptive applications. In this case it is sufficient to convert the content from the intermediate format to the adaptive application format and vice versa. Standards can also help in this process [5].

One of the first attempts to use information content in several applications was a conversion from the Interbook to AHA! [11]. In this case the

Interbook system serves as authoring tool and the AHA! system for adaptive presentation. Another approach described in [9] defines a conversion from the MOT system (My Online Teacher, [7]) to adaptive applications of AHA! and WHRULE systems. MOT is used as an authoring tool, where it is possible to define the content of adaptive application and adaptation strategy that specifies personalization of educational content according changing user characteristics. The conversion from MOT to AHA! uses intermediate format CAF (Common Adaptation Format) that defines a hierarchy of concepts and corresponding attributes of concepts using an XML.

3 Development of Reusable Content

We use the ontology as a mean for representing and maintaining data related to an application domain. Figure 1 depicts our proposal of domain model schema as it is modeled using the ontology. Considering adaptive Web-based applications we define also domain dependent part of the user model that is automatically generated from the domain model.

The domain model defines concepts and their content (for the content of concepts we use term *concept attributes*), domain attributes of concepts and typed *relations* to other concepts. Domain attributes represent information important for adapting but they are not necessarily viewable in the content presentation.

The user model defines for each concept a *User View* and a set of *User View Attributes*. The user view makes a connection among specific user, concept and attributes, which frame together an overlay user model used in most of the adaptive Web-based applications and defined in both AHAM and LAOS

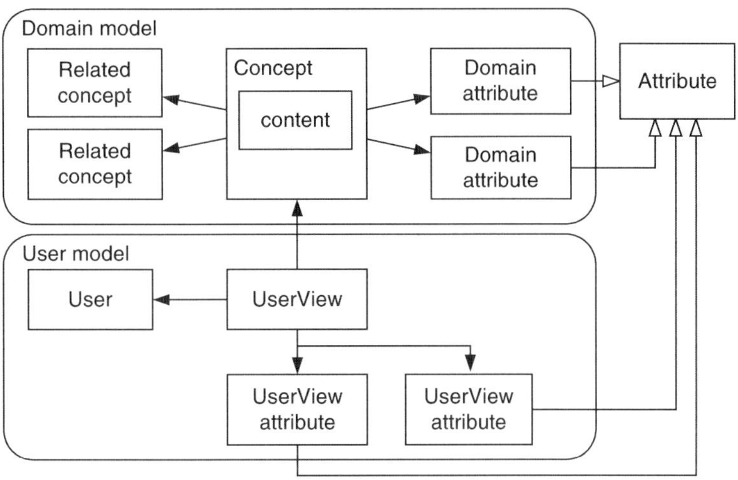

Fig. 1. Schema of adaptive application content model

reference models. Our approach is mainly oriented towards the content modeling. The user model is defined in such a way that can be semiautomatically derived from the domain model. Its actual representation strongly depends on the adaptive application and the means for user characteristics acquisition.

Boxes in Fig. 1 represent classes in the ontology (we represent it in OWL/RDF) and every connection (except those deriving the attributes) stands for a relation in the ontology. Specification of a structure of the adaptive application content is done by defining classes and their relations in the ontology. The content of adaptive application itself is represented by instances of classes defined in the ontology with connections specified by relations.

3.1 The Method for Domain Modeling

We proposed a method for modeling the content of adaptive application that uses *core ontology* designed for adaptive application content modeling. It defines steps essential for creating a new domain ontology, which is compliant with the core ontology, describes an export of the ontology into an intermediate format, which is used for importing into existing adaptive applications. We assume also a situation where the domain model already exists and provide a mapping into proposed domain model schema.

The method defines following steps for authoring adaptive application domain model represented by ontology:

Method 1 Modeling reusable content for adaptive applications

1. `Specify the concept classes and the concept class hierarchy`
2. `Define attributes for the concept classes`
   - `data attributes`
   - `domain attributes`
   - `user attributes`
3. `Specify and type relations between concepts of specified classes`
4. `Create concept instances as the class instances of the ontology`

Steps 1–3 are not as a rule performed in this order and in one iteration. Actually they are often mixed and reordered in several iterations with aim to develop consistent model of the content. Developing domain model is followed by the step of *delivering the ontology content into existing adaptive application.*

Step 1: Specify the Concept Classes and the Concept Class Hierarchy

In this step we classify information fragments which are supposed to be used in the adaptive application. Information fragments serve as a basis for determining the concepts and concept classes, which are identified for example by means of text analysis, i.e. subjects and nouns used in application context are considered (e.g., exercise, example, explanation in programming learning domain).

Concept classes are organized into a hierarchy based on their common properties. In this step the concept class hierarchy is not final and is often changed when discovered duplicate attributes or relations among the concepts. Entry point to the adaptive application is a concept of special type *root* (base).

Step 2: Define Attributes for the Concept Classes

Concept class defines its information fragments, attributes, which have defined their values in the concept instances. For each attribute we specify a type and a flag indicating whether the attribute has single or multiple values. Data or domain attribute value is assigned to the concept instance while the user attribute value is assigned to the concept and the user instance.

If there is a need in the adaptive application to monitor, for example the level of user knowledge about particular data attribute, good practise would be to transform the data attribute to particular concept class and define the user attribute *knowledge* for that class. In this way we can model a domain using existing standards for meta-data, e.g., IEEE LOM (Learning Object Metadata) or SCORM (Shareable Content Object Reference Model).

Step 3: Specify and Type Relations Between Concepts of Specified Classes

There are various ways of concept interconnections and dependencies with significant impact on the adaptive application behavior. We specify a relation between concept classes by defining a name of the relation, sets of source and target concept classes and type of the relation. In defining generic structure we use several obvious relations (such as those depicted on Fig. 2. Rectangles represent concept classes, lines represent named relations. Relation of type *fragment* is drawn without the lines. Relations can be specialized according specific need in particular application domain.

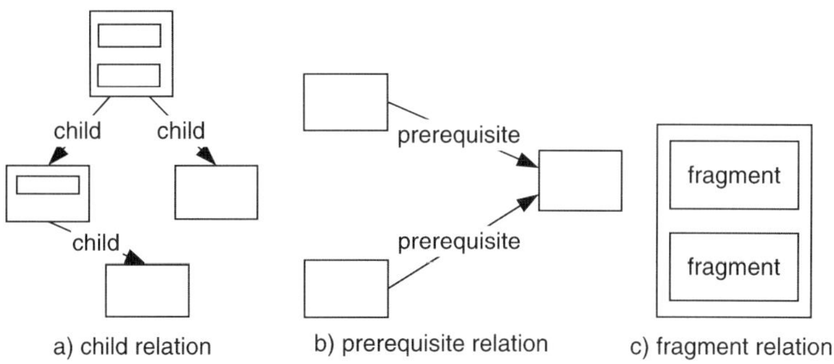

Fig. 2. Examples of types of relationships between concepts

Concept classes relations produce a structure of the content. The parent of all concepts is a single concept of the type *root*. If the content is modeled by more than one concept, at least one *child* relation where the parent is the root concept should be defined.

Step 4: Create Concept Instances as the Class Instances of the Ontology

Performing steps 1–3 leads to a structure of domain ontology. Step 4 is devoted to filling the information content into the ontology, i.e. information fragments are defined. This is needed for later content exporting into existing adaptive application.

Good practice is first to create an instance of the root concept and thereafter instances of concept classes in relation to the root concept class. Process is similarly performed for other concept classes. Creating a concept instance requires specifying the concept name, filling the values of concept attributes and assigning the other related concept instances.

Our method is not aimed at creating actual content of the adaptive application in the ontology. This can be done by various ways (e.g., manually using an ontology editor, automatically by transformation learning objects knowing core ontology and the source format using an existing OWL API). We concentrate on creating a model of the adaptive application not the application itself, therefore we define only what content should be present in the ontology in order to be able to perform delivering the ontology content into existing adaptive application.

3.2 Importing Existing Domain Model Ontologies

Assume that a class in an existing ontology represents a concept class. Then all the data properties of the class represent data attributes and all object properties represent relationships with other concept classes. All the classes in existing ontology including their properties represent the domain model structure of the adaptive application. Instances of all the classes in existing ontology represent the content of domain model of the adaptive application.

This approach helps to use existing sources in form of the ontology as domain models in adaptive applications. All original classes and object properties need to be additionally derived from the core ontology components, which keeps original ontology structure. All instances, data properties and restrictions remain unchanged.

When existing ontology is used as a domain model above described method for reusable domain model development is replaced by the following method 2.

When transforming ontology classes to concept classes in step 1 all data properties of ontology classes become data attributes of concept classes. Additionally we can define other attributes in step 3.

We assume that the existing ontology includes also classes' instances which are treated as concept instances after performing steps 1 and 2. Consequently

Method 2 Importing existing ontology domain model

```
1. Transform all classes in the ontology to concept classes
2. Transform all relations in the ontology to relations between concepts
3. Define attributes for the concept classes
   •  domain attributes
   •  user attributes
4. Specify types of relations between concepts
5. Specify root concept class
```

we have omitted the step of creating concept instances from the original method. In the case the ontology does not contain instances, we follow the original step 4 after specifying the *root* concept class.

Existing ontology is treated as the pure domain model which is extended by other properties of an adaptive application content by performing steps 3–5. Relation types in step 4 are needed to be refined in order to accurate interpretation of concept relations. The *root* concept class in step 5 is necessary for specifying the entry point of the adaptive application.

3.3 Delivering Ontology Content into Existing Adaptive Application

Our aim is to reuse the domain model across several adaptive Web-based applications. Creating reusable domain model is used for delivering the content into existing adaptive application. The process results in creating the intermediate format for adaptive application content and importing the content into particular adaptive application. The intermediate format contains concept instances, attribute values and concept relationship bindings. Next step is specific for each adaptive application as a transformation should be developed depending on actual domain model representation. In general following steps are needed:

- Create concepts and generate concept content from data and domain attributes.
- Create concept user attributes in user model.
- Link concepts according to concept relationships.

The intermediate format has to be efficient for reuse and delivering to adaptive systems, consistent and general enough to store the content of adaptive application. Our proposal for an intermediate format was inspired by the already mentioned CAF (Common Adaptation Format) used for a conversion from the MOT system do adaptive applications AHA! and WHRULE [9]. We have extended CAF to the CAFE (CAF Extended) format, which simplifies CAF format and introduces possibility of complete and consistent definition of concept attributes and relations.

The CAFE format defines the domain model by listing concepts and their data elements, domain attributes, user model attributes and relationships with other concepts. The relations among concepts may be more complex than only parent–child relation type represented in original CAF format. The lesson model from the CAF format is no longer needed, while the concept hierarchy can be derived from concept relations.

The attributes in CAFE format are explicitly divided into concept (data), domain and user model attribute sets. Each attribute has its value type defined. Each concept in CAFE lists its relationships to other concepts with defined relation type and the set of target concepts (*relationLink*).

An example bellow illustrates main parts of the concept description.

```
<cafe>
<domainmodel>
 <concept id="LispA1418" name="Lesson exercise">
    <conceptattributes>
       <attribute type="string" name="hint">
         <contents type="individual">Use the template ...</contents>
       </attribute>
    </conceptattributes>
    <domainattributes>...</domainattributes>
    <userattributes>...</userattributes>
    <conceptrelations>
       <relation type="linked" name="template">
          <relationlink conceptid="T_PROJECTION" />
       </relation>
    </conceptrelations>
 </concept>
</domainmodel>
</cafe>
```

Our method can be used with other intermediate formats. However, number of various formats obviously leads into explosion of possibilities of transformations that should be specified manually (with little help of software tools developed just for this purpose similarly as wrapper specification tools are being developed).

Export of the content into the intermediate format is performed in a cycle where all concept instances are read and parsed to retrieve values of attributes and relations to other concepts. Importing the content into existing adaptive application is specific for each application based on its domain model representation. One way wrapper for each particular model should be written.

4 Ontology Models of the Content

4.1 Core Ontology of Content Model

The core ontology is depicted in Fig. 3. It defines generic terms of adaptive application content and relations between them. We concentrate on domain

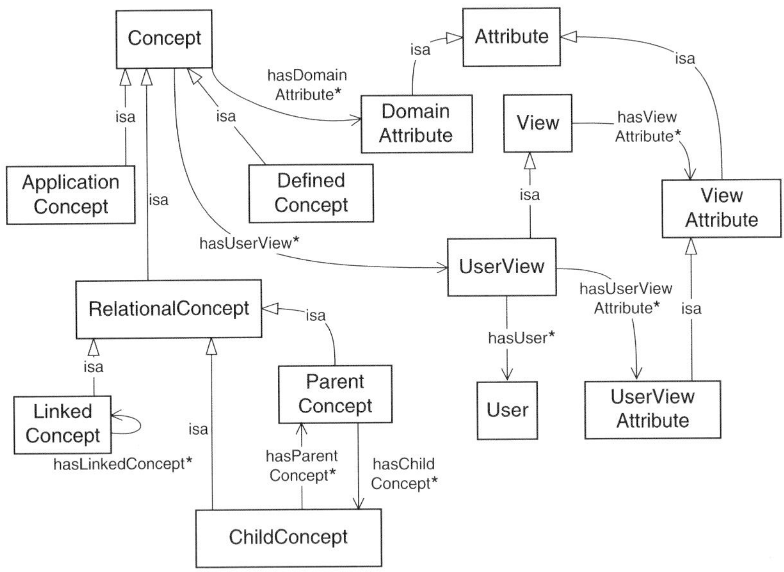

Fig. 3. Core ontology of adaptive application content model

modeling and recognize user attributes only on the level of automatic generation of domain dependent part of the user model. Interconnection of both domain and user models is realized using attributes and views.

Concepts

The *Concept* class is the base class for all concept classes. It presents abstract information unit (e.g., knowledge in educational application) while concept instance presents concrete information fragment (the content). Concept class defines basic data which include *conceptName*, *conceptDescription* and *conceptText*. We assume bidirectional relations between concepts which are supported and simple to maintain in OWL.

We specify three types of concepts on the highest level (classes derived directly from the *Concept* class):

- *DefinedConcept*: is the base class for all user defined concept classes. By deriving classes from the *DefinedConcept* class we differentiate user defined classes from standard classes from the core ontology.
- *ApplicationConcept*: stands for the root concept class, which is an entry point to the adaptive application. The *ApplicationConcept* class is also derived from the *ParentConcept* class, which defines its relation of type parent–child with some other concepts (application content).
- *RelationalConcept*: is the base class for classes in a relation. It indicates that the concept is in a relation with other concept. Type of the relation is defined by source and target relational concept classes, which are derived from the *RelationalConcept* class.

Relations

Hierarchy of *RelationalConcept* classes is depicted on Fig. 4. It demonstrates a technique of defining relation types using the ontology. It is possible to define new types of relations by adding new classes and relations. Each relation is complemented by an inverse relation that is used as reference and is treated automatically. Inverse relations are not depicted in Fig. 4.

For each relation type in the ontology there is a source and target relation class which are connected by object relation. Figure 5 demonstrates a creation of relationships between the concept classes. When we are creating a relationship between two concept classes we need to name the source and target relation classes and the relation itself. Base relation concept class is the *RelationalConcept* class and the base relation is the *hasRelatedConcept* object relation. They are used to derive more specific relations and relation classes like the *ParentConcept*, *ChildConcept* classes and the *hasChildConcept* object relation on Fig. 5.

To define a relationship of type parent–child we need to derive the source concept class from the *ParentConcept* class, derive the target concept class from the *ChildConcept* class and create the object relation derived from the *hasChildConcept* relation, which interconnects the source and the target concept classes. For each relationship we consider also an inverse relationship, which is considered only due to completeness of the model and for assisting reasoning in the ontology.

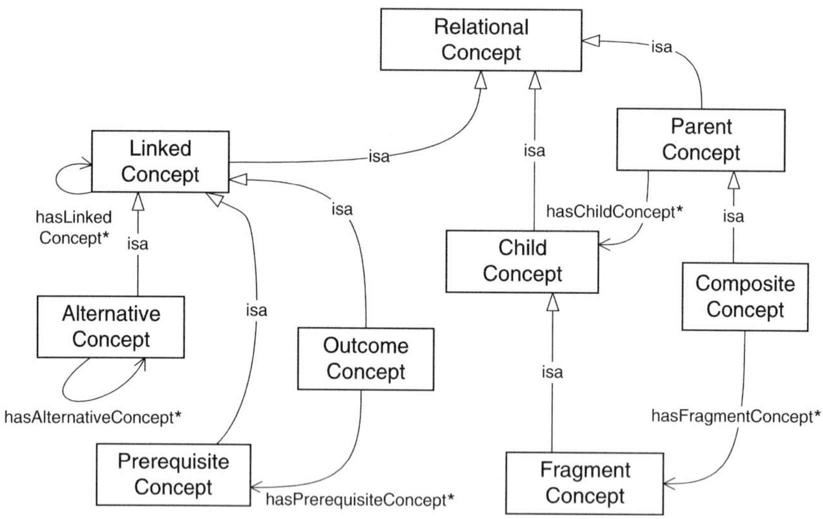

Fig. 4. Hierarchy of relational concept classes

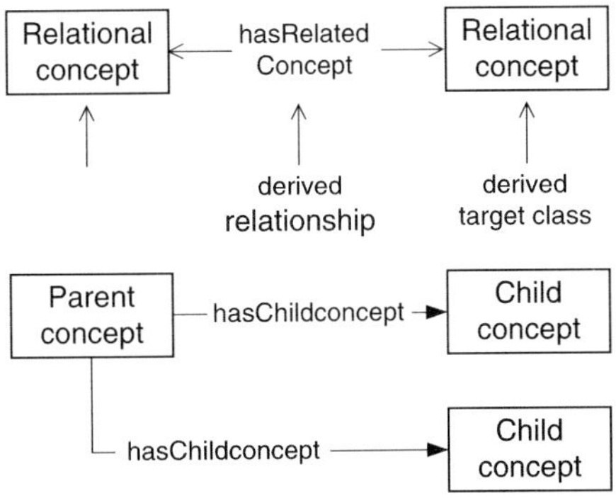

Fig. 5. Creation of relationships between the concept classes

Views

The *View* class represents an abstract view on the concept. *View* defines a set of attributes concerning the adaptive application entity, which can be used for adaptation purposes. For basic modeling we consider one type of view, the view of a user on the concept represented by the *UserView* class.

Views (represented as defined attributes sets) are defined to conform the consistence of modeling. In one of our drafts we considered to insert attributes directly into concept classes. This approach would define only the user meta-model and would not allow to store actual user attribute values. Moreover, the problems with determining attribute sets by ontological restrictions would not be trivial. By defining views we separated the definition of attributes from the content, which simplifies the ontology modeling, the ontology itself becomes usable as a model of adaptive application content for direct use and it is possible to employ actual standards of meta-data representations for particular domain.

User view on the concept implements the overlay user model in the ontology. *UserView* defines the set of attributes for the user model by relational property *hasUserViewAttribute*. The *hasUser* relation connects the view with factual user or with the set of users (in this case stereotype user model). *User view* defines the *ConceptVisited* attribute, which is common for all concepts. It specifies information whether the user has already visited the concept. This attribute is a standard in adaptive Web-based applications. More such often used attributes (e.g., showability, suitability, knowledge) according existing adaptive applications can be defined by extending the *UserView* with additional attributes.

Attributes

Attribute is a base class for all attributes except concept data attributes. It presents information processed by adaptive application with purpose to perform the adaptation. Attribute class represents identification and definition of information fragment for particular concept class, user or other entity from a domain model. For example, in the domain of learning programming the attribute can represent the information about solving the programming exercise or understanding exercised programming concepts. Detaching the attributes into separate classes we have enforced an abstraction from the attribute source and processing attribute values at the level of instances.

We mention three attribute types represented in core ontology:

- *DomainAttribute*: differentiates domain attributes from other types of attributes. Typical usage of domain attribute is meta-data related to a concept, which is not displayed in presentation but influences the adaptation process (e.g., difficulty of the programming exercise concept).
- *ViewAttribute*: deriving the attribute class from the *ViewAttribute* specifies the attribute of a view. Deriving is not straight but through additional attribute class specifying the type of the view, e.g., *UserViewAttribute* for the *UserView* view.
- *UserViewAttribute*: specifies that derived attribute class is valid for particular view class derived from the *UserView* class. Default user model attribute valid for all concept classes is the *ConceptVisited* attribute.

Domain attributes are not contained directly in the concepts, but rather assigned to concepts using object relations between the concept classes and the domain attributes classes. The *Concept* class is in relation *hasDomainAttribute* with instances of the *DomainAttribute* class. To define new domain attribute for the concept class we create a new attribute class derived from the *DomainAttribute* class and assign it to the concept class with new object relation derived from *hasDomainAttribute* relation (see Fig. 6).

Users

Instance of the *User* class represents a user of adaptive application. The *User* is referenced by multiple user views for multiple concepts. Creating view instances, setting view attribute values and their modification is the role of the adaptive application. Domain model ontology defines default user attribute values for particular concepts. Creating the instances of users is out

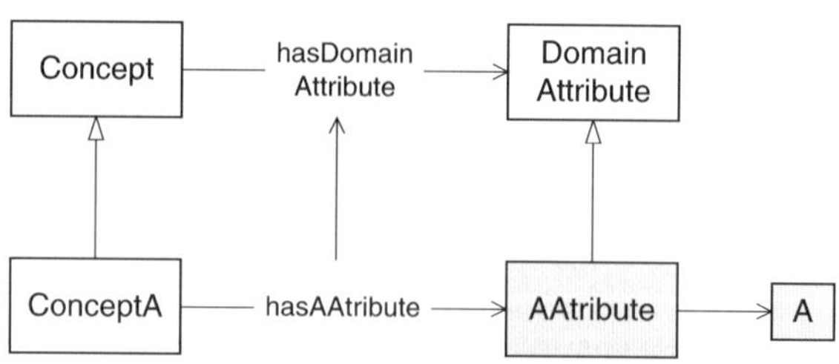

Fig. 6. Definition of a domain attribute

of the scope of domain modeling. It is in common provided manually (filling forms by the user) or (semi)automatically employing knowledge discovery techniques [18].

4.2 Domain and User Models Interconnection

Domain model defines concepts including their information content, relations between concepts and domain attributes for the concepts. User model consists of user views for each concept class from the domain model. The user view defines a set of attributes for domain concepts which are stored in the user model. We do not consider the domain independent user model while we concentrate on domain modeling. We model the user only to such extent that can be automatically gathered from developed domain model. Domain independent part of the user model is often defined in separated ontology or it can be accessed from shared source [2].

Interconnection between a domain model and a user model is depicted on Fig. 7. The *Concept*, *UserView*, *User* and *ConceptVisited* classes demonstrate the basic interconnection. The *DefinedConcept* does not define any attributes for the user model. As an example we consider e-learning domain where the *EducationalConcept* is specified. It defines two user model attributes *EducationalKnowledge* and *EducationalInterest* for which we have derived new view class *EducationalUserView*, which defines relations for connecting these attributes.

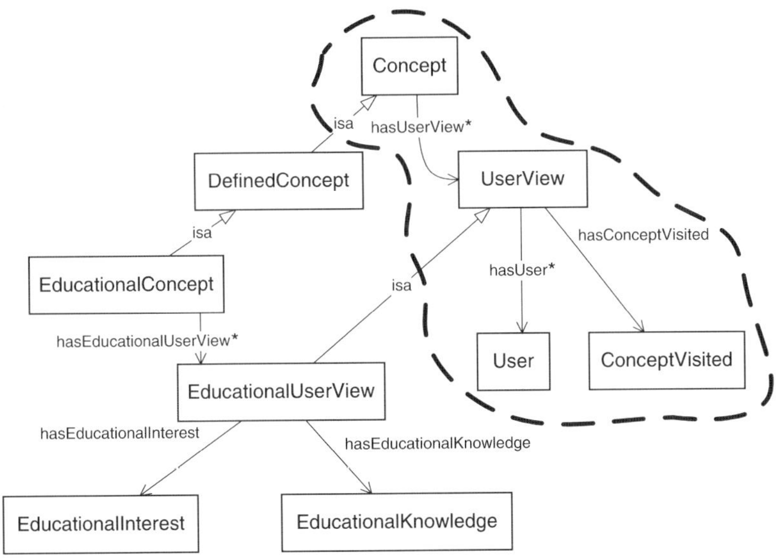

Fig. 7. Interconnection between domain and user models

5 Meta-Model of the Adaptive Application Content

In ontology it is possible to define data and object properties as functional or non-functional (multiple). Non-functional properties are represented and handled as sets with no regard on ordering the elements contained. This can result into unsuitable attributes order after delivering the content ontology into adaptive application where sequences of concepts or information fragments present important knowledge on the presentation (beside dynamically generated sequences). Explicit specification of elements ordering in the ontology would uselessly embarrass the domain model. We have proposed a meta-model, which contains a list of all concept classes and corresponding attributes and relationships. Each attribute and relationship in the meta-model has defined an order index and a flag of its visibility in the presentation.

After generating the meta-model from the ontology the ordering of attributes and relationships conforms their physical ordering in the ontology which can be random in general. Our method for delivering content into existing adaptive application uses the meta-model in the process of exporting the adaptive application context from the ontology into the intermediate format where the attributes and relations are stated in such order as will be presented after importing into the adaptive application.

Bellow is an example of meta-model for sequencing information fragments for a programming exercise in e-learning domain.

```
<meta_caf>
   <meta_concept name="ExerciseConcept">
      <meta_concept_attributes>
         <meta_attribute type="concept" show="true" order="1"
                          name="exerciseHint"/>
         <meta_attribute type="domain" show="false" order="2"
                          name="exerciseDifficulty"/>
      </meta_concept_attributes>
      <meta_concept_relations>
         <meta_relation type="fragment" show="true" order="1"
                         name="hasExerciseDefinition"/>
      </meta_concept_relations>
  </meta_concept> ... </meta_caf>
```

6 Evaluation of Proposed Method

In our experiments we focused on adaptive Web-based system AHA! [10] and authoring tool for adaptive applications MOT [9]. The AHA! system is partially compliant with the AHAM model and the MOT system is based on the LAOS model. Both provide authoring tools for creating adaptive applications. Level of authoring is superior in MOT, which provides simple and powerful definition of domain concept maps, lesson model (defined in the LAOS model) and adaptation strategies (LAG programs, [8]). On the other side, AHA! provides superior adaptive techniques for defining an adaptive presentation. This resulted to a proposal to use MOT as an authoring system and AHA! as a delivery system supported by transforming the content between these two systems (MOT2AHA, [9]). Transformation is either straightforward, or consists of two steps where the intermediate format (CAF, Common Adaptation Format) extended to CAFE (Common Adaptation Format Extended) is used.

We have used the core ontology for authoring described in this chapter together with existing converting tools for delivering the content into the AHA! application (CAF2AHA! tool developed at Eindhoven University).

Evaluation was realized in three stages:

1. Definition of prototype ontology in domain of learning programming by examples.
2. Export the prototype ontology into the adaptive application.
3. Developing software support for authoring the adaptive application content ontology.

In the first stage we developed the ontology describing a programming course using program exercises for languages Lisp and Prolog (based on the adaptive Web-based system ALEA used for learning programming in Functional and Logic programming course at the Slovak University of Technology in Bratislava [4]). At the same time we provided manual transformation of a part of the ALEA content into the AHA! application [3] in order to verify developed ontology.

Structure of the domain model is shown in Fig. 8. For simplicity we omit here inverse relations between the concepts. *ProgrammingExercises* represents the root concept of the application, which can include a set of program schemata (*TemplateConcept*) and a set of exercises (*ExerciseConcept*). Program schemata include concepts describing the program schema usage (*TemplateUsage*) and concepts containing exercises (*ExerciseConcept*). Each exercise contains one or more concepts defining the exercise (*ExerciseDefinition*) and its solution (*ExerciseSolution*). The *hasSubTemplate* relation enables to build a hierarchy of the template concept instances.

After defining a structure of ontology for learning programming domain, we filled the ontology up with instances of programming templates and exercises to enable further evaluation.

During the second phase in order to solve a problem of unwanted misplacing concept attributes in generated presentation we developed an ontology meta-model, which defines concept classes and their attributes and relations (see Sect. 5). By editing the meta-model and applying it during exporting to CAF format concept attributes and relations are resorted as specified in the meta-model.

Second and third stages of the evaluation are supported by software tools developed in order to make proposed sequence of steps defined by the method for modeling the content of adaptive application practicable. We developed

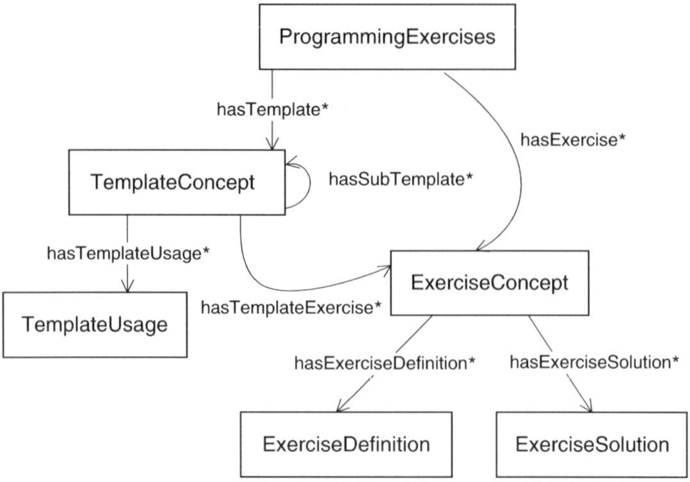

Fig. 8. Programming course domain model

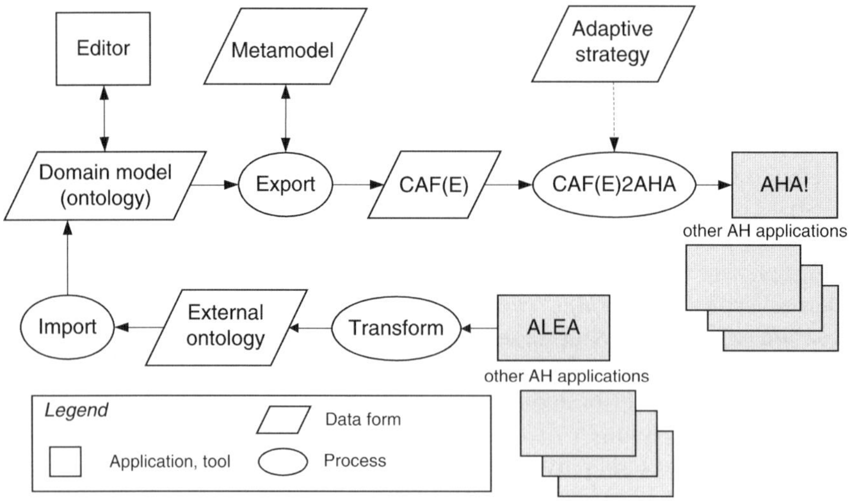

Fig. 9. Content transformation possibilities

tools for editing, importing and exporting the ontology defining the content of adaptive application. Figure 9 visualizes a process of transforming content represented by the domain model in ontology to the AHA! system. Moreover we added importing the content from the ALEA system, which was realized by a set of developed software tools for transformation XML representation of the content in ALEA into ontological representation.

7 Conclusions

In this chapter we described a method for adaptive application content modeling using ontology that allows the content reuse between applications. It is based on designed core ontology that is open and can be used for integrating other aspects of adaptive behavior and other layers of reference models. Adaptive applications benefit also from generated domain dependent part of user model.

Our long term goal is a use of ontology as knowledge representation in adaptive Web-based applications. It allows building both closed and open corpus systems using the same processes. Knowledge representation using the ontology reveals possibilities of reasoning and thus supports defining adaptive behavior. Moreover, automatic means for sequences of content generation (e.g., using knowledge on user behavior [19]) or checking its consistency could be employed.

Our work was aimed at delivering adaptive application content into existing adaptive systems or content conversion between content oriented Web-based applications. We have demonstrated authoring of adaptive application content using the ontology and its delivering into existing system in terms of "authoring once, delivering many", which is promising real utilization of value-added possibilities for modeling adaptive applications with means of the Semantic Web. Described approach of modeling adaptive applications content employing the ontology is a step to support the Semantic Web technologies in adaptive Web-based systems.

Acknowledgements

This work was partially supported by the Scientific Grant Agency of Slovak Republic, grant VG1/3102/06 and by the Cultural and Educational Grant Agency of the Slovak Republic, grant No. KEGA 3/5187/07.

The authors thank members of PeWe group (`www.fiit.stuba.sk/research/pewe`) and participants of the SMAP 2006 workshop (`www.smap2006.org`) for fruitful discussions and feedback on work presented in this chapter.

References

1. Andrejko A, Barla M, Bieliková M, Tvarožek M (2006) Ontology-based user modeling for web-based information systems. In: Proceedings of International Conference on Information Systems Development, ISD'06, Budapest, Hungaria
2. Bieliková M, Kuruc J (2005) Sharing user models for adaptive hypermedia applications. In: Proceedings of 5th International Conference on Intelligent Systems Design and Applications, ISDA'05, Wroclaw, Poland, ACM, New York 506–511

3. Bieliková M, Kuruc J, Andrejko A (2005) Learning programming with adaptive web-based hypermedia system AHA! In: Jakab F, et al. (eds) Proceedings of International Conference on Emerging e-Learning Technologies and Applications, ICETA'05, Košice, Slovakia, 251–256
4. Bieliková M (2006) An adaptive web-based system for learning programming. Int. J. Cont. Eng. Educ. Life-Long Learn., 16(1/2):122–136
5. Bureš M, Jelínek I (2005) Reusable Adaptive Hypermedia E-learning Content Using AICC. In: Proceedings of the IADIS International Conference on WWW/Internet'05, Lisboa, Spain, IADIS, vol. I, 376–378
6. Ceri S, Fraternali P, Matera M (2002) Conceptual modeling of data-intensive web applications. IEEE Internet Comput., 6(4):20–30
7. Cristea A I, De Mooij A (2003) LAOS: Layered WWW AHS authoring model and their corresponding algebraic operators. In: Proceeedings of the 12th International World Wide Web Conference, WWW'03, Budapest, Hungary, ACM, New York
8. Cristea AI, Verschoor M (2004) The LAG grammar for authoring adaptive web. In: Proceedings of the International Conference on Information Technology: Coding and Computing, ITCC'04, IEEE Computer Society Press, USA 382–386
9. Cristea AI, Smits D, De Bra P (2005) Writing MOT, reading AHA! – converting between an authoring and a delivery system for adaptive educational hypermedia. In: Proceedings of 3rd International Workshop on Authoring of Adaptive and Adaptable Educational Hypermedia, A3EH'05 at AIED'05, Amsterdam, The Netherlands
10. De Bra P, et al. (2003) AHA! – The adaptive hypermedia architecture. In: Proceedings of the of the ACM Hypertext Conference, Nottingham, UK, 81–84
11. De Bra P, Santic T, Brusilovsky P (2003) AHA! meets Interbook, and more. In: Proceedings of the AACE ELearn'03 Conference, Phoenix, AZ, 57–64
12. Gruber TR (1993) Towards principles for the design of ontologies used for knowledge sharing. In: Guarino N, Poli R (eds) Formal Ontology in Conceptual Analysis and Knowledge Representation, Kluwer, Dordretch
13. Houben GJ, Barna P, Frasincar F (2003) HERA: development of semantic web information systems. In: Proceedings of International Conference on Web Engineering, ICWE'03, LNCS 2722, Springer, Berlin Heidelberg New York, 529–538
14. Koch N, Wirsing M (2002) The Munich reference model for adaptive hypermedia applications. In: De Bra P, Brusilovsky P, Conejo R (eds) Proceedings of International Conference on Adaptive Hypermedia and Adaptive Web-based Systems, AH'02, LNCS 2347, Springer, Berlin Heidelberg New York, 213–222
15. Seefelder PA, Schwabe D (2004) A semantic meta-model for adaptive hypermedia systems. In: Proceedings of 3rd International Conference on Adaptive Hypermedia and Adaptive Web-Based Systems, AH'04, Eindhoven, The Netherlands, LNCS 3137, Springer, Berlin Heidelberg, New York, 360–365
16. Stewart C, Cristea AI, Brailsford T (2005) Authoring once, delivering many: creating reusable adaptive courseware. In: Proceedings of 4th IASTED International Conference on Web-Based Education, WBE'05, Grindelwald, Switzerland
17. Schwabe D, Rossi G: An object-oriented approach to web-based application design. Theory Pract. Object Syst., Special issue on the Internet, 4(4):207–225
18. Tvarožek M, Barla M, Bieliková M (2007) Personalized Presentation in Web-Based Information Systems. In: Proceeedings of SOFSEM 2007, J. van Leeuwen et al. (ed), LNCS 4362, Springer, Berlin Heidelberg New York, 796–807

19. Velart Z, Šaloun P (2006) User behavior patterns in the course of programming in C++. In: Proceedings of the International Workshop on Adaptivity, personalization and the Semantic Web, Odense, Denmark, 41–44
20. Wu H, Houben GJ, De Bra P (1998) AHAM: A reference model to support adaptive hypermedia authoring. In: Proceedings of the Conference on Information Science, Antwerp, 51–76

Indexing and Browsing of Color Images: Design Considerations

Christos Diou, Nikos Batalas, and Anastasios Delopoulos

Multimedia Understanding Group, Department of Electrical and Computer Engineering, Aristotle University of Thessaloniki, Greece diou@mug.ee.auth.gr, amv@ee.auth.gr, adelo@eng.auth.gr

Summary. This chapter deals with the various problems and decisions associated with the design of a content based image retrieval system. Image descriptors and descriptor similarity measures, indexing data structures and navigation approaches are examined through the evaluation of a set representative methods. Insight is provided regarding their efficiency and applicability. Furthermore the accuracy of using low dimensional FastMap point configurations for indexing is extensively evaluated through a set of experiments. While it is out of the scope of this chapter to offer a review of state of the art techniques in the problems above, the results presented aim at assisting in the design and development of practical, usable and possibly large scale image databases.

1 Introduction

The main goal of content based image retrieval research is to devise suitable representations of images in order to allow query and retrieval based on the visual properties of images instead of manually inserted user annotations. Often the queries themselves are images and the user expects similar images to be retrieved.

Significant research has been performed on image retrieval systems in the past few years and the promising results contributed to the development of the MPEG-7 standard [1, 2]. The ultimate goal of automatic semantic characterization of images based on their visual content remains largely unsolved (even though there are partially successful approaches under controlled environments e.g., [3–5]). Still, descriptions of images based on color, textures, shapes etc. provide adequate results for a user to begin a search.

Practical, real world image retrieval applications, however, have additional requirements. Queries must be answered fast, while at the same time the design and implementation must be scalable, allowing searches within large image datasets. This requirement implies that an efficient indexing mechanism must be employed for storing image descriptions in the database. Several data

C. Diou et al.: *Indexing and Browsing of Color Images: Design Considerations*, Studies in Computational Intelligence (SCI) **93**, 329–346 (2008)
www.springerlink.com

structures have been proposed to this end in the literature. Examples include a family of methods based on *kd*-trees [6, 7], hashing (e.g., [8]), P-Sphere trees [9] and others. While these approaches offer an improvement over the exhaustive search for results, they are less efficient than the corresponding indexing structures used for text. Furthermore, they cannot always be used in conjunction with high dimensional image descriptors. In fact, this is one of the major limiting factors preventing the creation of content based image databases at large scales (and ultimately, the Internet).

One may summarize the most important design issues that have to be resolved during the development of a content based image database system into the following:

1. The description of images.
2. The similarity measure for the selected description. Several image descriptors can be compared with more than one measure. One must choose the measure that provides the best retrieval results for the application at hand.
3. The indexing mechanism to be employed.
4. The proper number of dimensions for the target descriptors for practical retrieval times.
5. If needed, a method that will provide low dimensional feature vectors, given the original image descriptors.
6. The visualization and browsing interface.

This chapter discusses the above issues by providing evaluations of known methods, thus providing an overview of the design of an image retrieval system. For the description of images in Sect. 2, only global color descriptors are considered, the histogram and dominant color. For the dominant color a comparison between two distance metrics is performed that provides indications concerning the choice of a similarity measure.

The *kd*-tree is considered regarding the indexing of image descriptors in Sect. 3. It is illustrated that data structures of this kind are not efficient when the number of dimensions used by the image descriptors exceed a certain limit. Additionally, such indexing structures index points in a k-dimensional space but the dominant color descriptor does not define points in such a space.

Section 4 presents a solution to these problems, based on point configurations provided by methods such as Multidimensional Scaling and FastMap. Thus one can can derive low dimensional feature vectors from the original image descriptors, allowing the efficient use of indexing structures. In addition, point configurations allow the implementation of intuitive browsing interfaces by visualization of results on the 2- or 3-dimensional space. But the benefits of using compact, low dimensional image representations come with a certain cost in retrieval accuracy. Through a set of experiments, Sect. 5 attempts to quantify the deviation of retrieval results obtained through point configurations with respect to the ones obtained through the initial image descriptions. Finally, Sect. 6 summarizes the conclusions drawn from the presented evaluations.

2 Color Descriptors and Similarity Measures

The first step in the design of a content based image retrieval system involves the representation of images using a set of descriptors. These provide a compact description of visual cues (color, texture, shape) or interest points (e.g., SIFT [10]) that allows the definition of similarity measures between images. The following choices need to be made:

1. The visual cues or types of interest points that will be utilized (e.g., color and texture).
2. The descriptors for each cue.
3. The similarity measure to be used with each descriptor.
4. If more than one descriptors are used, a fusion strategy that will combine them into a single descriptor (e.g., [11]), or alternatively, combine the results of each similarity measure into a single value (e.g., [12]).

In this section it is assumed that a single global color descriptor will be used, thus an evaluation example is provided for issues 2 and 3 above.

2.1 Histogram and Dominant Color

Global color descriptors are used to describe color properties of an image independent of spatial color distribution. The most important descriptors of this form are the well-known histogram and dominant color descriptors.

Both these descriptors can be given by

$$D = \{(c_i, p_i)\,, \quad i = 1, \ldots, N\}\,, \tag{1}$$

where c_i is a color from a predefined colorspace and p_i is the percentage of image pixels having that color. Of course, having a 3-channel 8-bits per channel image described by (1) is very impractical, since (1) $2N = 2^{25}$ values are used to describe a single image and (2) this level of granularity is not informative for the purpose of image retrieval. Therefore, images are quantized prior to extraction of D.

In histograms, the given colorspace (e.g., RGB) is usually quantized to a predefined number of "bins" independent of the images. While this approach can reduce the number of values required for D (the colors c_i are essentially predefined for a given colorspace), the description does not adapt to each image. Consider, for example, a 24-bit RGB image with only 64 different colors all at the same color region. Then, if the colorspace is uniformly quantized at $N = 27$ levels (three levels for each color channel), all these colors will be concentrated at a single bin; the rest of the 26 color-value pairs are left unused.

The dominant color descriptor, on the other hand, overcomes this issue by allowing the use of the more general form of (1) where the colors c_i and their number N can be different for each individual image. Naturally, a method for

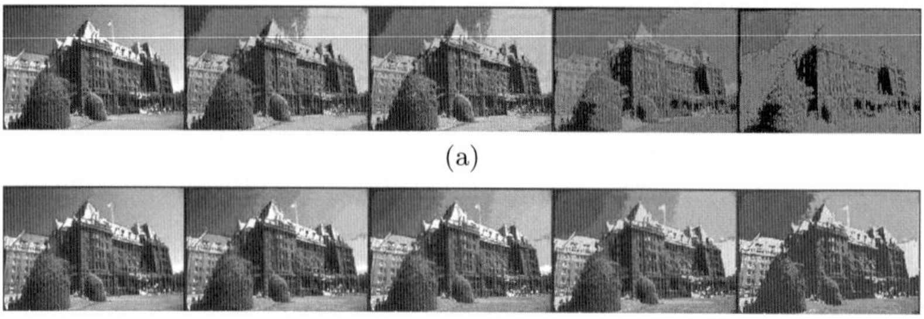

(a)

(b)

Fig. 1. Color reduction using 64, 32, 16 and 8 colors: (**a**) using uniform color space quantization and (**b**) octrees

selecting the appropriate dominant colors c_i for each image must be defined. In [2] regarding the MPEG-7 standard color descriptor the use of the Generalized Lloyd Algorithm is proposed. In this work, a different approach utilizing octrees for color reduction [13] was used.

Figure 1 illustrates an example of colorspace quantization (histogram) and quantization adapted to the image (dominant color) using octrees. In the latter case a more accurate description of the image is derived for image retrieval purposes. It is therefore reasonable to select the dominant color over the histogram descriptor for global color representation.The next section deals with the definition of similarity measures for these descriptors.

2.2 Distance Metrics

Given a predefined set of colors c_i, $i = 1, \ldots, N$, a $N \times N$ matrix A with $a_{jk} = 1 - d_{jk}/\max(d_{jk})$ where d_{jk} is the distance between c_j and c_k in their colorspace and two vectors h_1 and h_2 with the percentages of each color c_i, the quadratic histogram distance is given by

$$d_h(h_1, h_2) = (h_1 - h_2)^T A (h_1 - h_2). \tag{2}$$

However, (2) cannot be used if the colors c_i and their number N are different for each image.

Deng et al. proposed a similar quadratic metric in [14], for the dominant color descriptor. If $D_1 = \{(c_i, p_i), i = 1, \ldots, N_1\}$ and $D_2 = \{(b_j, q_j), j = 1, \ldots, N_2\}$ are two dominant color descriptors, then the distance between D_1 and D_2 is defined to be

$$d_q(D_1, D_2) = \sum_{i=1}^{N_1} p_i^2 + \sum_{j=1}^{N_2} q_j^2 - \sum_{i=1}^{N_1} \sum_{j=1}^{N_2} 2a_{ij} p_i q_j, \tag{3}$$

where the similarity coefficient a_{ij} is

$$a_{ij} = \begin{cases} 1 - d_{ij}/d_{\max}, & d_{ij} \leq T_d, \\ 0, & d_{ij} > T_d, \end{cases} \tag{4}$$

$d_{ij} = \|c_i - b_j\|$ is the euclidean distance between c_i and b_j, $d_{\max} = \alpha T_d$, α is an arbitrary value and T_d is the maximum distance for two colors to be considered similar.

Another metric that has been proposed for comparing two dominant color descriptors D_1 and D_2 is the Earth Mover's Distance (EMD). In simple terms, the EMD is a dissimilarity measure between two images indicating the amount of "work" required to "move" from the descriptor D_1 of the first image to D_2 of the second. Imagine the colors c_i in the first descriptor as locations in a field with piles of p_i mass of earth each. The colors b_j are also locations, but they consist of holes with capacity q_j earth each. EMD denotes the minimum work required to distribute the piles of earth at c_i to the holes in b_j. Computation of the EMD is based on a solution of the transportation problem and is covered in [15].

A simple experiment was set up in order to evaluate the performance of each distance metric in a realistic image database scenario. A total of 5,022 images from the corel dataset were used, where each of the images belongs to a predefined category C. All images were indexed using the dominant color descriptor with 16 color – percentage pairs. Each image was successively used as a query and a ranked list of results was retrieved. The performance of a distance metric was evaluated based on the semantic correspondence of the results, using the following precision measure:

$$precision_C = \frac{\sum_{n=1}^{N_C} A_n}{\sum_{n=1}^{N_C} 1/n}, \tag{5}$$

where N_C is the number of images in the query image category C, $A_n = 1/n$ if the nth result belongs in C and zero otherwise. Maximum precision is achieved when the first N_C results for a query $I \in C$ belong to C as well. Note that it is too optimistic to expect retrieval of the best results at the semantic level (same category) using only a global color descriptor; however the two distance measures can be compared in this manner.

A graph of the results is given in Fig. 2, where the horizontal axis corresponds to categories C and the vertical axis is the precision (average for all images in a category).The results are clearly in favor of the EMD, that consistently achieved higher precision compared to the quadratic distance.

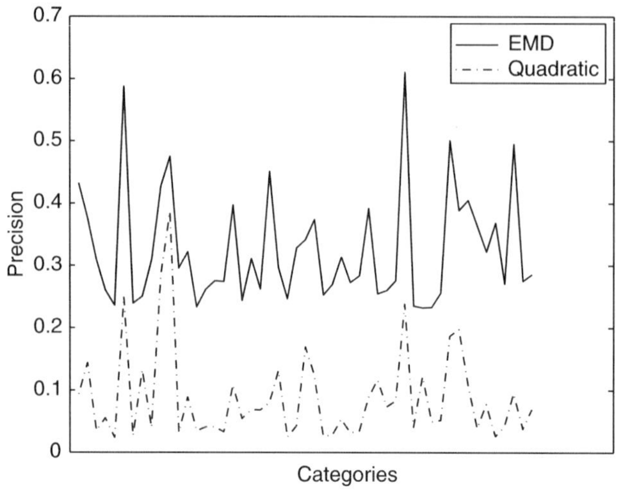

Fig. 2. Results for retrieval based on dominant color descriptors with quadratic distance and EMD. Precision (5) vs. category

3 Indexing and Dimensionality

Assume that the image descriptors and associated distance metrics have been determined and also that the descriptors can be expressed as feature vectors in a k-dimensional space S. In order to construct an image database, a method for solving the "Nearest Neighbor" problem must be selected: Given a set of points P (descriptors) in S (k-dimensional descriptor space) and a query point $q \in S$, find the closest point to q in P.

The simplest solution is to compare q against all images in the database, P. This approach, however, poses strict limitations to the size of the image database due to its high computational cost. To enable implementation of image databases at larger scales, indexing data structures have to be used that solve the Nearest Neighbor problem without visiting the entire database.

One of the most popular data structures proposed is the *kd*-tree [6, 7]. The idea is to construct a binary tree by successively using elements of the dataset as pivot points to partition the k-dimensional space into hyperrectangles, each containing at most one point. When searching, an initial estimate of the nearest neighbor is provided (by finding the hyperrectangle that contains the query) and then only hyperrectangles and pivot points that are possible to contain a point closer to the query than the initial estimate are visited. Thus, with *kd*-trees only a subset of the indexed points (i.e., database images) are visited, compared to the exhaustive search where the query is compared against all points in the dataset. Searching in n images with the *kd*-tree requires at least $\mathcal{O}(\log n)$ visits and $\mathcal{O}(n)$ at the worst case (same as the exhaustive search). The drawback is that the complexity for each point visit is increased (since branching conditions, etc. have to be evaluated).

Still, a very important problem remains, known as "the curse of dimensionality" that affects *kd*-tree efficiency. As the number of dimensions increases, an exponentially increasing number of hyperrectangles (thus points in the dataset) will have to be visited to find the nearest neighbor of a query point.

3.1 Limits of *kd*-Tree Effectiveness

In order to examine the behavior of *kd*-trees with respect to the dimensionality of the space considered, the *kd*-tree data structure and associated algorithms were implemented and a dataset of 10^5 uniformly distributed random points was created for various dimensions. The number of nodes visited per dimension was measured and the results are given in Fig. 3a.

The number of points visited for a number of dimensions close to 30 is practically the complete dataset and the *kd*-tree has no advantage over the exhaustive search. In fact, the upper limit of dimensions that the *kd*-tree is useful is lower, since each visit has additional costs in terms of CPU time. Figure 3b provides the time (in ms) required per dimension for indexing performed on the same dataset using *kd*-trees and exhaustive search on an average personal computer.

These results indicate that for the test computer and implementation the *kd*-tree keeps an advantage in terms of computational time for eight dimensions or less. Clearly, an optimized implementation of the algorithm would increase this limit, but practically it cannot exceed 15 dimensions. Furthermore, through the experiments conducted, it was observed that the efficiency of the *kd*-tree search is largely dependent on the size of the dataset to be searched. Larger datasets allow for the *kd*-tree to be more efficient in even higher dimensions, compared to exhaustive searching. Also, note that a uniform dataset (used in the empirical evaluation above) is the worst case scenario; *kd*-tree searches are significantly faster within distinctively clustered datasets.

4 Point Configurations

Two major problems can be identified with the use of *kd*-trees and similar indexing structures in image databases:

1. The number of dimensions used by image descriptors is prohibitive for efficient indexing.
2. Descriptors do not always define points in a k-dimensional space and thus *kd*-trees cannot be employed for indexing. The dominant color presented in Sect. 2.1 is an example of such a descriptor.

Low-dimensional embeddings of descriptors such as those produced by PCA is a possible solution to the high dimensionality problem. Again, however, this approach is only applicable to k-dimensional points. Both the above issues

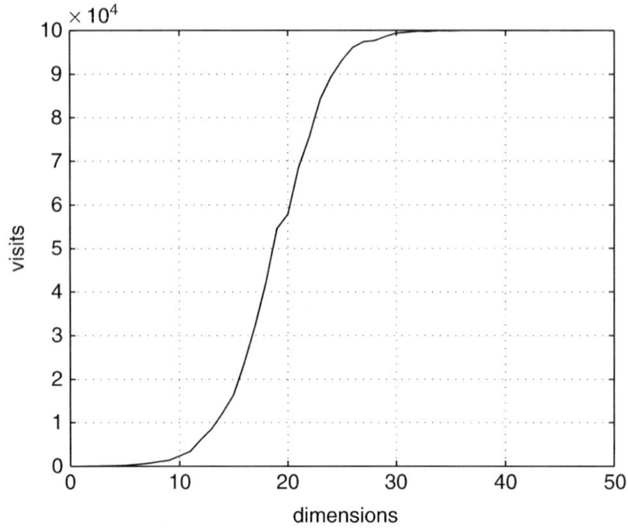

(a)

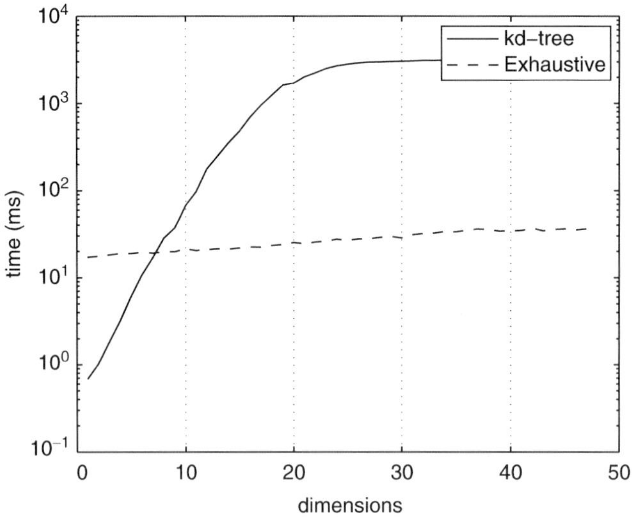

(b)

Fig. 3. (**a**) Number of visits for nearest neighbor search in the *kd*-tree for a dataset of 10^5 uniformly distributed random points. (**b**) Time for nearest neighbor search in ms for the *kd*-tree and exhaustive search for various dimensions on an average personal computer can be tackled if a low-dimensional point configuration is produced from the original descriptors.

This problem can be formulated as follows: "Given the observed distances d_{ij} between any two objects (images in the database), produce a configuration of points in the k-dimensional space, such that the new (euclidean) distances d'_{ij} are as close as possible to the original d_{ij} for all the points". A measure of effectiveness for a solution is Kruskal's *stress* function (6) [16].

$$\text{stress} = \left[\frac{\sum_{i,j}(d'_{ij} - d_{ij})^2}{\sum_{i,j} d_{ij}^2}\right]^{1/2} . \tag{6}$$

In the case of the dominant color descriptor, the original distances are given by the EMD, while the derived points are in an k-dimensional space with euclidean distances d'_{ij}.

4.1 MDS and FastMap

Two techniques that produce point configurations were evaluated, namely metric Multidimensional Scaling (MDS) [17] and the FastMap algorithm [18]. Metric MDS is a technique that receives as input the observed dissimilarities d_{ij} between objects and produces a configuration P' of points in the k-dimensional space through an iterative optimization process. Roughly, each object is assigned an k-d point (e.g., randomly) and then every point is examined by computing its distance from all the $N-1$ other points. The point is moved so that it optimizes the stress function. The complexity of performing MDS is $\mathcal{O}(N^2)$, where N is the number of objects.

FastMap, on the other hand, solves the same problem by starting from $1-d$ and recursively determining the coordinates of the N objects on a new axis, up to k–d. Computationally, this process is much more efficient than MDS, with its complexity being $\mathcal{O}(kN)$, where k is the number of dimensions of the target configuration. MDS and FastMap were compared with respect to the stress function (6) using the corel dataset. It can be seen in Fig. 4 that MDS achieved better results. But the main strength of FastMap is the $\mathcal{O}(1)$ complexity required for the insertion of a new point, contrary to $\mathcal{O}(N)$ required by MDS. In practice, answering a query with MDS (thus embedding the query object in the k-d space and finding its nearest neighbors) requires as much time as the exhaustive search at best. MDS is therefore not suitable for retrieval applications.

4.2 Browsing

An important aspect of any image retrieval system is the visualization and browsing interface. In the simplest case, thumbnails of result images can be provided in a list for the user to browse. In the case of images, however, a

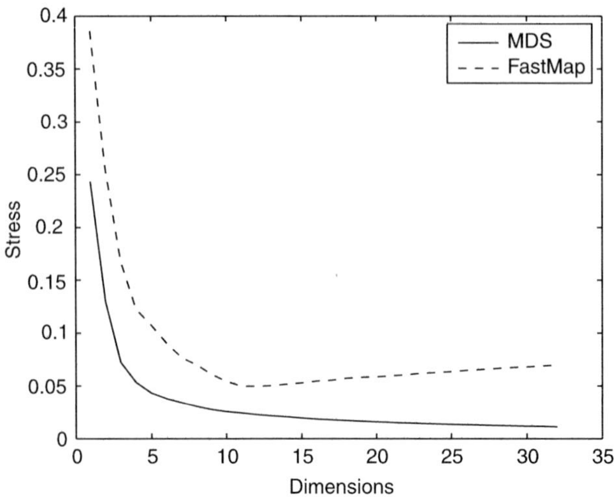

Fig. 4. MDS vs. FastMap stress performance for various dimensions. Experiment for 500 images

more efficient visualization and browsing interface would provide a grouping of similar results. Using MDS or FastMap to produce a configuration of points at two or three dimensions from the result dataset allows its visualization, as illustrated in Fig. 5. The user is able to view the entire result dataset, zoom to specific regions and select images. The examples in the figure are derived using FastMap. If the number of results is relatively small, MDS can be employed as well.

5 Efficiency of FastMap Configurations

Given a dominant color descriptor D_i for each image I_i in the database, the EMD measures d_{ij} between I_i and I_j are computed for all i, j. Subsequently, FastMap is applied to create a configuration P_k of k-dimensional points, one for each image. This allows the efficient use of kd-trees for indexing. The questions that naturally arise have to do (1) with the quality of the retrieval results and (2) how these results are affected by the choice of k.

In order to evaluate the performance of FastMap configurations for image retrieval, the ranking results obtained from EMD-based queries (Sect. 2) were used as ground truth with the same collection of 5,022 images from the corel dataset. For the same query, the difference in rank between the EMD and mapped results was used, as follows.

Initially, point configurations of the entire dataset were constructed for dimensions $k = 1, \ldots, 32$. For each k, each image I_j, $j = 1, \ldots, 5022$ was submitted as a query, and returned a ranking $r_{jk} = \left[I_1^{jk} \ldots I_{5022}^{jk} \right]$ of all images

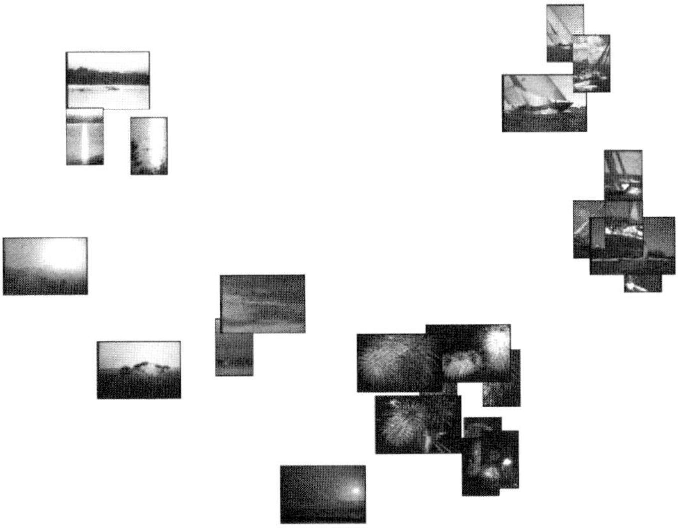

(a)

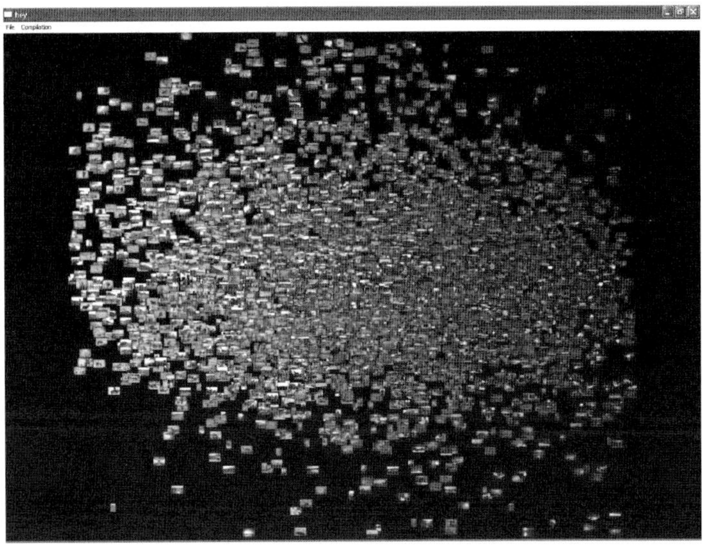

(b)

Fig. 5. (**a**) Result of applying FastMap for two dimensions on a set of images from three categories of the corel dataset. (**b**) Navigation on a larger dataset. Users can zoom in specific areas, modify all distances by a factor and select specific images

extracted from search at the k-d space. This ranking was compared with the corresponding EMD rank r_j^{EMD} for the same image, yielding a set of pairs $L_{jk} = \{(p_{j1}, p_{j1}^k), \ldots, (p_{j5022}, p_{j5022}^k)\}$. These pairs indicate the rank in EMD and FastMap for the same result image. For example, a query image from the

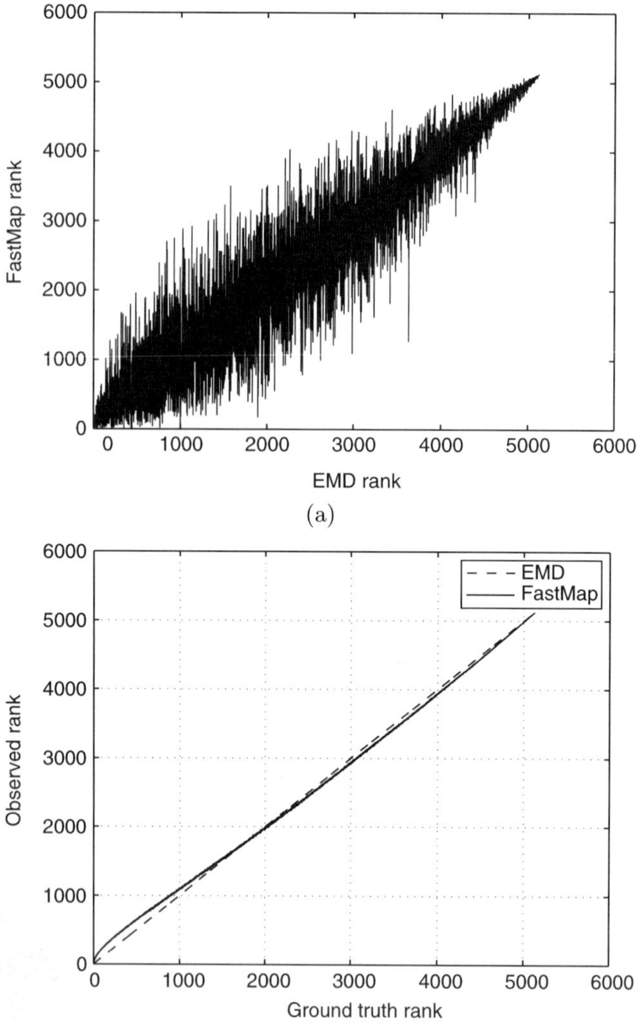

Fig. 6. Example rank results for $k = 6$ dimensions. The EMD rank (ground truth) is the $y = x$ line. (**a**) Results for a random image. (**b**) The average for all images

"action sailing" category gives $(2, 4)$ which reads "the 2nd result of EMD was ranked 4th using FastMap". Figure 6 shows an example of the rank results for a random image and the average for all images at $k = 6$ dimensions.

It can be seen that even though the results of EMD ranking and those of FastMap configurations are not identical, they are averagely very close to each other at six dimensions. The similarity is practically not improved if more than 10 dimensions are used. Hence this experiment is an indication that *by using FastMap configurations for indexing, the retrieval does not deviate significantly from the EMD results.* It is therefore highly unlikely to find images ranked in the first results by EMD at the lower ranks (i.e., least similar results) of FastMap based retrieval.

To quantify this observation, another experiment was conducted. Images were selected at random from all categories (resulting in 100 images in total) and the retrieval results for these images were observed for EMD and FastMap, as in the previous. Then, the distribution (pdf) of the random variable $\mathbf{d}_{I_q,k} = p_{jr} - p^k_{jr}$ was calculated for $k = 1, \ldots, 32$ dimensions, with p_{jr} and p^k_{jr} as defined above. This random variable quantifies the difference in rank of a result for a query image I_q at k dimensions, compared to EMD. The distribution (histogram) of $\mathbf{d}_{I_q,k}$ for a random query image I_q at $k = 6$ dimensions is shown in Fig. 7.

In order to remove the dependence of the estimated distribution of $\mathbf{d}_{I_q,k}$ on the query image I_q, a number of Monte Carlo experiments were performed, yielding the ensemble average over 100 randomly selected query images I_q. The resulting Monte Carlo mean $\mathbf{d}_k = E_I\{\mathbf{d}_{I_q,k}\}$ is depicted in in Figs. 8a–c for $k = 3$, 6 and 12 FastMap dimensions. The Monte Carlo variance for all ordinates of the estimate of $\mathbf{d}_{I_q,k}$ was pretty low for fixed k (e.g., Fig. 8d for $k = 6$), thus ensuring that the adopted distribution estimators are meaningful and representative.

These distributions allow the computation of the probability $P(-d < \mathbf{d}_k \leq d)$ that the difference of results between EMD and FastMap will be less than d. Figure 9 displays how these probabilities are affected by the number of dimensions k for various d. It can be deduced that the FastMap results are, in fact, expected to deviate from the original EMD ranking (for $d = 25$, P

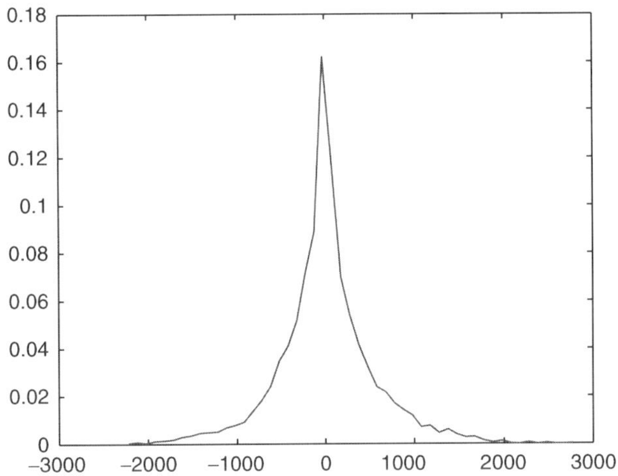

Fig. 7. The distribution of $\mathbf{d}_{I_q,k}$ for a random image at six dimensions

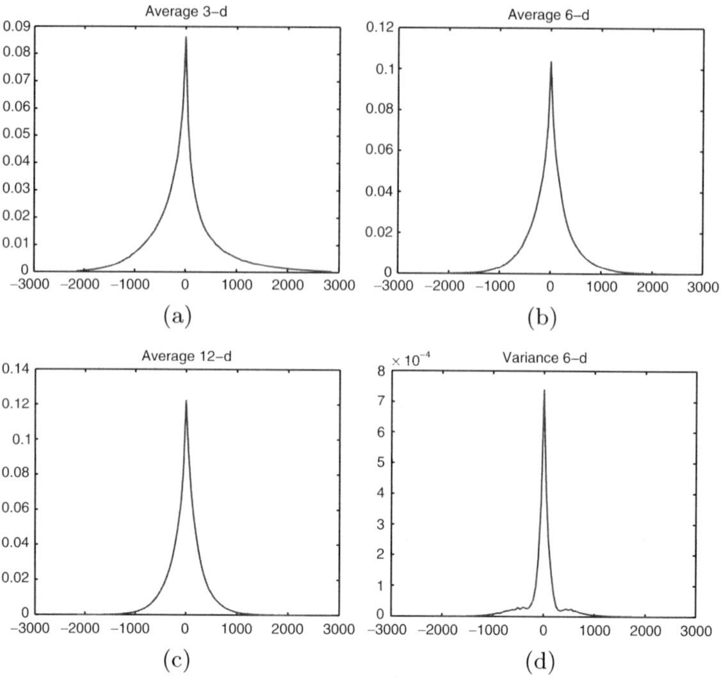

Fig. 8. (**a**), (**b**), (**c**) The distribution of $\mathbf{d}_k$ for $k = 3, 6, 12$ dimensions, respectively. (**d**) The variance of $\mathbf{d}_{I_q,k}$ for $k = 6$ across different query images. Notice that peak variance is 7.4×10^{-4}

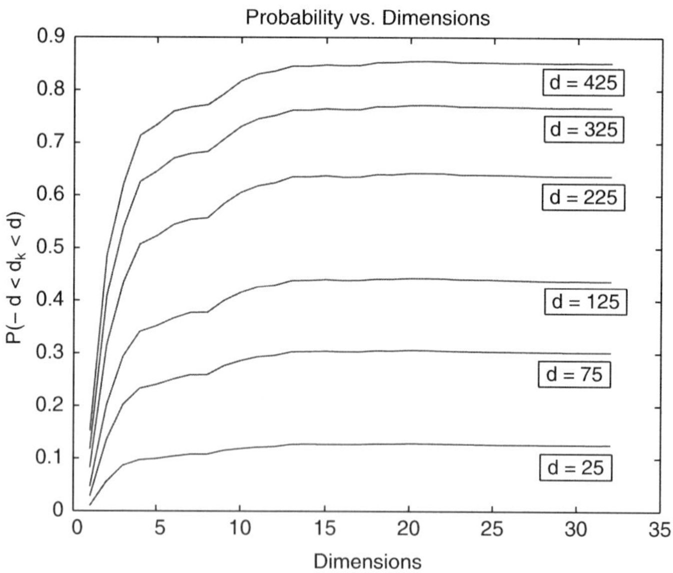

Fig. 9. The probability $P(-d < \mathbf{d}_k \leq d)$ vs. dimension k for various d

is low), however it is highly unlikely for the best EMD results to be ranked lower than 400 using FastMap.

Whether these results can support the practical use of FastMap in image database retrieval largely depends on the application. More specifically, it depends on the browsing and retrieval interface used and the number of results that are originally visualized. There is a tradeoff between the number of results returned (larger result datasets possibly require more browsing effort from the user) and query response times.

Assume the design requirement: "For a query image I_q, the first r results of the corresponding EMD results should be returned". One must find the number of results r_{fk} that must be returned to the user if the dataset is mapped to k dimensions using FastMap. These r_{fk} results should contain the first r results of EMD with a high probability. Again a set of Monte Carlo experiments are performed that provide the ensemble average over 100 randomly selected query images I_q. The resulting Monte Carlo mean $p(r, r_{fk})$ indicates the estimated percentage of the first r results of EMD present in the first r_{fk} of FastMap at k dimensions. Its value is depicted for various r_{fk} in Fig. 10. For example, if $k = 8$ and $r_{f8} = 100$ then it is estimated that 92% of the first $r = 10$ EMD results will appear. In other words, the probability for any of the first 10 EMD results will be found if 100 results are returned using FastMap at eight dimensions, is estimated to be 0.92.

6 Conclusions

From the results of the previous sections several useful conclusions can be drawn.

For the global color representation and specifically the dominant color descriptor, the Earth Mover's Distance appeared to be the most accurate distance metric in the conducted experiments.

For the indexing problem, *kd*-trees were more effective than exhaustive search, but their practical application is limited to a relatively low number of dimensions. Moreover, *kd*-tree structures index points in a k-dimensional space that dominant color descriptors do not provide.

For visualization and navigation purposes, the MDS approach proved more precise than FastMap in terms of the stress measure. However its practical use is limited due its high computational cost. FastMap is better suited to large-scale image databases.

FastMap was also used to provide low dimensional point configurations for efficient indexing. A series of experiments illustrated how the EMD ranking results are affected by the number of dimensions used in the FastMap configurations. Using a configuration with relatively low dimensionality (e.g., 6–8 dimensions) in conjunction with a browsing interface that allows the concurrent visualization of a large result set seems to be a reasonable compromise between retrieval accuracy and fast response times.

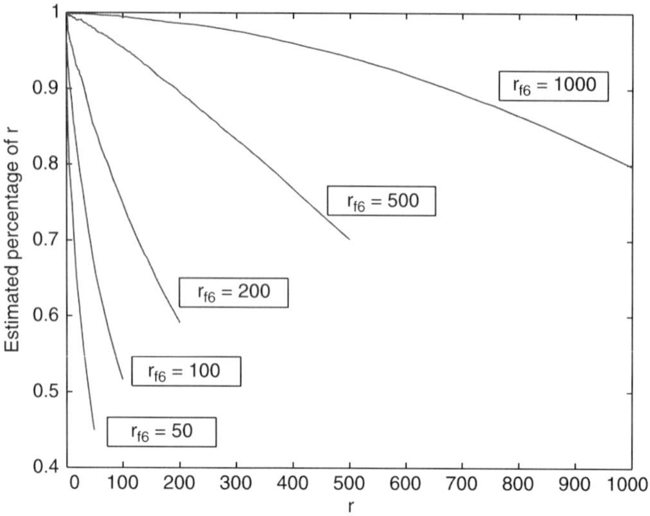

(a)

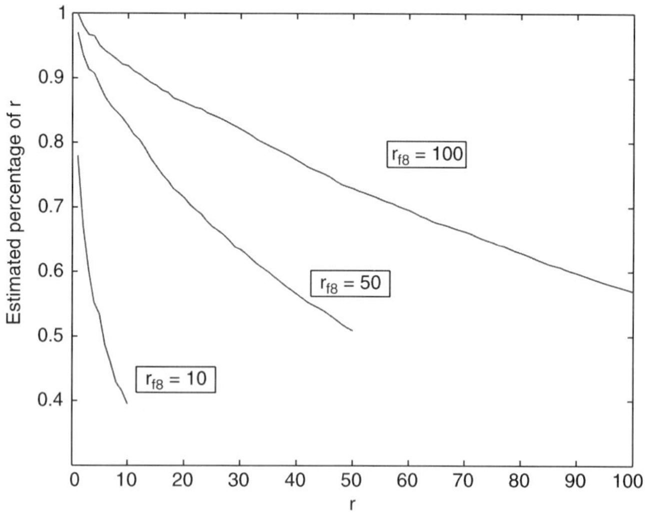

(b)

Fig. 10. (**a**) $p(r, r_{f6})$ for various result numbers and $k = 6$. (**b**) More detailed view for small number of results and $k = 8$

These evaluations outline the problems that must be tackled and the decisions that need to be made for the various stages of a content based image retrieval system design. For the example evaluations above, an image database system implementation can be proposed. The operations related to the construction of the database as well as query answering are outlined in Figs. 11a,b, respectively.

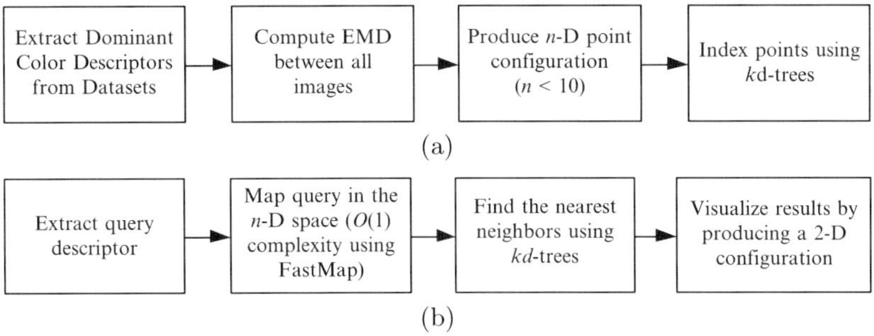

Fig. 11. (**a**) Steps for constructing an image database. (**b**) The process of query answering

Acknowledgment

Christos Diou is supported by the Greek State Scholarships Foundation.

References

1. T. Sikora. The mpeg-7 visual standard for content description – an overview. *IEEE Transactions on Circuits and Systems for Video Technology*, 11(6):696–702, 2001.
2. B. S. Manjunath, P. Salembier, and T. Sikora. *Introduction to MPEG-7: Multimedia Content Description Interface*. Wiley, New York, 2002.
3. A. Dorado and E. Izquierdo. Semantic labeling of images combining color, texture and keywords. In *IEEE International Conference on Image Processing, ICIP*, volume 3, pages 9–12, Barcelona, Spain, September 2003.
4. A. Yavlinsky, E. Schofield, and S. Ruger. Automated image annotation using global features and robust nonparametric density estimation. In *Proceedings of the International Conference on Image and Video Retrieval (CIVR'05)*, 2005.
5. J. Winn, A. Criminisi, and T. Minka. Object categorization by learned universal visual dictionary. In *Proceedings of the Tenth IEEE International Conference on Computer Vision (ICCV 2005)*, 2005.
6. J. L. Bentley. Multidimensional binary search trees used for associative searching. *Communications of ACM*, 18(9):509–517, 1975.
7. A. W. Moore. An introductory tutorial on kd-trees. Technical Report 209, Robotics Institute, Carnegie Mellon University, Pittsburgh, PA, Computer Laboratory, University of Cambridge, 1991.
8. A. Gionis, P. Indyk, and R. Motwani. Similarity search in high dimensions via hashing. In *The VLDB Journal*, 518–529, 1999.
9. Jonathan Goldstein and Raghu Ramakrishnan. Contrast plots and p-sphere trees: Space vs. time in nearest neighbour searches. In *VLDB 2000, Proceedings of 26th International Conference on Very Large Data Bases, September 10–14, 2000, Cairo, Egypt*, pages 429–440, 2000.
10. D. G. Lowe. Distinctive image features from scale-invariant keypoints. *International Journal of Computer Vision*, 60(2):91–110, 2004.

11. J. Yang, J. Yang, D. Zhang, and J. Lu. Feature fusion: parallel strategy vs. serial strategy. *Pattern Recognition*, 36(6):1369–1381, 2003.
12. A. Kushki, P. Androutsos, K. N. Plataniotis, and A. N. Venetsanopoulos. Retrieval of images from artistic repositories using a decision fusion framework. *IEEE Transactions on Image Processing*, 13(3):277–292, 2004.
13. M. Gervautz and W. Purgathofer. A simple method for color quantization: Octree quantization. In *New Trends in Computer Graphics*. Springer, Berlin Heidelberg New York, 1988.
14. Y. Deng, B. S. Manjunath, C. Kenney, M. S. Moore, and H. Shin. An efficient color representation for image retrieval. *IEEE Transactions on Image Processing*, 10(1):140–147, January 2001.
15. Y. Rubner, C. Tomasi, and L. J. Guibas. The earth mover's distance as a metric for image retrieval. *International Journal of Computer Vision*, 40(2):99–121, November 2000.
16. J. B. Kruskal. Multi-dimensional scaling by optimizing goodness-of-fit to a nonmetric hypothesis. *Psychometrica*, 29:1–27, 1964.
17. Y. Rubner, C. Tomasi, and L. J. Guibas. Adaptive color-image embeddings for database navigation. In *Proceedings of the 1998 IEEE Asian Conference on Computer Vision*, Hong Kong, 1998.
18. C. Faloutsos and K.-I. Lin. FastMap: A fast algorithm for indexing, data-mining and visualization of traditional and multimedia datasets. In *Proceedings of the 1995 ACM SIGMOD International Conference on Management of Data*, pages 163–174, San Jose, CA, 1995.

Generic Personal Assistance Architecture for Mobile Environments

Gerald Bieber

Fraunhofer-Institute for Computer Graphics, Rostock, Germany

Summary. The computing power of mobile computers is comparable to the PCs of some years ago. In addition, PDAs or mobile phones are equipped with a high connectivity and provide a multimodal interface by a high resolution display, vibration feedback and sound functionality. This feature enables the use of mobile phones as personal information manager and personal assistance for business and the everyday life. The progress of personal assistance is supported by the inclusion of external environment information and user related data. By using new interfaces and sensors, an additional and comprehensive understanding of the user's situation can be determined. Hereby a planning of the user's task is supported and a scheduling of the user's activities will be available. This chapter shows an improving approach of the generic architecture for mobile personal situation aware assistance and describes the modules by sample applications and gives an outlook on upcoming challenges of research in activity and assistance models.

1 Introduction

PDAs or mobile phones are equipped with a software package which helps the user to organize his business or everyday life. The given assistant functionality helps to manage contact information, calendar dates, to-do lists, etc., for business or private activities.

The everyday usage of PDAs shows that the physical limitations of display or interaction possibilities lead to a restricted application field. To open the limited application area, new technological human–computer-interaction approaches can be developed by display/visualization technologies (e.g. high resolution displays, SVG, fisheye view, etc.) or interaction technologies (e.g. letter recognition for pen interaction support, speech recognition, etc.). The technology helps to mitigate the problem partly. Another approach is to support the user by providing the right information, at the right time, at the right place. The context information minimizes the overhead information given to the user and therefore helps to work with the limitations very efficiently. The

G. Bieber: *Generic Personal Assistance Architecture for Mobile Environments*, Studies in Computational Intelligence (SCI) **93**, 347–368 (2008)
www.springerlink.com

understanding of the user needs is an important requirement for the assistance functionality. Hereby the aim of the user has to be transformed into a computer understandable goal and executable actions. The optimal sequence of the actions is determined by a scheduler who requires sufficient context information of environment and user. Beside the context information of the external environment, a consideration of inner conditions of a user (e.g. current task state, mood, attention, etc.) is needed for an extended assistance model.

The following section of the chapter gives a brief summary of related work. After presenting the overall definition of the user's goal, a general concept of generation of tasks is giving. In the next section of the chapter, the generic approach of personal assistance architecture is described and finally a compilation of current concerns and future work is shown in the last section.

2 Related Work

At the beginning of using handheld computers with PIM (personal information management) functionality, the needed context information was given by user input. Electronic calendars or to-do lists became a powerful tool and are still one of the core applications for PDAs. The advantage of processing and browsing through information on mobile devices as well as a context related presentation of graphical, textual, hierarchic and time-related information opened a wide range of applications. The research focused on mobile interaction technology. This lead to new solutions and applications, e.g. mobile electronic books, exhibition or tourist guides, mobile maintenance systems, etc.

The next generation of mobile computers was able to achieve external environment information by sensors. The common hardware of the PDAs supported infrared interfaces and so IrDA-Beacons were used to determine position information. This lead to location aware museum guides (MOBIS) [1] and challenging research projects on office guides (Active Badge) [2], remembrance agents (MIT) [3] and others. The emphasis of regarded sensor types covered mainly position information. The situation model consisted only by location but it could cover location based services (LBS) to support the user. Currently the car navigation tools, using PDAs and GPS receivers are very successful.

The latest technology generation allows the inclusion of new sensors which measure not only environment but user specific and biometric information. Sensors do measure EEG pattern, blood pressure, skin resistance, physical activity, etc. The enriched situation information has to be integrated into a generic mobile personal assistant model.

The major focus of personal assistance is based of providing task and task relevant information to the user. Many tasks can only executed by the inclusion of constraints, like environment conditions, physical or psychological parameters [4].

The aim of a computer based mobile personal assistance has been addressed by many visions and research projects. Hereby two major concepts of mobile personal assistance are established as follows:

- Interaction with mobile devices (e.g. PDA, mobile phone, etc.). Hereby the user receives permanently proactive support by his own interaction device.
- Mobile user in intelligent environment (mobile ambient intelligence). The user will be supported by ambient intelligence to interact with ad hoc networks and appliance ensembles.

While Ambient Intelligence (AmI) enables a connected and collaborative ensemble of electronic devices which are sensitive and responsible to people, mobile agents without a constant physical representation might become available. Anind Dey also supposes this development by his Context Toolkit in 2001.

The assistance concepts of agent technology could be applied in location based architectural models [5] and could be enhanced by Microsoft Research to a prototype architecture for ubiquitous computing, the EasyLiving system. The Oxygen/Metaglue-Project by MIT provide an architecture that organizes agents into societies. The PHENOM project of Philips is a long-term research project, which aims to be capable of butler-like behavior. The common ground and major challenges to support the user in ambient environments are Perception, Analysis, Decision, and Action/User-Interaction. The approaches of a general structure for ambient environments or smart spaces mainly addressed the aspects of the intelligence, devices and ad hoc networks. Because the focus of mobile personal assistance is set on the user and not on actuators or devices, only parts of the architecture of agents or ambient networks are relevant for this work.

The project Specter (2003–2005, funded by BMBF, Germany) developed a Conceptual Base Architecture which allowed developing a system which keeps track of user's actions and affective states. Specter created a personal journal which, together with a user model learned on the basis of it, supported the generation of context-appropriate recommendations and actions. The framework was designed especially for instrumented environments.

The joined research project SAiMotion, (situation ware assistance in motion, 2001–2004, funded by BMBF, Germany) discussed the special needs of the user in limited application fields and worked out architectural requirements. One project focus lied on the assistance functionality of mobile computers and the special need to support the user while he is performing his tasks. Hereby it's necessary to use a suitable task model. The basic work on task models was done by F. Paterno who described "A task defines how the user can reach a goal in a specific application domain" and worked out this consideration to the development of Concurrent Task Trees (CTT) [6]. The CTT are able to model hierarchal and temporal constraints and pick up the initial proposal of GOMS (goals, operators, methods, selection rules). The later concept of Diane+ (Balbo et al. 1998) represents tasks in a graphical notation explicitly stating who performs the task (end-user, system, or interaction of the two). The underlying motivation of these models is a descriptive presentation of the task, task description, task flow and their topology and constraints.

Alternative approaches described workflow processes and defined tasks. In 1991, SAP and Saarland University Germany developed the concept of EPK (Ereignisgesteuerte Prozesskette). EPK bases of the Petri-Net-Theory and connection elements, which are known by netplan applications. The core of EPK works with rules engines and using core elements like event, condition, actions.

This concept is one of the most common, because it is used by SAP R/3 as well as within ARIS. Users of EPK are large companies, e.g. BASF, DaimlerChrysler, Siemens, etc., are one of the customers.

The experience of the modeling of business processes can be adapted for modeling personal tasks. In [7], it is shown that general concepts of visualization of personal tasks with their state and conditions could be solved. The aim of creating a generic Personal tasks model leads to the inclusion of ontology and application relevant approaches. Research projects of MAP (Multimedia-Workspace of tomorrow, BMBF) or FIRB-VICOM (Italy) used a context manager to receive a computerized consciousness about the task history and task state. This assures that personal tasks consist of a suitable structure. Schilit [8] included the context into the task description. It could be shown by [9] that user relevant task models are basing on the concept that the user is a physical object, which is presented in space and time. Every task is also an object of space and time dimensions [10] and even for time irrelevant personal tasks it is very helpful to integrated actions in a homogenous space/time ontology.

2.1 Personal Task Model

Personal Task descriptions are used to define task states and to describe actions to achieve the user's goal. The aim of using the Personal Tasks Model is to design a personal assistance system for mobile computers. To support mobile personal assistance, some pre-assumptions can be done [11]. Hereby a personal task

- Is an atomic task, which is the basis of compound tasks or subtasks
- Has a definite start and end
- Is not interruptible
- Has to be performed only at one place
- Has effects
- Is not spread over time or location

Suitable Task descriptions where given by the concepts of personal agents [12], remembrance assistance (RA), mobile buddy alert (ZGDV) or

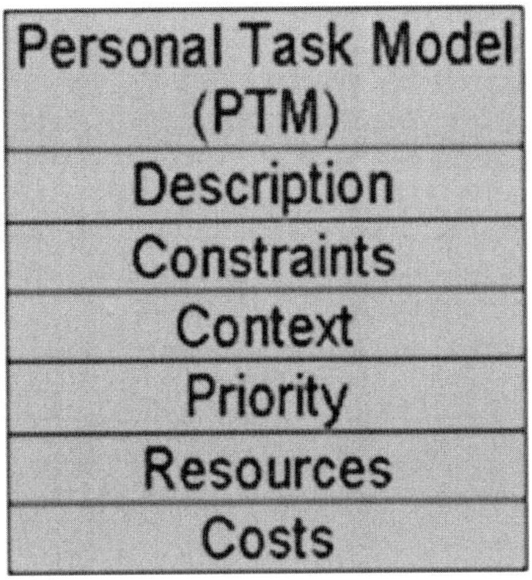

Fig. 1. Personal task model (PTM)

MEMOS [13], etc. Some applications are basing on the informal CommonKads methodology but field applications, e.g. mobile fair guides on PDAs (e.g. TouristGuide, etc.), uses the following basic task model (Fig. 1) for a computerized computation.

Every task does contain a general description which is readable by man and machine. The description block includes the identification and a short description. The next component contains constraints, e.g. when or where the task can be performed. The constraints provide additional information about the space/time domain. The context describes the status information such as if the task is performed or not and include the control about subtask and topology. The context may contain frequency pattern (how often the task is performed). The priority describes the ranking and importance of the task. The resource block defines preconditions before the task is performed, postconditions which conditions are existing after executing the task and what resources are needed for the execution of the task (e.g. expense of money, etc.). For an adequate scheduling of task, a cost estimation component is part of the task model. It assumes which expense will be generated by performing the task. Costs are usually a fictive currency with single or multidimensional values.

Field applications might raise the need to expand the elements of the task definition like durability, creator, etc., but the given modules can be regarded as the core elements of the personal task model.

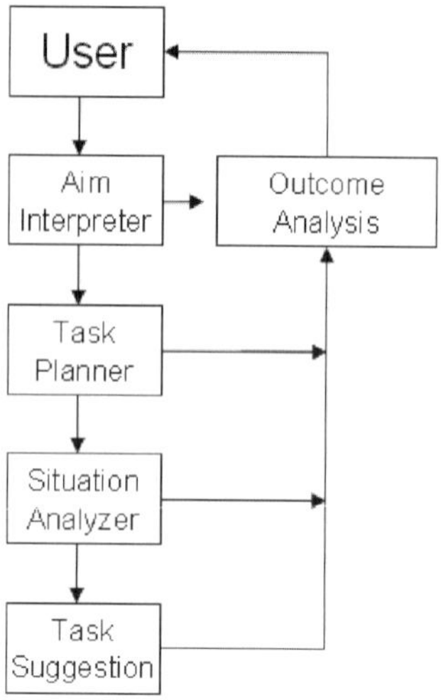

Fig. 2. Personal assistance

3 Personal Assistant

The objective of a personal assistant is to help the user to achieve his personal tasks. To fulfill this aim, the personal assistant has to understand and to know about the user's goal and desires. This can be performed by an aim interpreter.

The personal assistance (Fig. 2) can be performed within two methods. The first possibility is the mobile personal assistance in appearance of a mobile device. The user is observed by this device and can interact and control it and the environment like interacting with a remote control. The other method of personal assistance can be found within intelligent environments. Hereby the user is surrounded by a connected, interactive device ensemble and he receives assistance by the Ambient Intelligence (AmI) [14] itself. The intelligent environment can provide assistance functionality by itself.

To achieve the user's goal, several actions have to be performed. The task planner identifies the needed actions. Hereby some tasks do have constraints, e.g. spatial or time restrictions.

By inclusion of the current situation the right tasks can be suggested to achieve the user's goals.

Hereby the following steps are to be addressed:

- Identification of user needs
- Providing execution plan to achieve users goal
- Supporting user by providing suitable tasks for current situation
- Keeping track of activities or task state
- Evaluation of user activities

3.1 Goal Description

Within the existing domain, the goal has to be identified. The user has a goal in mind but his utterances are describing something different. The inner state concerning the goal is mostly represented different in an external computer based assistance system (Fig. 3). Another reason of a different representation of the user's goal is that the representation ability of the reception computer is limited [15]. To give an example, a travel assistance system which tries to suggest the route information from start to the destination can only display the closed available information.

Currently two common approaches are available for determining the user's goal. The user is active and selects a goal out of the set of predefined possibilities. The alternative way is passive; the user's utterances are mapped to goals by interpretation (Fig. 4).

The interpretation is depending from user preferences, user situation and many other parameters. Even planning systems and methods of artificial intelligence can be used, the probability might improved by, e.g. Markov Models,

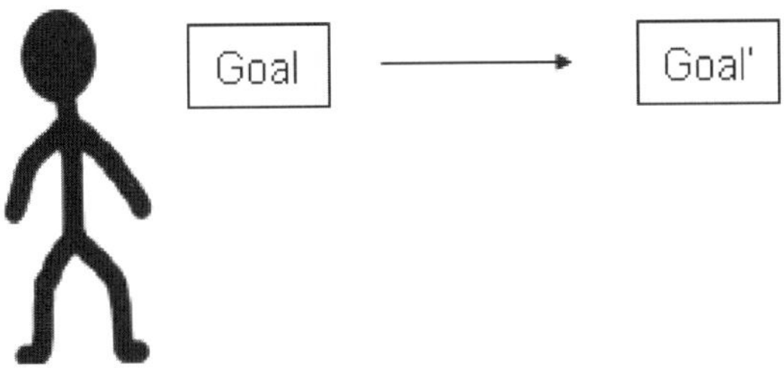

Fig. 3. Inner and expressed goal of a user

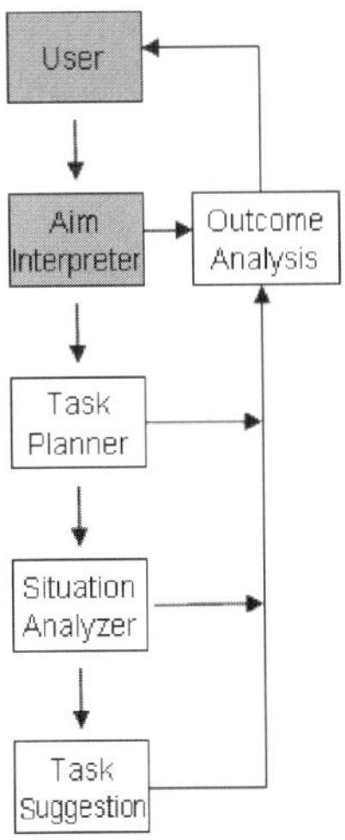

Fig. 4. Personal assistance structure

Fig. 5. Aim interpretation by avatars

but still a high chance of misinterpretation is given. Independent of the quality of the identification of the user's goal, a computerized goal processing and a generation of strategies to achieve the desired conditions will be possible.

The project Multimodal Assistance for Infotainment and Service Infrastructures EMBASSI (BMBF) worked on several areas, one was the aim interpretation of the user.

The aim interpreter by avatars (Fig. 5) includes multimodal information and provides AR/VR interfaces. Embassi was a lead project of the federal ministry of education and research, Germany (2000–2004). Its focus was the development new paradigms and architectures for the intuitive interaction with technical infrastructures of the everyday life, such as home entertainment and control, public terminals, and car infotainment systems [16].

The Embassi structure (Fig. 6) is continued by further research projects, e.g. Dynamite (Germany) but its focus is on personal support for interacting with real world devices.

3.2 Situation Influence by Task Execution

As described in the concept of GOMS (Goals, Operators, Methods and Selection rules by Card, Moran and Newell), or TAG (Task Action Grammars by Payne and Green) and UAN (User Action Notation by Hartson, Siochi and Hix), the task execution is regarded as an operation to achieve the user's goal. Hereby it is very reasonable that a task is atomic, not able to be divided in further subtasks. The operations can be external (e.g. changing conditions in the environment), mental operations (e.g. decision making) or primitive or clustered in high-level operations. Like a machine needs energy for operation, the task performance might consume resources (e.g. money, calories, etc.) or need resources like catalytic converters. The catalytic converters can

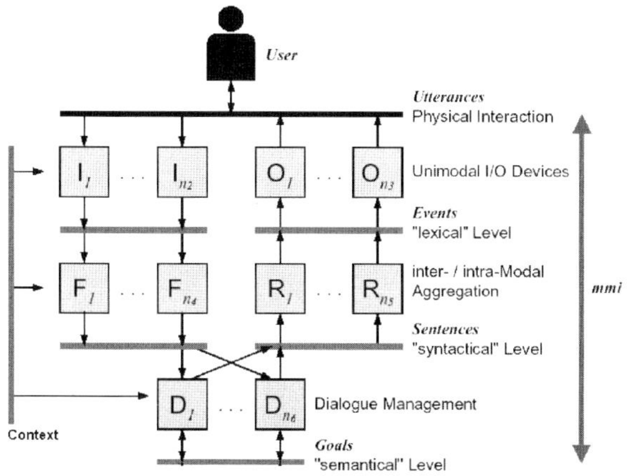

Fig. 6. Embassi layer structure

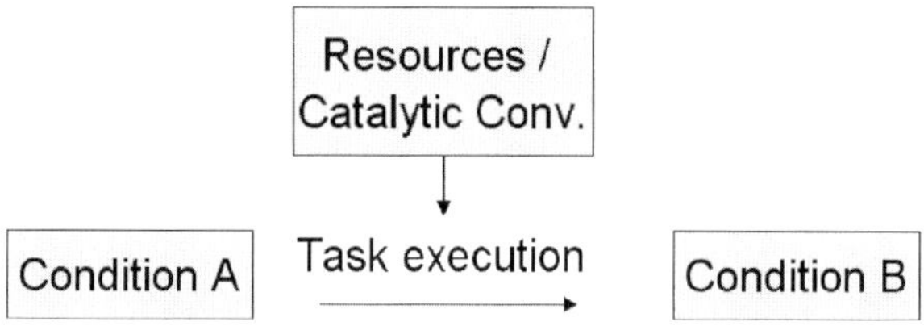

Fig. 7. Influence of task on environment

be needed but no consumption occurs. Figure 7 illustrates that the task performance influences the condition from the origin (condition A) to the present (condition B) by using resources. The origin situation will be changed to a new one; the final state of the task is completed.

3.3 Task Generation

It is necessary to perform tasks to achieve the user's goal. It could be shown in [17] that three major methods do exist to generate possible tasks by knowing the goal. This methods can be in real-time or offline, computerized, computer aided or manually. The concepts to generate personal tasks to achieve the user's goal are classified in following main groups:

- Planning (by planning systems)
- Template (by predefined task lists)
- User input

The task generation is an integral module of the personal task planning architecture. This assumption leads to the following structure (Fig. 8).

A task planner generates task by the given goals and constraints. Hereby planning systems are usable and known by production planning systems, e.g.

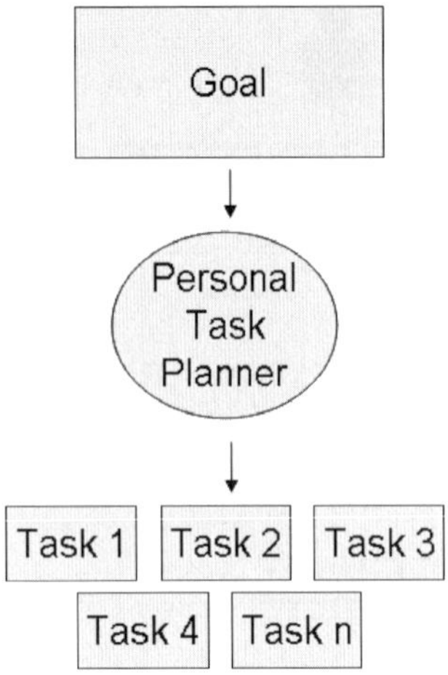

Fig. 8. Task composition by a planning module

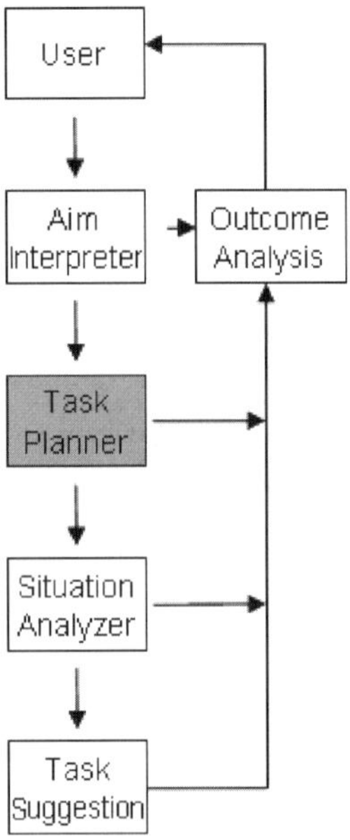

Fig. 9. Task generation

OPS5, Fuzzy Systems, Clips/JESS (Java expert shell) or by ART and Expert Systems. For a personal assistance, the knowledge about the task structure of an unknown problem or goal might be very fuzzy. For the user it might be more efficient to use templates for the task generation (Fig. 9).

Templates can be used whenever a task list is known for the solution of achieving a current goal. The usage of templates is very reasonable when the same goal will be addressed often. This can be generated by computers or users as well.

Beside computerized task generating or templates, the user might compose talks directly. This manual talk-list can be reordered by the personal assistant and might provide additional information whenever the situation requires.

The origin of the techniques of task generation is founded by workflow management, Computer Supported Cooperative Work (CSCW) and operation research (OR).

3.4 Situation Dependent Task Suggestion

A certain strategy, which consists of many operations, is needed to achieve a given goal. The right selection of the tasks and the sequence of the operations (task execution list) are relevant. Hereby the efficiency can be determined by the final result and the over all costs of the performed operations. The sequence of the task execution is defined within the execution plan; the strategy defines the task selection and the order of the tasks.

Even a set of tasks can be generated by the task planner; the order of the task can be varying during the task executions. The execution plan can be re-ordered after each task execution. This re-ordering will be done by a tasks scheduler (Fig. 10).

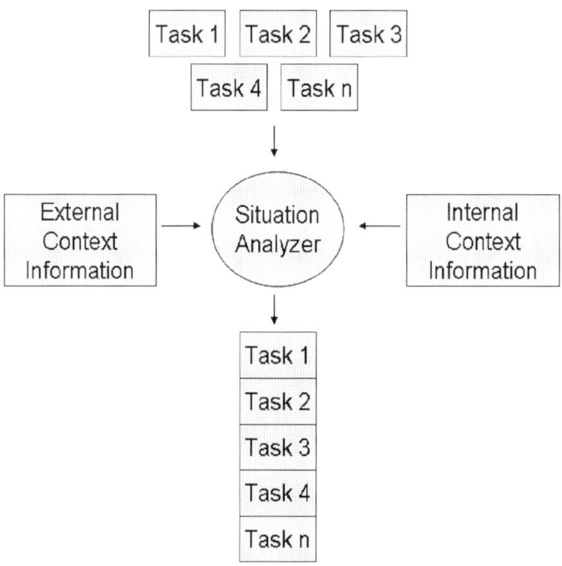

Fig. 10. Personal task scheduling

The technology of evaluating the efficiency, calculation costs and underlying dependencies are broadly evaluated by the operation research (OR) and can be applied to personal assistance.

The schedule of a plan describes the order of the execution of the tasks without inclusion of the current situation the user is in. For an optimal assistance of the user and a fitting of the tasks for the present situation, a situation analyzer (SA) has to be integrated into the concept of the personal task assistant [18]. By regarding the current situation, the SA reschedules the task list and generates a new plan under consideration of the given strategy. This might be done by rule-engines (e.g. OPS5, Jess, etc.) or task scheduler (e.g. heuristically scheduling systems). Hereby the external situations as well as internal context information (e.g. task state, system connectivity, etc.) are integrated into the plan generation. Because every tasks changes conditions within the environment, a feedback loop is necessary to integrate sudden, unexpected or unknown influences.

3.5 Control Circuit

As described, the mobile personal assistance is able to suggest tasks to the user which are suitable for execution. In [19] it could be worked out that the user often interrupts the current task, changes to another operation and returns afterwards to the primary task. It seems that a significant factor is not only the external interrupt but also internal reasons and mental states for this behavior. Another irregular human behavior is that the suggested task will not executed – even the execution is more convenient, more logical or produces only a minimum of costs. Instead of the suggested task, another will be executed. This leads to take additional requirements into account for implementation of a mobile personal assistance. Once the execution plan is generated, it is not stabile and the system permanently has to compare current situation with the estimated situation. A robust technology of comparing nominal and actual condition is known by control engineering. This enables the use of a feedback loop and allows a constantly rescheduling and generation of a current execution plan of the tasks within personal assistance systems.

4 Personal Assistance Architecture

The concept of task management and scheduling could be proofed by application prototypes. The early MONAD System [20] (MObile Network ADministration assistant) enables the assistant of a network administrator by supporting simultaneous execution of task on spatial distributed and complex systems. The supported actions are computer and telephone installation, setup routines or wiring in patch rooms. For this type of activity, an exemplary strategy could be developed which uses a suitable segmentation of the actions into elementary and atomic tasks. The scheduling was performed by an operation system using OPS5 and single actions could be formulated as production rules. The rules engine could act as a reasoning engine that analyzed the current situation and reschedules the execution plan whenever is was necessary. Because only the current situation was regarded into the consideration (opportunistic scheduling), it could be shown that a scheduling with total cost consideration would be more effective.

The inclusion of current situation and the total costs leads to the mobile personal assistance architecture.

Figure 11 shows the modules of the aim interpreter, which generates the fittest goal (B) by the user utterances (A). To achieve this goal, certain actions have to be executed. The task planner module defines these. The situation analyzer regards the current situation of inner and external conditions and schedules the possible tasks under consideration of the best cost ratio. The user executes the suggested task (or others) and the assistance system evaluates the post-condition with the aim of the user.

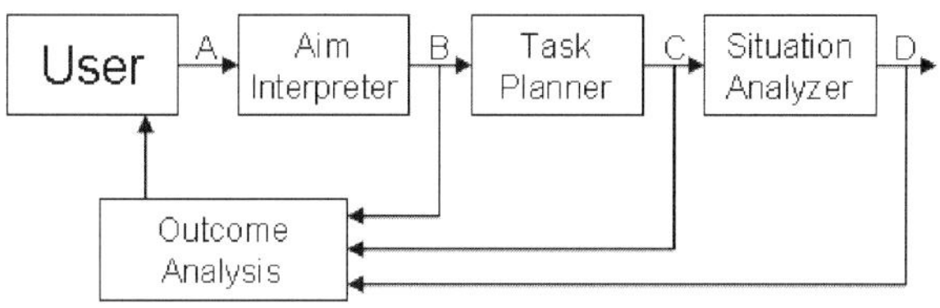

Fig. 11. Personal assistance

Fig. 12. Task scheduling on a PDA

5 Proof of Concept-Applications

5.1 xGuide: Electronic Exhibition Guide

The general approach of the evaluation of a system is the real use with objective users within a closed scenario. Hereby it is desirable to have an application field with a short software life-cycle. The scenario of using a mobile guide on mobile phones or PDAs for exhibitions is very useful. The average user is high educated, the user group is very homogeneous and in general the business people (users) own a high standard device.

The scenario spans a working day and the spatial constraints are given by the physical limitations of the fair ground. The first widely and real used mobile exhibition guide was established by eGuide of the Fraunhofer-Institute IGD Rostock at the world largest exhibition fair CeBIT, Hannover, Germany in 2000 [19]. The mobile guide was an electronic list of every exhibitor with additional information, e.g. number of the hall and booth. At this time, the system could only provide a search on the list of exhibitors and sorting function to support a personalized tour. This guide did not use any external sensors. Today, the next generation of the exhibition and convention xGuide is able to locate the own position by sensors (e.g. infrared/Bluetooth/GSM). This application enables a situation analyzing and initiates by the inclusion of the know position a rescheduling (Fig. 13).

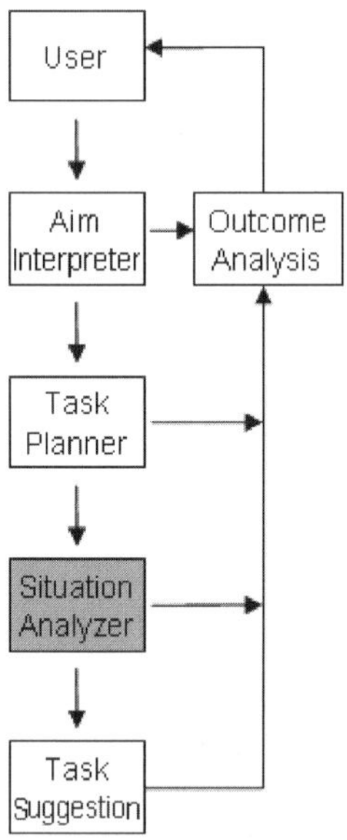

Fig. 13. Situation analyzing

5.2 Applied Task Definition

The general visitor of a trade show likes to visit booths, a lecture as well as ambling around of the fairground. Hereby the actions of the visitor can be defined as time and space related tasks, e.g. to visit an exhibitor (without time relevance), to visit a talk (time and place relevance), to have an appointment (e.g. without place), to do something without time/location relevance (doing telephone calls, etc.). The place of every booth or lecture is known within the fairground and the scheduling module is able to suggest the best task for the current position of the visitor. This assumption leads to a very suitable task planning system.

Figure 12 shows a suggestion of the execution of the tasks under consideration of the costs which are estimated by the shortest path and the constraints of the tasks (e.g. talks which have a defined start and ending).

The scheduling functionality calculates the shortest path (traveling salesman problem) and for time depended tasks it uses the common algorithm for the delivery problem in combination with heuristically search (using, e.g. Lin–Kernighan). The personal assistant tool xGuide minimized the detour and suggests an optimized task-list for the current situation. In application, we could figure out that the visitor of a trade-show is wasting a high amount of time just by walking. As another outcome, it could be identified that the task list of the visitor is in many cases longer as the list of the performed task [21]. The visitor is also often performing tasks spontaneously and he is acting in inspiration. Hereby the need of an extensive user model does exist because this behavior is correlated with the individual user. A very interesting outcome within the application scenario is that user doesn't like a total optimizing tool because he needs a special amount of redundant information to be able to receive a good impression of the event.

5.3 StepMan

The department Mobile Multimedia-Technologies and Entertainment Technologies of the Fraunhofer IGD-R developed the StepMan Technology. StepMan has been designed as an add-on application for MP3 players, smartphones and PDAs (personal digital assistants) which can also play music. Joggers wear sensors that deliver data regarding pulse and speed to the device. A software program continuously adjusts the tempo of the music to the runner's style according to jogging speed and body stress. The music pitch remains unchanged. Conversely, users can set up their own training program with intermittent breaks and performance limits, forcing the runner to adjust his speed to the music. Mostly any song has a rhythm. The speed of the music depends on the genre of the song and is about 70–140 beats per minute. The StepMan does an adaptation of the music speed to the walking speed of the runner (Figs. 14 and 15).

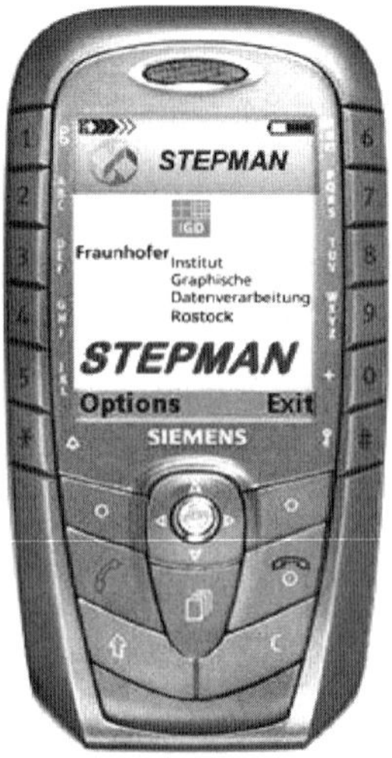

Fig. 14. Situation analyzing

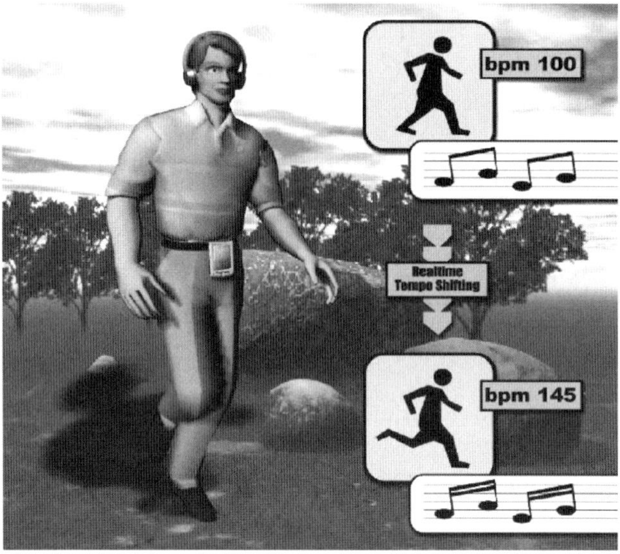

Fig. 15. StepMan music adaptation

StepMan is a mixture of entertainment-, interaction- and mobile multimedia- technologies. The first prototype based on a pocketPC, Compaq iPAQ h3970 with a 400 MHz Intel PXA250 processor, using an electronic acceleration sensor for step detection and a ChipOx sensor from EnviteC for oxygen saturation and pulse rates. All implementation was done in native code, C++. The current implementation on a mobile phone, first version demonstrated as the application "Runster" on the CeBIT exhibition for Siemens, showed that StepMan is running on a non performant calculation platform (a mobile phone with 120 MHz).

Within the system all sensor data are analyzed and can be used to control the music or to give situation relevant feedback to the jogger.

Even without a step sensor it is possible to create a training profile as an important functionality of StepMan. In this case, the tempo of music will change after a predefined profile in order to give a jogger a basic beat to follow. This simple interaction allows a first training activity for joggers, but we depend on the good will of joggers. If they do not follow the given tempo of music we cannot force them to do so. Within our tests this problem did not occur.

5.4 Personal Assistant DiaTrace

Using an activity sensor to obtain additional situation information about the user might be also helpful in application areas like wellness, fitness and health. One relevant issue concerning assistance is the monitoring of patients, real-time processing of the data and proactive feedback to the user.

Some diseases are caused by the lack of physical activity, e.g. metabolic diseases like diabetes. Hereby the diabetes-assistant reacts as a non-obtrusive interface because the patient achieves situation related feedback in real-time. The assistant consists of a sensor unit (MoSeBo) for activity measurement and a calculation, a mobile phone, which is connected to the MotionSensor-Board (Fig. 18). The java-phone performs the activity classifying algorithms and provides multimedia-feedback to the wearer in order to patient specific settings.

DiaTrace provides activity information as easy as a scale for weight. The user stays on a scale and knows if he is overweighed or not – and it belongs to him to change this condition. The aim of DiaTrace is to be a digital assistant which gives advises and it also displays the current and average activity intensity plus qualitative information of activity (e.g. cycling, car, bus riding, sitting, walking, etc.) (Figs. 16 and 17). In addition it reminds the user – when he is inactive while sitting behind his t.v. – to perform some sports, or congratulates that the daily amount is reached.

The major part of diabetes patients do suffer on disadjustment. Hereby the unofficial numbers says that 75% are not well adjusted. The amount of food, the time of meals and the medicine is considered in the adjustment but not the energy consumption mainly caused by physical activity. DiaTrace includes this in the treatment of the patients. The concept "DiaTrace" was awarded by the Mecklenburg–Vorpommern innovation contest in 2005.

The first tests of DiaTrace could work out that a 24 h support by mobile assistants is possible and users agreed to use it (Fig. 19). The conditions of everyday life are leading to a new understanding of the system requirement. The system works well with defined sensor fixations (e.g. belt, hip, pocket),

Fig. 16. Current activity

Fig. 17. History overview

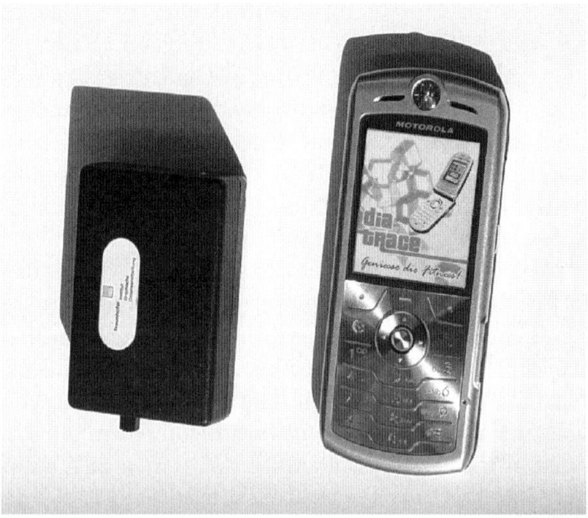

Fig. 18. DiaTrace, sensor and phone

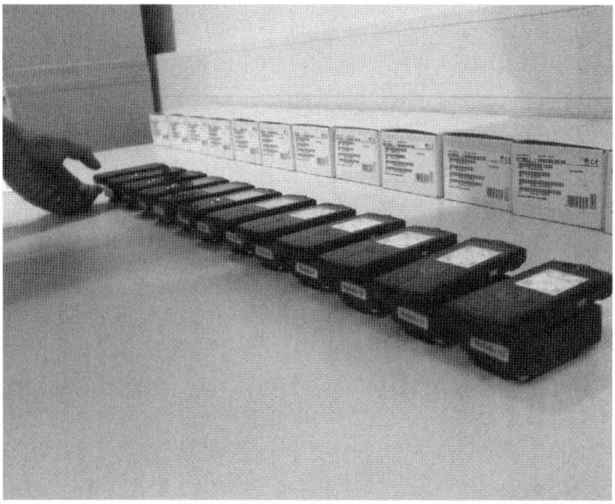

Fig. 19. Evaluation-systems DiaTrace

but in real life conditions, mostly women had difficulties for an adequate sensor fixation because they often wearing skirts without pockets. Here DiaTrace was carried in a hand-bag which implies affected sensor data.

5.5 Mobile Personal Trainer

In addition to the described use of DiaTrace and StepMan, the technology of a real time analysis of physical activity can be integrated into a system called Mobile Personal Trainer. The overall goal of a personal trainer is to strengthen or at least to remain the physical condition of the user. Hereby an analysis of the present physical condition of the user will be done, the personal constraints regarded and an individual training plan can be presented.

The virtual trainer uses additional sensors to track the condition of sports people while doing the exercise. It motivates during the training and later on gives a feedback about the effects of the sporting unit. Even normal movement activities can be regarded as sporting units for patient or handicapped people. By inclusion of user specific information, e.g. step length, weight, etc., the personal trainer can easily calculate the speed and total length of the run, calorie consumption and can integrated the data to an activity profile of the user. The training functionality can be used for various application fields, e.g. health care, fitness, running, swimming, etc.

For non professional athletes, inner reasons but also outstanding conditions (e.g. weather condition, family, illness, etc.) are leading to the fact that an optimal trainings plan will be always changed. The Mobile Personal Trainer also works as a sport organizer and can automatic reschedule a training plan. Each training unit is treated as a personal task and can be optimized by integration into a personal task management system. The user will be informed about the next trainings units, alternative training possibilities, etc., with the same notifying instruments as personal task, e.g. to-do lists and organizers.

6 Outcome Analysis

The personal assistance system needs a validation of its suggestions and the user reactions. Hereby it is very interesting to know the user response regarding assistance.

To develop a method of user response of a mobile exhibition guide, a tagging of exhibits by 2D-Codes can be performed. Relevant exhibits provide an optical 2D-barcode with an information link. Hereby the visitor of an exhibition can collect a list of exhibits, exhibitors, time and date by performing a 2D-barcode data collection with a mobile phone (Fig. 20). The visitor takes a photo with the 2D-barcode and the software on the mobile phone is able to decode the information. This information will be saved on the mobile phone. If a visitor would like to receive all tagged information via email, he sends his personal list to a host-server. On this server, another application extracts all information and provides an email response which sends relevant information to the user and calculates statistical information such access information, emphasis time, number of exhibits and exhibitors. In future, it will be possible to match the proposed task list of the user for an exhibition day with the performed visits.

Fig. 20. Feedback retrieval by 2D barcodes

7 Conclusion and Further Work

The chapter describes the general concept from task generation to situation-aware task scheduling on mobile personal assistance systems. We could see that a task scheduling is very helpful in closed application domains. Hereby a generic task model for the description of tasks enables a situation analyzer to identify the fittest task for the current situation.

The concepts of the personal mobile assistance architecture could be already used in application fields of tourist- and exhibition guides, fitness trainer and health care assistants and will span spare time and business applications in future.

Therefore the research on the measurement of the quality of a mobile assistant should be advanced. This feedback retrieval must be performed almost without any interference and disturbance to the user. The concept on the outcome analysis will show if it will be accepted by the user.

In the application area of exhibitions, it can be determined that the absolute optimization of every action is not necessary – the user likes assistance but not a leading string. The exhibition scenario includes ambling over the fair ground and passive receiving information as well as actively searching and examination the exhibitors. The mobile exhibition system xGuide surely well supports the users but it doesn't cover the total need of a comprehensive assistant.

Concerning exhibition guides, the further work will cover the examination of activity periods, walking distances, breaks, duration on booth (stops) with one possible outcome to improve the usage of shuttle services.

Further improvement of mobile situation aware personal assistances can be done by enhancement of comprehensive situation modeling. Upcoming research is on physical activity distribution (e.g. using pedometers or motion sensor boards) within an exhibition day. Furthermore, the inclusion of inner conditions like fatigue, feed ache, mood, etc., could be integrated by new sensors (skin resistance, etc.) into a mobile personal assistance system. The additional sensors should be non obtrusive.

In future, personal assistance systems might be able to predict the outcome of various task alternatives reasonable to use multiple mobile assistance systems with ambient interfaces.

References

1. Chavez E., Kirste T., Mobile visitor information system, CG topics 3/99, Darmstadt, Germany, ISSN 0936-2770, 1999
2. Want R., Hopper A., Falcão V., Gibbons J., The active badge location system, Olivetti Research Ltd. (ORL)/Xerox PARC research laboratories, ACM Transactions on Information Systems, 10:91–102, Jan. 1992
3. Rhodes B., Using physical context for just-in-time information retrieval, IEEE Transactions on Computers, 52(8):1011–1014, Aug. 2003

4. Chávez E., Ide R., Kirste T., Interactive applications of personal situation-aware assistants. Computers & Graphics, 23(6):903–915, 1999
5. Satoh I., Location-Based Services in Ubiquitous Computing Environments, Lecture Notes in Computer Science, vol. 2910, Springer, Berlin, Germany, 2003
6. Paternò F., Mancini C., Meniconi S., ConcurTaskTrees: A diagrammatic notation for specifying task models. In Proc. of IFIP Int. Conf. on Human-Computer Interaction Interact '97 (Sydney, July 1997). Chapman & Hall, London, 1997, pp. 362–369
7. Bieber G., Tominski C., Visualization techniques for personal tasks on mobile computers, In Proceedings of the HCII2003, vol. 3, Lawrence Erlbaum, Crete, Greece, 2003, ISBN 0-8058-4932-7
8. Schilit W.N., A system architecture for context-aware mobile computing, Ph.D. Thesis, Columbia University, New York, 1995
9. Bieber G., The approach of a personal task model for mobile computing, In MOST International Conference 2002, Warsaw, Poland, ISBN 83-87091-32-4
10. Garlan D., Schmerl B., The RADAR Architecture for personal cognitive assistance, International Journal of Software Engineering and Knowledge Engineering, 17(2), Apr. 2007
11. Mcgraw K., Herbison-Briggs K., Knowledge Acquisition, Principles and Guidelines, International Editions, Prentice Hall, London, 1989
12. Pereira J., Englmeier K., Rojas C., A model for personal assistance in complex information spaces, In Proceedings of the American Society for Information Science and Technology, vol. 39(1), 2005
13. Voinikonis A., Irmscher K., Schulze H., Distributed processing of reminding tasks within the mobile memory aid system, MEMOS, Personal and Ubiquitous Computing, Springer, London, DOI 10.1007/s00779-004-0332-5, 2005
14. ISTAG, Involving users in the Development of Ambient Intelligence, ISTAG Report on Experience and Application Research, 2004
15. Heider T., Kirste T., Architecture considerations for interoperable multi-modal assistant systems, In Proc. DSV-IS 2002, Rostock, Germany, 2002
16. Hildebrand A., Sa V., EMBASSI: Electronic Multimedia and Service Assistance, In Proc. IMC 2000, Rostock, Germany, 2000
17. Iqbal S.T., MeWS-IT: A mental workload based system for interruption timing. In Proceedings of the ACM Symposium on User Interface Software and Technology, Doctoral Symposium, Seattle, WA, Oct. 2005
18. Chen G., Kotz D., A survey of context-aware mobile computing research, Dartmouth Computer Science Technical Report TR2000-381, 2000
19. Giersich M., Bieber G., Personal mobile navigational systems – design considerations and experiences, Computer & Graphics, 25(4):563–570, Elsevier Science, UK, 2001, ISSN 0097-8493
20. Kirste T., Rieck A., A mobile network administration system: Conception and realization. In Proc. AKIDA'98, Aachen, Germany, June 3–4, 1998
21. Oppermann R., Specht M., A context-sensitive nomadic exhibition guide, In Handheld and Ubiquitous Computing (Proc. 2nd Int. Symp., Bristol, UK, Sep. 2000), P. Thomas and H.W. Gellersen (eds.), Springer, Berlin, pp. 127–142, 2000

Printing: Krips bv, Meppel, The Netherlands
Binding: Stürtz, Würzburg, Germany